MAJID, KAMAL IZELDEEN
THEORY OF STRUCTURES, WIT
000303689

624.04 M23

0303689-X

Theory of Structures

Theory of Structures
with matrix notation

K. I. MAJID,
B.Sc., Ph.D., D.Sc., C.Eng., F.I.C.E., F.I.Struct.E.

*Professor of Civil Engineering,
University of Aston in Birmingham*

Newnes — Butterworths

LONDON — BOSTON

Sydney — Wellington — Durban — Toronto

The Butterworth Group

United Kingdom	**Butterworth and Co (Publishers) Ltd** London: 88 Kingsway, WC2B 6AB
Australia	**Butterworths Pty Ltd** Sydney: 586 Pacific Highway, Chatswood, NSW 2067 Also at Melbourne, Brisbane, Adelaide and Perth
Canada	**Butterworth and Co (Canada) Ltd** Toronto: 2265 Midland Avenue, Scarborough, Ontario M1P 4S1
New Zealand	**Butterworths of New Zealand Ltd** Wellington: 77-85 Customhouse Quay, 1 CPO Box 472
South Africa	**Butterworth and Co (South Africa) (Pty) Ltd** Durban: 152-154 Gale Street
USA	**Butterworth (Publishers) Inc** Boston: 19 Cummings Park, Woburn, Mass. 01801

First published 1978

©Butterworth & Co. (Publishers) Ltd, 1978

All rights reserved. No part of this publication may be reproduced or transmitted in any form or by any means, including photocopying and recording, without the written permission of the copyright holder, application for which should be addressed to the Publishers. Such written permission must also be obtained before any part of this publication is stored in a retrieval system of any nature.

This book is sold subject to the Standard Conditions of Sale of Net Books and may not be re-sold in the UK below the net price given by the Publishers in their current price list.

ISBN 0 408 00303 0 Cased
0 408 00323 5 Limp

British Library Cataloguing in Publication Data

Majid, Kamal Izeldeen
　Theory of Structures.
　1. Structures, Theory of
　I. Title
　624'.17　　TA654　　77-30307

ISBN 0-408-00303-0
0-408-00323-5 Pbk

Typeset By
Reproduction Drawings Ltd
Sutton Surrey

Printed in Scotland by Thomson Litho Ltd.,
East Kilbride

Preface

The use of computers in structural engineering is now well established. The methods at present used in structural analyses are therefore based on matrix algebra which is more suitable to computers. However, these matrix methods are often taught too late during degree courses to the extent that the student, well versed in matrix algebra from school, does not appreciate the advantages of them. On the other hand, present day courses in the theory of structures suffer from duplication which causes congestion as well as confusion as to which of the many methods prescribed is the most useful.

This book is therefore written to fulfill two main objectives. First, to introduce the modern matrix methods gradually from an early stage and in a simple manner. Second, to remove duplication by omitting topics that are out of use so that those currently practised are covered thoroughly. Prominence is given to the fundamental aspects of structural analysis. Thus whenever possible the equations of equilibrium and the stress-strain relationships of the material are emphasised. Such structural properties as stiffness, flexibility, energy, elasticity, plasticity etc. are not only described in detail but also used extensively. To be consistent, a single set of sign conventions is adopted throughout. It is considered that the right hand screw rule is more advantageous than the left particularly because students become familiar with this rule at school.

This volume treats simple structures and covers the course given in the first two years of undergraduate study. Chapter 1 covers the analysis of pin jointed isostatic plane frames. Once the member forces are calculated, these are expressed in terms of the external forces in the form of a matrix equation. This expression later forms the basis of the matrix force method. Chapter 2 introduces the structural properties such as stress, strain, stiffness, flexibility and strain energy of axially loaded members. The relationship between the stiffness, the stress and the strain which forms the basis of the finite element method is introduced in this chapter with the aid of a single member.

Chapter 3 is devoted to laterally loaded beams while Chapter 4 covers bent and curved members in which bending moments and shearing forces as well as axial forces are present. The local axes, as opposed to the global axes, of a member are defined in Chapter 4. Identification of members and their ends is introduced here with the aim of training the student for the future use of member numbering. This chapter is concluded with the analysis of three pinned archs as isostatic frames consisting of bent or curved members.

Both the matrix force and matrix displacement methods are based on the slope deflection equations. These equations are derived in Chapter 5 and used extensively to calculate the deflections of isostatic and simple hyperstatic members and frames. The slope deflection equations are derived from a general displacement function. In this manner the procedure used to present the finite element method is illustrated. The method of dividing uniformly distributed loads into two loading systems, as adopted by matrix methods for discrete problems, is also described.

Chapter 6 introduces the influence lines. The virtual work method is adopted to draw the influence lines for shearing forces, bending moments, etc. for isostatic members and frames. The dangerous section for design is defined in this chapter.

Chapter 7 defines the flexural properties of structures and members, and the strain energies of a member in pure bending and pure shear are introduced. The virtual work equation in its general form has been derived in this chapter and it is demonstrated that in fact this equation is another form of defining the state of equilibrium in a member.

Chapter 8 is devoted to the plastic theory. It has been emphasised that frames at incipient collapse are isostatic and the equations of static equilibrium are sufficient to calculate the failure load of such frames.

Chapter 9 introduces the force method for calculating deflections of pin jointed and rigidly jointed frames. It is considered that calculating deflections by the force method is simpler than the use of the virtual work equations, for, once the relationships between member forces and external loads are established, they can be used directly to calculate joint deflections. This chapter is concluded by extending the force method to cover the analysis of hyperstatic structures.

Chapter 10 is concerned with the analysis of columns. The modern method of using the stability functions in the analysis of columns is adopted throughout. In this chapter the significance of the elastic instability on the one hand and of the non-linear analysis of columns on the other hand are both explained in detail.

Chapter 11 is concerned with the analysis of pin jointed space frames. The method of tension coefficients is brought up to date and presented as the basis of the matrix force method. The chapter then deals with the matrix displacement method for the analysis of hyperstatic pin jointed space frames.

Chapter 12 covers the displacement method for the analysis of rigidly jointed plane frames. This covers the linear and the non-linear elastic analysis of frames including the state of elastic instability.

The volume contains more than 120 worked examples and each chapter is concluded with exercises with answers.

The author wishes to thank Mr David J. Newman, B.A. (Cantab.) of Taylor Woodrow International and Dr David Just of Aston University for reading the manuscript and resolving and correcting the examples as well as for their valuable suggestions and contributions; Dr David Anderson of the University of Warwick for reading the text and making final suggestions; Mrs Janet Allen for typing the script and Mrs Patricia Taber for producing the drawings.

Contents

1. **Pin jointed isostatic plane structures**   1

    1.1–Introduction, 1.2–Representation of supports, 1.3–Equilibrium of coplanar forces, 1.4–Structural equilibrium, 1.5–Calculation of reactions, 1.6–Member forces, 1.7–The method of sections, 1.8–Combination of methods of joint and section, 1.9–Unstable, isostatic and hyperstatic frames, 1.10–Consideration of members and supports, 1.11–Examples, Exercises on chapter 1.

2. **Structural properties**   23

    2.1–Direct stress in a member, 2.2–Direct strains, 2.3–The stress-strain curve, 2.4–Hooke's law, 2.5–Stiffness, 2.6–Flexibility, 2.7–Examples, 2.8–Temperature stresses, 2.9–Example, 2.10–The principle of superposition, 2.11–Strain energy, 2.12–The work equation, 2.13–Relationships among stress, strain and stiffness, 2.14–Principles of structural analysis, 2.15–Examples, 2.16–Properties of symmetrical structures. 2.17–The equivalent half frame, 2.18–The equivalent loads, 2.19–An advantage of matrix methods, 2.20–The critical design load, 2.21–Example, Exercises on chapter 2.

3. **Isostatic beams**   58

    3.1–Introduction, 3.2–Shearing force and bending moments in beams, 3.2.1–Definitions, 3.2.2–A cantilever with an end load, 3.2.3–Sign conventions, 3.2.4–A cantilever carrying an inclined force, 3.2.5–A uniformly loaded cantilever, 3.2.6–A cantilever with several loads, 3.2.7–Simply supported beams, 3.2.8–Examples. 3.3–Relationships for the load, the shearing force and the bending moment, 3.4–Examples, 3.5–Bending stresses in beams, 3.6–Example, 3.7–Generalised Macaulay's method for shearing forces and bending moments, 3.7.1–A beam with point loads, 3.7.2–A beam with uniform loads, 3.8–Example, Exercises on chapter 3.

4. **Bent members and structures**   90

    4.1–Introduction, 4.2–Plane bent members, 4.3–Plane curved members, 4.4–Examples, 4.5–Bent member loaded out of its plane, 4.6–Curved member loaded out of its plane, 4.7–Three pinned arches, 4.8–The local axes of a member, 4.9–Examples, Exercises on chapter 4.

5. **Slopes and deflections**   111

    5.1–Introduction, 5.2–Elastic bending of beams, 5.3–The deflection function, 5.4–The slope-deflection equation, 5.5–Deflection and slope of a cantilever, 5.6–Deflection of a simply supported beam loaded at mid-span, 5.7–A beam carrying a point load unsymmetrically, 5.8–Beam with end moments and midspan load, 5.9–Fixed ended beam carrying a central load, 5.10–Uniformly loaded simply supported beam, 5.11–A uniformly loaded fixed ended beam, 5.12–Treatment of several loads, 5.13–Examples, 5.14–Deflection and moment at an unloaded specific point, 5.15–Examples, 5.16–Treatment of uniform loading, 5.17–Example, 5.18–Further examples, Exercises on chapter 5.

Contents

6 Influence lines   141

6.1–Introduction, 6.2–Influence line for support reactions, 6.3–Influence line for shearing force, 6.4– Example, 6.5–Influence line for bending moments, 6.6– Example, 6.7–The dangerous section, 6.8–The dangerous section for several loads, 6.9–The significance of influence lines, 6.10–Influence lines for pin jointed frames, 6.11–Examples, Exercises on chapter 6.

7 Flexural properties of structures   155

7.1–Introduction, 7.2–The flexural stiffness, 7.3–Examples, 7.4–The flexibility, 7.5–Examples, 7.6–Strain energy in pure bending, 7.7–Examples, 7.8–Strain energy in pure shear, 7.9–The effect of shear stress distribution, 7.10–Example, 7.11–The work equation, 7.12–Examples, 7.13–Plasticity, 7.14–Evaluation of the fully plastic moment, 7.15–Example, 7.16–The shape factor, 7.17–Examples, Exercises on chapter 7.

8 The plastic theory   183

8.1–Introduction, 8.2–The load factor, 8.3–Proportional loading, 8.4–The use of the virtual work equation, 8.5–Collapse of continuous beams, 8.6–Selection of mechanisms, 8.7–Example, 8.8–Design by rigid-plastic theory, 8.9–Example, 8.10–Effect of axial load on Mp, 8.11–Examples, 8.12–Limitation of the plastic theory, 8.13–The yield line theory in plates, 8.14–Rectangular reinforced concrete slabs, 8.15–Example, Exercises on chapter 8.

9 Analysis of structures by the force method   205

9.1–Deflections of structures. 9.2–Relationship between member and joint deflections, 9.3–Examples, 9.4–Deflection of unloaded joints, 9.5–Deflection of a structure with non-linear material characteristics, 9.6–Examples, 9.7–Analysis of hyperstatic frames by the force method, 9.8–Examples, 9.9–Deflection of bent members, 9.10–Examples, 9.11–Assembly and thermal forces in hyperstatic frames, Exercises on chapter 9.

10 Analysis of columns   239

10.1–Introduction, 10.2–The deformed shape of a column, 10.3–Method 1: The use of energy and displacement functions, 10.4–Method 2: The use of stability functions, 10.5–The stability functions, 10.6–The critical load of a propped cantilever, 10.7–The force-rotation curve for a propped cantilever, 10.8–Example, 10.9–A column in double curvature, 10.10–A column in pure sway, 10.11–The general column, 10.12–Examples, 10.13–Deflection of an eccentrically loaded column, 10.14–The stresses in an eccentrically loaded column, 10.15–Example, 10.16–Initially curved pin-ended column, 10.17–The maximum stress in curved pin-ended column, 10.18–Design of pin-ended column, 10.19–Laterally loaded columns, 10.20–Example, Exercises on chapter 10.

11 Pin jointed space structures   266

11.1–Introduction, 11.2–Concurrent forces in space, 11.3–Equilibrium of isostatic space structures, 11.4–Stability of pin jointed space structures, 11.5–Direction cosines of a space member, 11.6– Example, 11.7–The cartesian components of a force, 11.8–Analysis of isostatic space frames, 11.9–Examples, 11.10–Deflections of isostatic space frames, 11.11–Example, 11.12– Relationship between member deformations and joint deflections, 11.13–The joint equilibrium equations, 11.14–Examples, Exercises on chapter 11.

| 12 | **The stiffness method for the analysis of rigidly jointed plane frames** | 294 |

12.1–Introduction, 12.2–The steps required in a linear analysis, 12.3–Relationships between member deformations and joint displacements, 12.4–Examples, 12.5–The components of the member forces, 12.6–Examples, 12.7–Non-linear elastic analysis of frames, 12.8–Elastic instability of frames, 12.9–Solution of homogeneous equations, 12.10–Examples, 12.11–Assembly and thermal forces, 12.12–Examples, Exercises on chapter 12.

Appendix 1 The stability functions 343

Index 353

1

Statically determinate

Pin jointed isostatic plane structures

1.1. Introduction

A structure consists of one or more members joined together and supported by its foundations. The members are made from structural material, such as steel, reinforced concrete, timber etc. A purpose of constructing a structure is to carry loads applied to it either at its joints or elsewhere along its members. These loads are transmitted to the foundations by the members; because of this forces are also developed in these members.

The members and joints as well as the supports of a plane structure all lie in one plane. The forces acting on a plane structure also lie in the same plane. The members of a pin jointed plane structure are connected together by means of frictionless pins. An isostatic pin jointed structure is one which can be analysed by means of the equations of static equilibrium. When analysing a structure, the purpose is to calculate the forces in the members and the reactions at the supports.

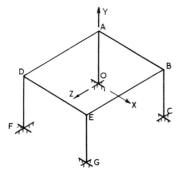

a. A rigidly jointed space structure

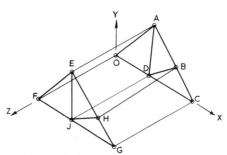

b. A pin jointed space structure

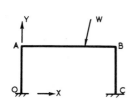

c. A rigidly jointed plane structure

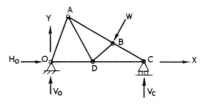

d. A pin jointed plane structure.

Figure 1.1

In reality structures are not plane but three dimensional. The members of a real structure are either welded together or connected by means of plates and bolts or rivets. In the case of reinforced concrete structures, the members are often monolithic with each other. Thus a plane structure is in fact an idealisation of a real three dimensional structure. In *Figure 1.1a* a three dimensional structure is shown. This consists of four columns rigidly jointed to four horizontal beams. *Figure 1.1b* shows a three dimensional pin jointed structure. To simplify the analysis of a structure it is often considered that a part of it is a two dimensional plane structure. For instance the part OABC of the structure in *Figure 1.1a* is shown as a plane structure in *Figure 1.1c*. Similarly the part OABCD of the structure in *Figure 1.1b* is shown as a plane structure in *Figure 1.1d*. Here the members are shown to be connected together by means of small circles. This is the way frictionless pins are represented. A load W is shown acting on this plane structure at joint B. This causes forces to develop in the members which in turn cause the development of reactive force at the supports. In analysing a structure it is often necessary to calculate the support reactions first.

1.2. Representation of supports

A structure is prevented from moving as a rigid body because it is constrained by its supports. There are various types of supports. A frictionless pinned support is represented by a small circle, as shown in *Figure 1.2*. Such a support restrains the structure from translation in any direction but permits rotation about the support. The force p_1, developed in member AB, tries to move end A of the structure upwards and to the right. The reactions H_A and V_A prevent these movements. However, if the support at C is disconnected, the force W at B can rotate the member AB about the support at A.

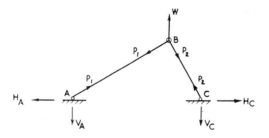

Figure 1.2

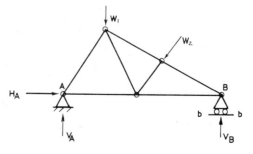

Figure 1.3

Pin Jointed Isostatic Plane Structures

A roller support also permits rotation about the support. It also permits translation in one direction but restrains translation in any other direction. In *Figure 1.3*, the roller at support B permits the structure to move along the plane bb and thus the support offers no reaction in this direction. The roller, however, prevents movement in a direction perpendicular to bb, thus giving rise to the reaction V_B. The support at A is another way of representing a frictionless pin.

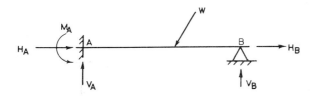

Figure 1.4

Other types of support are shown in *Figure 1.4*. The support at B represents a knife edge which acts exactly as a frictionless pin. The support at A shows that the beam is here built into the support. The beam is therefore prevented from rotation and translation. Three reactions are thus developed at A. H_A and V_A prevent horizontal and vertical translation while the reactive moment M_A prevents rotation of the beam at the support. For a structure to be prevented from moving as a rigid body, it must be constrained by at least three reaction components.

1.3. Equilibrium of coplanar forces

A set of forces acting in a plane are in equilibrium if:
1. The algebraic sum of the projections of all the forces on two perpendicular axes, in the plane, vanish.
2. The algebraic sum of the moments of the same forces about any point in the plane also vanishes. These statements are expressed mathematically as:

$$\left. \begin{array}{l} \Sigma p_{xi} = 0 \\ \Sigma p_{yi} = 0 \\ \text{and} \quad \Sigma M_i = 0 \end{array} \right\} \quad (1.1)$$

where p_{xi} and p_{yi} are the projections of a force p_i on two perpendicular axes such as the cartesian X-Y axes and M_i is the moment of the same force about a chosen point in the plane. The projection of a force is also called its component. In *Figure 1.5* the force p_i, represented by the vector OA, is resolved into two components.

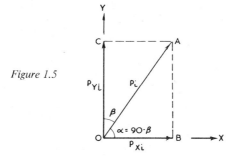

Figure 1.5

p_{xi}, represented by the vector OB, is the component of p_i in the X direction, while p_{yi}, represented by the vector OC, is the component of the force in the Y direction. Force p_i is called the resultant of the two components.

It is evident from *Figure 1.5* that

$$p_{xi} = p_i \cos \alpha$$
and
$$p_{yi} = p_i \cos \beta = p_i \sin \alpha \quad (1.2)$$

If a particle, such as B in *Figure 1.6*, is in equilibrium under the forces p_1, p_2, p_3 etc. then the first two of the equation of equilibrium (1.1) can be used to relate the components of these forces, thus

$$\Sigma p_{xi} = p_1 \cos \alpha_1 + p_2 \cos \alpha_2 + p_3 \cos \alpha_3 = 0$$
$$\Sigma p_{yi} = p_1 \cos \beta_1 + p_2 \cos \beta_2 + p_3 \cos \beta_3 = 0 \quad (1.3)$$

Notice the angles α and β between a force and the positive X and Y axes are always measured in an anticlockwise direction, starting from an axis and finishing on the force. These positive angular directions are indicated in *Figures 1.6 a* and *b* by arrows.

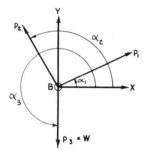

a. Forces at B and their positive angles with X axis

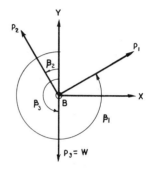

b. Positive angles between Y axis and the forces at B

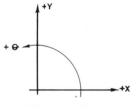

c. Sign convention

Figure 1.6

The last equation of equilibrium (1.1) ensures that the resultant of a set of forces in equilibrium does not give rise to a couple. The forces p_1, p_2, p_3 in *Figure 1.6* all pass through point B and thus the moment of each force about point B is equal to zero. However, to ensure that these forces are in equilibrium, the algebraic sum of their moments about any other point, such as C in *Figure 1.7*, must also equal zero,

i.e. $\quad p_1 \times Cf + p_2 \times Ce - p_3 \times Cd = 0$

Here Cf, Ce and Cd are the perpendicular distances between the point C and the lines of action of the forces p_1, p_2 and p_3 respectively. It is also noticed that the moments of p_1 and p_2 about C are both anticlockwise and taken as positive. The moment of p_3 about C is clockwise and is negative.

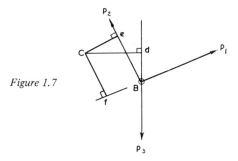

Figure 1.7

1.4. Structural equilibrium

When a structure is in equilibrium, the following conditions are satisfied.
1. The forces applied externally to the structure are in equilibrium with the reactions developed at the supports. Here the structure as a whole is considered as a rigid body.
2. Each part of the structure is in equilibrium.
3. Each joint in the structure is in equilibrium.
4. Each member of the structure is also in equilibrium.

These rules are used when a structure is being analysed.

1.5. Calculation of reactions

To calculate the reaction components at the supports of a pin jointed plane structure, the equilibrium of either a part or the whole of the structure is considered. If a sufficient number of such equations can be developed to calculate all the reaction components then the structure is said to be externally isostatic.

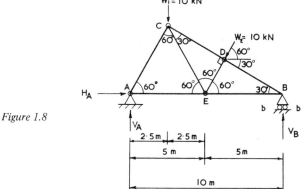

Figure 1.8

As an example, consider the pin jointed plane structure shown in *Figure 1.8* which has three unknown reactions H_A, V_A and V_B. The vertical load W_1 has no horizontal component while the horizontal component of the load W_2 is $W_2 \cos 60$ acting to the left. Thus when resolving the forces and reactions horizontally, only H_A and W_2 contribute to the equation of equilibrium, thus

$$H_A - W_2 \cos 60 = 0$$

which gives $\quad H_A = W_2 \cos 60 = 10 \times 0.5 = 5 \text{kN}$

A second equation of equilibrium, for the whole structure, is developed by taking moments of the external forces and reactions about point B. The sum of these moments must vanish. Now H_A and V_B pass through point B and therefore have no moments about B. The moment of V_A about B is $-V_A \times 10$ and is clockwise and that of W_1 is $W_1 \times 7.5$ which is anticlockwise. The moment of W_2 about B is also anticlockwise. The perpendicular distance from B to the line of action of W_2 is equal to BD which is equal to EB sin 60 = $5 \times 0.5\sqrt{3}$. Thus the moment of W_2 about B is $5 \times 0.5\sqrt{3} \times W_2$. Substituting these values in the last of equations (1.1) we obtain

$$-V_A \times 10 + W_1 \times 7.5 + 5 \times 0.5\sqrt{3} \times W_2 = 0$$

i.e. $\quad -10V_A + 10 \times 7.5 + 10 \times 5 \times 0.5\sqrt{3} = 0$

∴ $\quad V_A = 11.83 \text{ kN}$

Finally resolving the forces vertically and using the second of equations (1.1)

$$V_A + V_B - W_1 - W_2 \times 0.5\sqrt{3} = 0$$

i.e. $\quad 11.83 + V_B - 10 - 10 \times 0.5\sqrt{3} = 0$

∴ $\quad V_B = 6.83 \text{ kN}$

All the reaction components of the plane structure were obtained by using the three equations of equilibrium. This structure is therefore externally isostatic. In general, because there are only three equations of equilibrium of coplanar forces, it is only possible to calculate the values of three reaction components. For structures with more than three reactions, other methods have to be used to determine the extra unknowns.

1.6. Member forces

As stated earlier, by applying external forces to a structure, forces are developed in the members. For instance, when the force L is applied to joint B of the structure in Figure 1.9, forces p_1 and p_2 are developed in members 1 and 2. The component of L in the direction of AB tries to pull member AB in the direction away from A. A tensile force p_1 is developed in the member that balances this component of L at B. Similar the component of the load L resolved in the direction of BC presses against member BC and a compressive force is developed in this member.

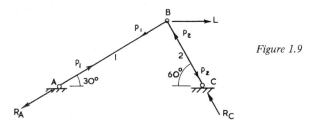

Figure 1.9

The equations of static equilibrium can be used to calculate the forces in the members of an isostatic pin jointed structure. Before doing this it is often necessary to calculate the support reactions first. One method for calculating member forces is known as the method of joints. This applies the equations of equilibrium to the joints of the structure; thus making use of the rule that once a structure is in equilibrium then each joint in it is also in equilibrium. The procedure is thus to resolve the forces acting on a joint in two perpendicular directions and develop two

Pin Jointed Isostatic Plane Structures

equations of static equilibrium for the joint. These two equations can only be used to calculate two unknown forces. Now the forces applied externally to the joints are known and thus only two unknown member forces can be calculated at each joint.

To analyse an isostatic pin jointed structure in a systematic manner, all the unknown reactions and member forces are assumed to be positive. Thus the reaction components are assumed to be acting in the same sense as the positive X and Y axes. Member forces are assumed to be tensile and positive. The result of the analysis will decide the validity of these initial assumptions. Thus a reaction with a negative sign acts in a direction opposite to its assumed direction. A negative force in a member indicates that the force in that member is in fact compressive.

As an example, consider the analysis of the pin jointed plane structure shown in *Figure 1.10*. All the reactions and member forces are first assumed to be positive as shown in *Figure 1.10a*. To begin with, the support reactions are calculated.

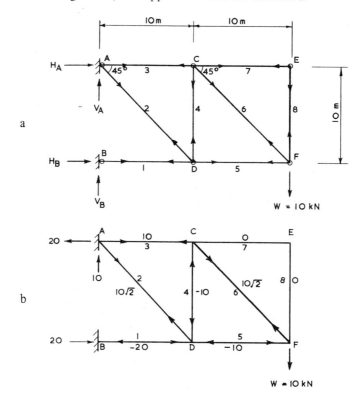

Figure 1.10

For the structure as a shole, taking moments of the external force W and the reactions at A and B about point A, it is noticed that reactions V_A, H_A and V_B pass through A and therefore have no moments about this point. The only other forces left are therfore W and H_B; thus

$$H_B \times 10 - W \times 20 = 0$$

Hence $\qquad H_B = 2W = 20 \text{ kN}$

Resolving the forces horizontally

$$H_A + H_B = 0$$

Thus $\qquad H_A = -H_B = -20 \text{ kN}$

This result indicates that the horizontal reaction H_A acts to the left.

Resolving the forces vertically

$$V_A + V_B - W = 0$$

i.e. $\quad V_A + V_B = W = 10 \text{ kN} \quad (1.4)$

Here both V_A and V_B are unknowns. It is therefore necessary to derive a further equilibrium equation. However, the equilibrium of the structure as a whole does not give any further information about the structure. For this reason a different equilibrium rule must be used to derive a further equation. Now since each joint in the structure is in equilibrium, joint B is selected. Resolving the forces acting at B vertically, it is found that $V_B = 0$. Substituting this value in equation (1.4) gives

$$V_A = W = 10 \text{ kN}$$

The sign of V_A indicates that it is acting upwards.

To calculate the member forces, the equilibrium of each joint is now considered. Joints with not more than two unknowns are considered first. Joint B has only one unknown force which is the force p_1 in member 1.

Resolving the forces at B horizontally

$$H_B + p_1 = 0$$

Thus $\quad p_1 = -H_B = -20 \text{ kN}$

The negative sign indicates that p_1 is a compressive force.

Joint A is considered next. The unknown forces acting at A are p_2 and p_3. Resolving the forces at A vertically

$$V_A - p_2 \times 1/\sqrt{2} = 0$$

Thus $\quad p_2 = \sqrt{2} \times V_A = 10\sqrt{2} \text{ kN}$

Resolving the forces at A horizontally

$$H_A + p_3 + p_2 \times 1/\sqrt{2} = 0$$

Thus $\quad p_3 = -H_A - p_2/\sqrt{2} = 20 - 10\sqrt{2}/\sqrt{2} = 10 \text{ kN}$

At joint C there are three unknown forces p_4, p_6 and p_7, while at joint D only p_4 and p_5 are unknown. Hence moving to joint D and resolving the forces vertically

$$p_4 + p_2 \times 1/\sqrt{2} = 0$$

∴ $\quad p_4 = -p_2/\sqrt{2} = -10\sqrt{2}/\sqrt{2} = -10 \text{ kN}$

Resolving the forces at D horizontally

$$p_1 - p_5 + p_2 \times 1/\sqrt{2} = 0$$

∴ $\quad p_5 = p_1 + p_2/\sqrt{2} = -20 + 10\sqrt{2}/\sqrt{2} = -10 \text{ kN}$

Now moving to C and resolving the forces there vertically

$$p_4 + p_6 \times 1/\sqrt{2} = 0$$

∴ $\quad p_6 = -\sqrt{2}\, p_4 = -\sqrt{2} \times (-10) = 10\sqrt{2} \text{ kN}$

Resolving the forces horizontally at C

$$p_7 - p_3 + p_6 \times 1/\sqrt{2} = 0$$

∴ $\quad p_7 = p_3 - p_6/\sqrt{2} = 10 - 10\sqrt{2}/\sqrt{2} = 0$

Moving to E and resolving the forces there vertically, we obtain $p_8 = 0$
All the forces are now calculated. These are shown in *Figure 1.10b* with their correct signs.

To understand modern methods of structural analysis it is useful to train the reader to express the member forces in terms of the external forces and in matrix form. In the case of the structure in *Figure 1.10* this takes the form

$$\begin{bmatrix} p_1 \\ p_2 \\ p_3 \\ p_4 \\ p_5 \\ p_6 \\ p_7 \\ p_8 \end{bmatrix} = \begin{bmatrix} -2 \\ \sqrt{2} \\ 1 \\ -1 \\ -1 \\ \sqrt{2} \\ 0 \\ 0 \end{bmatrix} [W] \qquad (1.5)$$

Here $\{p_1 \; p_2 \; p_3 \; p_4 \; p_5 \; p_6 \; p_7 \; p_8\} = \mathbf{P}$ is a column vector that lists all the member forces. This vector can be simply written as $\mathbf{P}$. The column vector $\{-2 \; \sqrt{2} \; 1 \; -1 \; -1 \; \sqrt{2} \; 0 \; 0\}$ is known as the load transformation matrix $\mathbf{B}$. Finally the column vector $\{W\} = \mathbf{W}$ is the external load matrix. Here because there is only one external load acting this matrix consists of only one element. The matrix equation (1.5) can be written as

$$\mathbf{P} = \mathbf{B} \mathbf{W} \qquad (1.6)$$

[In this book matrices and vectors are printed in bold type.]

1.7. The method of sections

Another method for the analysis of a pin jointed structure is known as the method of sections. This makes use of the rule that when a structure is in equilibrium then each part of the structure is also in equilibrium. This method is useful especially when only a few of the member forces are required to be determined. The procedure is to cut the structure into two parts by an imaginary section and then consider the equilibrium of the resulting parts of the structure. Because there are only three equations of equilibrium, cutting a structure by a section can only determine three unknown forces. If more than three forces are required to be determined then the structure is cut by more than one section.

As an example, consider the pin jointed plane frame shown in *Figure 1.11* in which it is required to calculate the member forces $p_1, p_2, p_3, p_4, p_5, p_6$ and p_7.

The reactions at the supports are first calculated by considering the equilibrium of the whole frame. Resolving the forces horizontally we obtain $H_A = 0$.

Taking moments about point J

$$W \times 2 - V_A \times 4 = 0$$

∴
$$V_A = W/2$$

By resolving the forces vertically, it is found that V_J also equals $W/2$. To calculate the forces in members 1, 2 and 3 these are cut by an imaginary section a-a as shown in *Figure 1.11a*. The part of the frame to the left of this section is shown in *Figure 1.11b*. This part is in equilibrium under the reactive force V_A and the member forces $p_1, p_2,$ and p_3. These latter forces are now acting externally on the part ABCD, for which taking moments about D

$$p_3 \times 1.5 - V_A \times 1 = 0$$

Thus
$$p_3 = V_A/1.5 = W/3 \text{ (tensile)}$$

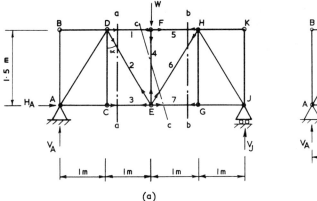
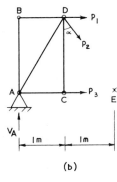
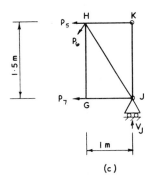
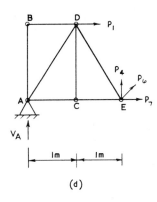

Figure 1.11

From the vertical equilibrium of ABCD

$$V_A - p_2 \cos \alpha = 0$$

∴ $\quad p_2 = V_A/\cos \alpha = 0.6W$ (tensile)

Both p_2 and p_3 pass through point E. Taking moments about this point gives

$$-V_A \times 2 - p_1 \times 1.5 = 0$$

∴ $\quad p_1 = -2V_A/1.5 = -W/1.5$ (compressive)

Similarly, the forces in members 5, 6 and 7 are calculated by cutting these members by section b-b as shown in *Figure 1.11a*. The part of the frame to the right of this section is shown in *Figure 1.11c*. Taking moments about point H, resolving the forces vertically and then taking moments about point E, it is found that

$$p_5 = -W/1.5 \text{ (compressive)}$$
$$p_6 = 0.6W \text{ (tensile)}$$

and $\quad p_7 = W/3$ (tensile)

Because the reaction $H_A = 0$, the support at A acts as a roller. Thus the frame is symmetrical and loaded symmetrically. For this reason the forces in the members are also symmetrical, and $p_1 = p_5$, $p_2 = p_6$ and $p_3 = p_7$.

To calculate the force in member 4, the frame is now cut by section c-c as shown in *Figure 1.11a*. The part of the frame to the left of this section is shown in *Figure 1.11d*. Resolving the forces acting on this part vertically

$$V_A + p_4 + p_6 \cos \alpha = 0$$

∴
$$p_4 = -V_A - p_6 \cos \alpha = -0.5W - 0.6W \times (1.5/1.8)$$

i.e.
$$p_4 = -W \text{ (compressive)}$$

1.8. Combination of methods of joint and section

Sometimes, it may be easier to use the method of section in conjunction with the method of joints. For instance, in the frame of *Figure 1.11*, while the forces p_1, p_2, p_3, p_5, p_6 and p_7 are calculated by the method of section, the force p_4 can be calculated by considering the vertical equilibrium of joint F. This directly gives $p_4 = -W$.

As another example consider the pin jointed frame shown in *Figure 1.12* where it is required to calculate the forces in members 1, 2, 3 and 4. By resolving the external forces and the reactions horizontally, it is found that $H_K = 0$. Taking moments about point J

$$V_K \times 4a - W \times a - W \times 2a - W \times 3a = 0$$

∴
$$V_K = 3W/2 = V_J$$

To calculate the forces in members 1, 2, 3 and 4, two sections and one joint are considered. The part of the structure to the right of section A-A is shown in *Figure 1.12b*. For this taking moments about point G

$$aV_K + p_1 \times 2h = 0$$

Thus
$$p_1 = -aV_K/2h = -3aW/4h \text{ (compressive)}$$

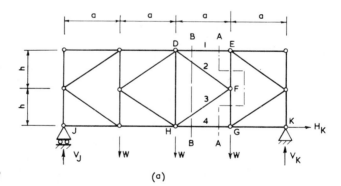

Figure 1.12   (a)

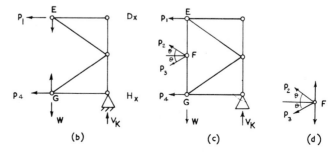

Similarly by taking moments about point E, it is found that $p_4 = 3aW/4h$ which is a tensile force.

Next consider the part of the structure to the right of section B-B, *Figure 1.12c*, and resolve the forces vertically

$$V_K + p_2 \sin \theta - p_3 \sin \theta - W = 0 \qquad (1.7)$$

Now using the method of joints at F, *Figure 1.12d*, and resolving the forces horizontally

$$-p_2 \cos \theta - p_3 \cos \theta = 0$$

i.e. $\qquad\qquad\qquad p_3 = -p_2 \qquad\qquad\qquad\qquad (1.8)$

Substituting for p_3 from equation (1.8) into equation (1.7), it is found that

$$p_2 = -0.25\ W/\sin \theta \quad \text{(compressive)}$$

Thus $\qquad\qquad p_3 = 0.25\ W/\sin \theta \quad \text{(tensile)}$

1.9. Unstable, isostatic and hyperstatic frames

So far it has been stated that isostatic plane frames can be analysed using the equations of static equilibrium. Before doing this it is necessary to find out whether the frame is isostatic. A plane frame is isostatic if it has a sufficient number of members and joints, arranged in a suitable manner so that it is just stable. For instance, the triangular frame shown in *Figure 1.13a* is just stable because its shape cannot be changed by a force without altering the length of its members. On the other hand the frame shown in *Figure 1.13b* is unstable because its shape can be changed from the position ACDB to AC'D'B with a very small force acting at C or D. This frame can be rendered stable and also isostatic if a member such as CB (*Figure 1.13c*) is added to it. Finally the frame shown in *Figure 1.13d* is 'over stiff' or 'over stable' because any one of its members can be removed without it becoming unstable. This frame is said to be hyperstatic. The equations of static equilibrium, on their own, are not sufficient to analyse such a frame.

A stable frame can be developed by starting with a triangle, with three members and three joints, and then successively connecting each additional joint by two new members. Thus starting with the stable triangular frame ABC shown in *Figure 1.13e*, the stable frame shown in *Figure 1.13c* is developed by connecting the new joint D by two members CD and BD. This process may be continued as shown in *Figure 1.13f*. Joint E is connected to ABCD by another two members DE and EB and then joint F is added by means of members EF and BF. Thus for a stable isostatic frame

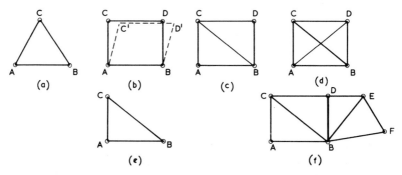

Figure 1.13

with a total of m members and j joints the first three members and joints belong to the initial triangular frame. Such a frame will therefore have $m - 3$ new members and $j - 3$ new joints. Now since the number of new members are twice as many as the number of new joints, it follows that

which gives
$$m - 3 = 2(j - 3)$$
$$m = 2j - 3 \tag{1.9}$$

In developing an isostatic frame, care must be taken to arrange the members and joints so that each part of the frame is stable. For instance, the frame in *Figure 1.14a* has nine members and six joints and satisfies equation (1.9). It is stable and internally isostatic (reactions are excluded here). The frame shown in *Figure 1.14b* also has

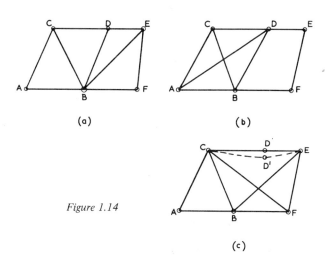

Figure 1.14

nine members and six joints but it is not internally isostatic. Here joint D is connected to the initial triangle ABC by three new members instead of two, part ABCD is therefore hyperstatic. On the other hand the two joints E and F are connected to ABCD by three members instead of four. Thus part DEFB is unstable. In *Figure 1.14c*, joints CDE are on a straight line and do not form a triangle. A small force can move joint D to D'. This part of the frame is therefore unstable. The rest of the frame on the other hand is stable.

1.10. Consideration of members and supports

It was pointed out that when using the equations of equilibrium to calculate the reactions and member forces by the method of joints, it is possible to derive two equations at each joint. Thus for a frame with j joints there are $2j$ available equations. If a frame is supported with r reaction components and has m members, then to calculate all r unknown components and m member forces, $m + r$ must be equal to $2j$

i.e.
$$m + r = 2j \tag{1.10}$$

It was stated earlier that to restrain a frame as a whole three reaction components are necessary, thus $r = 3$ at least. For instance, the frame shown in *Figure 1.15a* has six joints ($j = 6$), nine members ($m = 9$) and three reaction components ($r = 3$). For this frame equation (1.10) is satisfied and the structure is isostatic. The three reaction components can thus be calculated. Removing the supports results in the

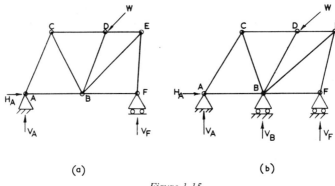

(a)   (b)

Figure 1.15

frame shown in *Figure 1.14a* and equation (1.9) will be satisfied, thus the member forces can be calculated and the frame is internally isostatic. The frame in *Figure 1.15b* has $m = 9$, $r = 4$ and $j = 6$. This frame does not satisfy equation (1.10). The reason is that the number of reactions is one too many and the frame is externally hyperstatic. Removing the supports once again results in the frame shown in *Figure 1.14a* which satisfies equation (1.9). Thus the frame is internally isostatic. Once the extra reaction (say V_B) is calculated, by methods given later, the member forces can be determined by equations of static equilibrium. The unknown force V_B is called a redundant force and for the time being the other reactions and the member forces can only be calculated in terms of V_B.

1.11. Examples

Example 1. Calculate the member forces in the frame shown in *Figure 1.16*. Hence write down equations **P = B W**.

Answer: In this frame $m = 9$, $r = 3$ and $j = 6$, thus equation (1.10) is satisfied. Removing the reaction components, it is found that equation (1.9) is also satisfied. The frame is thus isostatic both externally and internally.

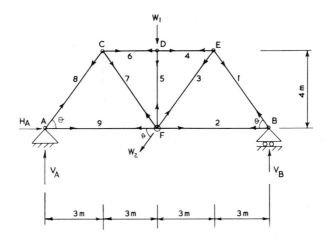

Figure 1.16

Pin Jointed Isostatic Plane Structures

From the horizontal equilibrium of the frame as a whole, we obtain

$$H_A - W_2 \cos \theta = 0$$
$$\therefore H_A = W_2 \cos \theta = 0.6 W_2$$

Taking moments of the external loads and the reactions about B

$$-V_A \times 12 + W_1 \times 6 + W_2 \sin \theta \times 6 = 0$$
$$\therefore V_A = 6W_1/12 + 6W_2 \times 0.8/12 = 0.5 W_1 + 0.4 W_2$$

Resolving the external loads and the reactions vertically

$$V_A + V_B - W_1 - W_2 \sin \theta = 0$$
$$\therefore V_B = W_1 + W_2 \times 0.8 - V_A = W_1 + 0.8 W_2 - 0.5 W_1 - 0.4 W_2$$
$$V_B = 0.5 W_1 + 0.4 W_2$$

Having calculated the reactions, the values of the member forces can now be calculated.

At B, resolving vertically

$$V_B + p_1 \sin \theta = 0$$
$$p_1 = -V_B/\sin \theta = -(0.5 W_1 + 0.4 W_2)/\sin \theta$$
$$p_1 = -0.5 W_1/0.8 - 0.4 W_2/0.8$$
$$p_1 = -0.625 W_1 - 0.5 W_2 \text{ (compression)}$$

At B, resolving horizontally

$$-p_2 - p_1 \cos \theta = 0$$
$$p_2 = -p_1 \times 0.6 = 0.625 W_1 \times 0.6 + 0.5 W_2 \times 0.6$$
$$p_2 = 0.375 W_1 + 0.3 W_2 \text{ (tension)}$$

At E, resolving vertically

$$-p_1 \sin \theta - p_3 \sin \theta = 0$$
$$\therefore p_3 = -p_1 = 0.625 W_1 + 0.5 W_2 \text{ (tension)}$$

At E, resolving horizontally

$$-p_4 - p_3 \cos \theta + p_1 \cos \theta = 0$$
$$\therefore p_4 = -p_3 \times 0.6 + p_1 \times 0.6 = -0.75 W_1 - 0.6 W_2 \text{ (compression)}$$

At D, resolving vertically

$$-W_1 - p_5 = 0$$
$$p_5 = -W_1 \text{ (compression)}$$

At D, resolving horizontally

$$p_4 - p_6 = 0$$
$$\therefore p_6 = p_4 = -0.75 W_1 - 0.6 W_2 \text{ (compression)}$$

At A, resolving vertically

$$V_A + p_8 \sin \theta = 0$$
$$\therefore p_8 = -V_A/\sin \theta = -0.625 W_1 - 0.5 W_2 \text{ (compression)}$$

At A, resolving horizontally

$$H_A + p_8 \cos \theta + p_9 = 0$$

$$\therefore \quad p_9 = -H_A - p_8 \cos \theta = -0.6W_2 + 0.625W_1 \times 0.6 + 0.5W_2 \times 0.6$$

$$p_9 = 0.375W_1 - 0.3W_2$$

Thus if $0.375W_1 > 0.3W_2$ i.e. if $W_1 > 0.3W_2/0.375$ the force in member 9 will be tensile. Otherwise it will be a compressive force.

At C, resolving the forces vertically

$$-p_8 \sin \theta - p_7 \sin \theta = 0$$

$$\therefore \quad p_7 = -p_8 = 0.625W_1 + 0.5W_2 \quad \text{(tension)}$$

Equations **P = B W** are

$$\begin{bmatrix} p_1 \\ p_2 \\ p_3 \\ p_4 \\ p_5 \\ p_6 \\ p_7 \\ p_8 \\ p_9 \end{bmatrix} = \begin{bmatrix} -0.625 & -0.5 \\ 0.375 & 0.3 \\ 0.625 & 0.5 \\ -0.75 & -0.6 \\ -1 & 0 \\ -0.75 & -0.6 \\ 0.625 & 0.5 \\ -0.625 & -0.5 \\ 0.375 & -0.3 \end{bmatrix} \begin{bmatrix} W_1 \\ W_2 \end{bmatrix}$$

Example 2. Is the frame shown in *Figure 1.17* isostatic? Calculate the forces in members 1, 2, 3 and 4.

Answer: $\quad m = 19, r = 3 \text{ and } j = 11$

$$m + r = 19 + 3 = 22 = 2j$$

Therefore equation (1.10) is satisfied and all the member forces and reaction components can be calculated. The frame is isostatic. Considering the part of the frame to the right of section A-A, as shown in *Figure 1.17b*, and taking moments about point H

$$p_1 \times 6 + W \times 4 \sin \theta = 0$$

The lever arm of the force W with respect to H is $HM = 4 \sin \theta$.

$$\therefore \quad p_1 = -4W \sin \theta / 6 = -4W \times 0.6/6 = -0.4W$$

The force p_1 can also be obtained as follows.

The horizontal component of the force W passes through point H and thus has no moment about this point. The vertical component of this force is $W \sin \theta$ and the lever arm of this component with respect to H is equal to $HL = 4m$. The moment of the vertical component of W about H is thus $W \sin \theta \times 4$. The moment of p_1 about H is $6p_1$, thus

$$6p_1 + W \sin \theta \times 4 = 0$$

$$\therefore \quad p_1 = -W \times 0.6 \times 4/6 = -0.4W.$$

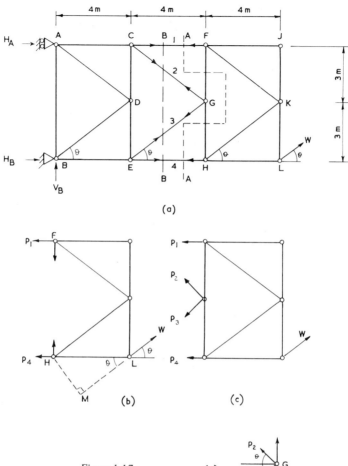

Figure 1.17

Similarly, taking moments about point F

$$-p_4 \times 6 + W \sin \theta \times 4 + W \cos \theta \times 6 = 0$$

$$\therefore \quad p_4 = 4W \times 0.6/6 + 6W \times 0.8/6 = 1.2W$$

Here both components of W have moments about point F.

Considering next the part of the frame to the right of section B-B, *Figure 1.17c*, and resolving the forces vertically

$$p_2 \sin \theta + W \sin \theta - p_3 \sin \theta = 0$$

i.e. $\qquad p_2 + W - p_3 = 0 \qquad (1.11)$

Finally, from the horizontal equilibrium at joint G, *Figure 1.17d*

$$p_2 \cos \theta + p_3 \cos \theta = 0$$

$$\therefore \quad p_3 = -p_2$$

Substituting for p_3 in equation (1.11)

$$p_2 + W + p_2 = 0$$

∴ $$p_2 = -0.5W$$

and $$p_3 = 0.5W$$

Example 3. Are the frames in *Figure 1.18* isostatic?

Answer: The frame shown in *Figure 1.18a* has $m = 5, j = 4$ and $r = 3$. These satisfy equation (1.10). When the reaction components are removed, the frame with $m = 5$ and $j = 4$ will satisfy equation (1.9). The frame is therefore stable and internally isostatic. With $2j = 8$ there are sufficient equations to calculate the reaction components and the member forces.

The frame shown in *Figure 1.18b* has $m = 4, r = 3$ and $j = 4$, thus $m + r = 7 < 2j$. This frame is unstable because it has not a sufficient number of members.

The stability of the frame shown in *Figure 1.18b* can be restored either by an extra reaction component as in *Figure 1.18c* or by an extra member as in *Figure 1.18 a* and *d*. The frame in *Figure 1.18c* satisfies equation (1.10) but not equation (1.9). Nevertheless with $2j = 8$, there are a sufficient number of equilibrium equations to solve for four unknown reaction components and four member forces. This frame is therefore isostatic. Frame d satisfies equation (1.9) and (1.10) and is therefore isostatic externally and internally.

The frame shown in *Figure 1.18e* has $m = 4, r = 5$ and $j = 4$ and $m + r = 9$ which is more than $2j$. There is an insufficient number of equations to solve for the nine unknowns. This frame is hyperstatic.

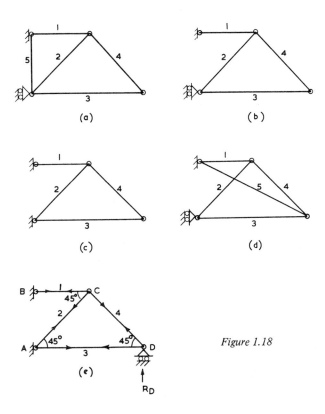

Figure 1.18

Example 4. In the hyperstatic frame shown in *Figure 1.18e* (a) select the vertical reaction R_D as the redundant force and calculate the member forces in terms of R_D; (b) select the force p_4 in member 4 as the redundant force and calculate the member forces in terms of p_4.

Note: Often when a structure is hyperstatic, the redundant forces can be selected in more than one way. In this question two such ways are suggested. Once a redundant force is selected it is often useful, as will be explained later, to express the member forces in terms of this force.

Answer: (a) R_D is redundant.

Resolve forces at D vertically

$$R_D + p_4 \times 1/\sqrt{2} = 0$$

$$\therefore p_4 = -\sqrt{2} \times R_D.$$

Resolve forces at D horizontally

$$-p_3 - p_4 \times 1/\sqrt{2} = 0$$

$$\therefore p_3 = -p_4/\sqrt{2} = R_D$$

Resolving forces at C vertically

$$(-p_2 \times 1/\sqrt{2}) - (p_4 \times 1/\sqrt{2}) = 0$$

$$\therefore p_2 = -p_4 = \sqrt{2} \times R_D$$

Resolving forces at C horizontally

$$-p_1 - (p_2 \times 1/\sqrt{2}) + (p_4 \times 1/\sqrt{2}) = 0$$

$$\therefore p_1 = -2R_D$$

The answers can be expressed in matrix form as

$$\begin{bmatrix} p_1 \\ p_2 \\ p_3 \\ p_4 \end{bmatrix} = \begin{bmatrix} -2 \\ \sqrt{2} \\ 1 \\ -\sqrt{2} \end{bmatrix} \begin{bmatrix} R_D \end{bmatrix}$$

(b) p_4 is redundant.

Resolving forces at D horizontally

$$-p_3 - p_4 \times 1/\sqrt{2} = 0$$

$$\therefore p_3 = -p_4/\sqrt{2}$$

Resolving forces at C vertically

$$(-p_2 \times 1/\sqrt{2}) - (p_4 \times 1/\sqrt{2}) = 0$$

$$\therefore p_2 = -p_4$$

Resolving forces at C horizontally

$$-p_1 - (p_2 \times 1/\sqrt{2}) + (p_4 \times 1/\sqrt{2}) = 0$$

$$\therefore p_1 = \sqrt{2} \times p_4$$

and

$$\begin{bmatrix} p_1 \\ p_2 \\ p_3 \\ p_4 \end{bmatrix} = \begin{bmatrix} \sqrt{2} \\ -1 \\ -1/\sqrt{2} \\ 1 \end{bmatrix} [p_4]$$

Exercises on Chapter 1

1. Calculate the forces in members AB, CB and AD in the pin jointed frame shown in *Figure 1.19*.

Ans. $p_{AB} = -4.46W$, $p_{CB} = 3.87W$, $p_{AD} = 2.45W$.

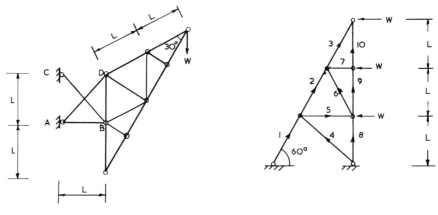

Figure 1.19   Figure 1.20

2. Calculate the forces in members 2, 3, 6, 7, 9 and 10 in the pin jointed frame shown in *Figure 1.20* (a) by the method of joint resolution, (b) by the method of sections when $W = 10$ kN.

Ans. $p_2 = -30, p_3 = -20, p_6 = 10, p_7 = -10, p_9 = 17.32, p_{10} = 17.32$ kN.

3. With $W = 10$ kN calculate the force in member AB of the pin jointed frame shown in *Figure 1.21*.

Ans. 31.38 kN.

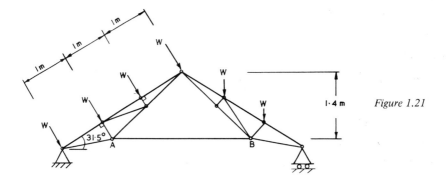

Figure 1.21

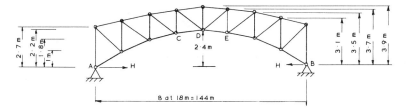

Figure 1.22

4. The horizontal thrust at A and B of the bridge shown in *Figure 1.22* is H. Calculate the member forces in terms of H and tabulate the results in the matrix from $\mathbf{P} = \mathbf{B}\mathbf{W}$. For $H = 10$ kN what is the largest compressive force in the bridge? Make use of the symmetry of the structure.

Ans. 24.7 kN in members CD and DE.

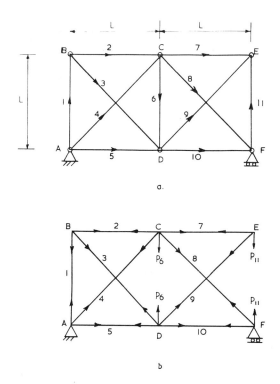

Figure 1.23

5. The forces in the members of the pin jointed frame shown in *Figure 1.23* can be expressed in terms of the forces in the redundant members p_6 and p_{11}. Tabulate these in the matrix form $\mathbf{P} = \mathbf{B}\mathbf{W}$. If $p_6 = 10$ kN and $p_{11} = 20$ kN what is the force in member 1?

Ans. $p_1 = -10$ kN.

Pin Jointed Isostatic Plane Structures

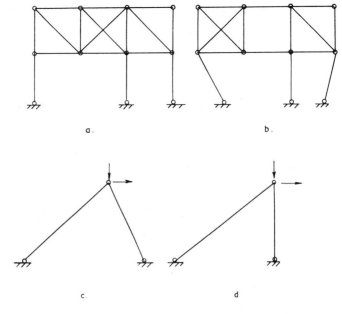

Figure 1.24

6. Indicate, giving reasons, whether the pin jointed frames in *Figure 1.24* are stable, isostatic or hyperstatic.

2
Structural properties

2.1. Direct stress in a member

Consider a pin ended member AB of length L and cross sectional area A as shown in *Figure 2.1*. When a tensile force p acts on the member, it is assumed that this force is evenly distributed over the cross sectional area of the member. If the member is cut somewhere such as D, then equal forces p are required to be applied at D, as shown in *Figure 2.1b*, so that the two parts AD and DB are kept in equilibrium. The member is thus under a tensile force p throughout. The intensity of the force acting on the unit cross sectional area of the member is known as the stress in the member. The tensile stress is thus given by

$$\sigma = p/A \qquad \text{kN/mm}^2 \qquad (2.1)$$

When the forces p, acting on the member, are reversed in direction at each end they tend to compress the member and this produces a compressive stress therein. Both tensile and compressive stresses are referred to as direct stresses.

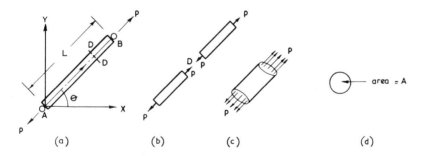

Figure 2.1

Consider a pin jointed structure with N members $1, 2, \ldots N$. Let the forces in these members be $p_1, p_2, \ldots p_i \ldots p_N$ and let the cross sectional areas of the members be $A_1, A_2, \ldots A_i \ldots A_N$. The stresses $\sigma_1, \sigma_2, \ldots \sigma_i \ldots \sigma_N$ in the members are given by equation (2.1) as

$$\left.\begin{array}{l} \sigma_1 = p_1/A_1 \\ \sigma_2 = p_2/A_2 \\ \quad\vdots \\ \sigma_i = p_i/A_i \\ \quad\vdots \\ \sigma_N = p_N/A_N \end{array}\right\} \qquad (2.2)$$

These equations can be written in matrix form as

$$\begin{bmatrix} \sigma_1 \\ \sigma_2 \\ \vdots \\ \sigma_i \\ \vdots \\ \sigma_N \end{bmatrix} = \begin{bmatrix} 1/A_1 & & & & 0 \\ & 1/A_2 & & & \\ & & \ddots & & \\ & & & 1/A_i & \\ & & & & \ddots \\ 0 & & & & & 1/A_N \end{bmatrix} \begin{bmatrix} p_1 \\ p_2 \\ \vdots \\ p_i \\ \vdots \\ p_N \end{bmatrix} \quad (2.3a)$$

or simply
$$\boldsymbol{\sigma} = \boldsymbol{\alpha} \mathbf{P} \quad (2.3b)$$

where $\boldsymbol{\sigma} = \{\sigma_1\ \sigma_2\ \ldots\ \sigma_i\ \ldots\ \sigma_N\}$ is a column vector of member stresses, $\mathbf{P} = \{p_1, p_2, \ldots p_i \ldots p_N\}$ is the column vector of member forces and the diagonal matrix

$$\boldsymbol{\alpha} = \begin{bmatrix} 1/A_1 & & & & 0 \\ & 1/A_2 & & & \\ & & \ddots & & \\ & & & 1/A_i & \\ & & & & \ddots \\ 0 & & & & & 1/A_N \end{bmatrix} \quad (2.4)$$

contains the reciprocal of the member cross sectional areas as elements on the leading diagonal. This matrix is square, i.e. it has the same number of rows and columns. Once the forces in the members are calculated, by the methods given in the last chapter, the stresses in the members are calculated using equation (2.3). Alternatively if the limiting stresses in the members are $\sigma_{w1}, \sigma_{w2}, \ldots \sigma_{wi} \ldots \sigma_{wN}$, then the required area of each member is obtained from equation (2.1) as

$$\left.\begin{array}{l} A_1 = p_1/\sigma_{w1} \\ A_2 = p_2/\sigma_{w2} \\ A_i = p_i/\sigma_{wi} \\ A_N = p_N/\sigma_{wN} \end{array}\right\} \quad (2.5)$$

Equations (2.5) can also be written in matrix form as

$$\begin{bmatrix} A_1 \\ A_2 \\ \vdots \\ A_i \\ \vdots \\ A_N \end{bmatrix} = \begin{bmatrix} 1/\sigma_{w1} & & & & 0 \\ & 1/\sigma_{w2} & & & \\ & & \ddots & & \\ & & & 1/\sigma_{wi} & \\ & & & & \ddots \\ 0 & & & & & 1/\sigma_{wN} \end{bmatrix} \begin{bmatrix} p_1 \\ p_2 \\ \vdots \\ p_i \\ \vdots \\ p_N \end{bmatrix} \quad (2.5.a)$$

or simply
$$\mathbf{A} = \mathbf{S}_w \mathbf{P} \quad (2.5.b)$$

The column vector $\mathbf{A} = \{A_1\ A_2\ \ldots A_i \ldots A_N\}$ contains the unknown member areas and the matrix

$$\mathbf{S}_w = \begin{bmatrix} 1/\sigma_{w1} & & & & 0 \\ & 1/\sigma_{w2} & & & \\ & & \ddots & & \\ & & & 1/\sigma_{wi} & \\ & & & & \ddots \\ 0 & & & & & 1/\sigma_{wN} \end{bmatrix} \quad (2.6)$$

is square and diagonal and called the limiting stress matrix.

2.2. Direct strains

When a tensile load *p* acts on a member, it elongates the member by a small, but measurable, amount. In *Figure 2.2* the member AB is shown to be pinned to its support at A and subject to a tensile axial force *p* at B. The original length of the member is *L*. The force *p* causes the member to elongate by an amount δ*L* and for this reason point B moves to B′ and the final length of the member becomes *L* + δ*L*.

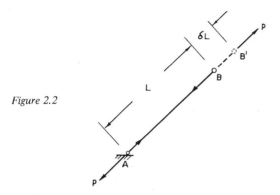

Figure 2.2

The strain ε in the member is the ratio of the elongation of the member and its original length, i.e.

$$\epsilon = \delta L / L \qquad (2.7)$$

The strain is thus the elongation of a unit length of the original member. Since both δ*L* and *L* are measured in units of length, e.g. in mm, the strain has no units. It is just a ratio.

Equations similar to (2.7) can be written for each member of a structure and these equations can be put in matrix form as

$$\varepsilon = l \delta \qquad (2.8)$$

where $\varepsilon = \{\epsilon_1, \epsilon_2, \ldots \epsilon_i \ldots \epsilon_N\}$ is the column vector of the strains in each member, $\delta = \{\delta L_1, \delta L_2, \ldots \delta L_i \ldots \delta L_N\}$ is the vector of the deformation of the members. The matrix *l* is square and diagonal with the reciprocal 1/*L* for each member appearing on the leading diagonal.

2.3. The stress-strain curve

For many engineering materials, such as steel, the graph of the stress in a member against the resulting strain is similar to that shown in *Figure 2.3*. Between the origin O and point A, the graph is nearly a straight line and the material is said to behave elastically. Point A is called the limit of proportionality. After A, the graph tends towards the horizontal axis and the increase in strain becomes large for small increases in the stress. Between points B and C, it is said that the material behaves plastically. The portion BC of the graph is nearly horizontal. After point C, the slope of the graph increases slightly and it is said that the material is strain hardened. At D the material fails suddenly.

If the member is loaded to a stress level below A and then unloaded, the stress-strain curve, for unloading, coincides with that for loading the member. However after A, if the member is unloaded, at a point such as F, then the unloading stress-strain curve follows the path FG. When the load is completely removed, a residual strain is left in the member as a permanent set. A material with a stress-strain curve

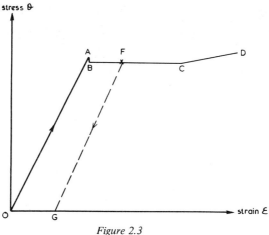
Figure 2.3

such as that shown in *Figure 2.3* is called a ductile material. A brittle material, such as concrete, has a different type of stress-strain curve. The linear part of the curve is much shorter and due to local failures, such as the cracking of the concrete at some parts, the ordinate of the curve may suddenly fall before total failure.

2.4. Hooke's law

It was stated in the last section that the part between points O and A of *Figure 2.3* is nearly linear. Engineers assume that for small loads applied to a member, the stress-strain curve is linear. That is to say the deformation of the member is directly proportional to the applied load. This assumption is known as Hooke's law. For a material that obeys Hooke's law the ratio

$$E = \sigma/\epsilon \tag{2.9}$$

between stress and strain is constant. This ratio is called Young's modulus of elasticity. The units of stress are in kN/mm² while strain has no unit. Thus the units of Young's modulus of elasticity are also in kN/mm². For structural steel the value of E is about 207 kN/mm². It is evident that E is the slope of the stress-strain curve for the linear portion OA.

Now $\sigma = p/A$ and $\epsilon = \delta L/L$, it follows, from equation (2.9) that

$$E = \sigma/\epsilon = (p/A)/(\delta L/L) = pL/A\delta L \tag{2.10}$$

2.5. Stiffness

Consider a force p applied to a member that obeys Hooke's law. The member has area A and length L. Let the deformation (elongation or contraction) of the member be δL. The equation (2.10) holds true for the member. Rearranging this equation

$$p = (EA/L)\delta L \tag{2.11}$$

or
$$p = k\delta L \tag{2.12}$$

where the property $k = EA/L$ (kN/mm) is known as the stiffness of the member and is defined as the load applied to the member which causes unit deformation of the member. This is because equation (2.12) can be rearranged to become

$$k = p/\delta L \tag{2.13}$$

Structural Properties 27

Thus the stiffness k is the force per unit deformation.

Equations similar to (2.11) and (2.12) can be written for each member of a pin jointed structure. For a structure with a total of N members,

$$\left.\begin{aligned} p_1 &= k_1 \delta L_1 = (E_1 A_1/L_1)\, \delta L_1 \\ p_2 &= k_2 \delta L_2 = (E_2 A_2/L_2)\, \delta L_2 \\ p_i &= k_i \delta L_i = (E_i A_i/L_i)\, \delta L_i \\ p_N &= k_N \delta L_N = (E_N A_N/L_N)\, \delta L_N \end{aligned}\right\} \tag{2.14}$$

These equations can be written in matrix form thus

$$\begin{bmatrix} p_1 \\ p_2 \\ p_i \\ p_N \end{bmatrix} = \begin{bmatrix} E_1 A_1/L_1 & & & 0 \\ & E_2 A_2/L_2 & & \\ & & E_i A_i/L_i & \\ 0 & & & E_N A_N/L_N \end{bmatrix} \begin{bmatrix} \delta L_1 \\ \delta L_2 \\ \delta L_i \\ \delta L_N \end{bmatrix} \tag{2.15a}$$

or simply $\quad\mathbf{P} = \mathbf{k}\boldsymbol{\delta}$ (2.15b)

The diagonal matrix **k** is known as the stiffness matrix of the members. The leading diagonal of the matrix consists of the stiffnesses $k = EA/L$ of the members.

2.6. Flexibility

Equation (2.11) can be rearranged in a different manner, thus

$$\delta L = (L/EA)p \tag{2.16}$$

or $\qquad \delta L = fp \tag{2.17}$

The property $f = L/EA$ (mm/kN) is known as the flexibility of the member. It is defined as the deformation caused in a member by a unit load. This is because equation (2.17) can be rearranged to read $f = \delta L/p$.

Equations similar to (2.16) and (2.17) can be written for every member of a pin jointed structure and in matrix form these become

$$\begin{bmatrix} \delta L_1 \\ \delta L_2 \\ \delta L_i \\ \delta L_N \end{bmatrix} = \begin{bmatrix} L_1/E_1 A_1 & & & 0 \\ & L_2/E_2 A_2 & & \\ & & L_i/E_i A_i & \\ 0 & & & L_N/E_N A_N \end{bmatrix} \begin{bmatrix} p_1 \\ p_2 \\ p_i \\ p_N \end{bmatrix} \tag{2.18a}$$

or simply $\quad\boldsymbol{\delta} = \mathbf{f}\mathbf{P}$ (2.18b)

The diagonal square matrix **f** is called the flexibility matrix of the members. Comparing equations (2.12) and (2.17) it is noticed that $f = 1/k$. This means that the flexibility of a member is the reciprocal of its stiffness. Similarly comparing equations (2.15) and (2.18), we deduce that

$$\mathbf{f} = \mathbf{k}^{-1} \qquad (2.19)$$

Thus the flexibility matrix is the inverse of the stiffness matrix.
Since k is a diagonal matrix, its inverse is given by taking the reciprocal of the elements EA/L of its leading diagonal.

In equations (2.11) and (2.16), the original length L of a member and its area A are constants. The value of E, for the linear part of the stress-strain curve, i.e. for OA in *Figure 2.3*, is also constant. Thus the stiffness and the flexibility of a linearly elastic member are constant properties of the member. However, the member has stiffness and flexibility even after the limit of proportionality, A in *Figure 2.3*, is exceeded. Here, because E is no longer constant, the stiffness and the flexibility of the member changes from point to point in the region AD of the stress-strain curve. The stiffness of the member is reduced while its flexibility is increased. Once the member fails, at D in *Figure 2.3*, its stiffness becomes zero and its flexibility becomes infinite. Thus failure of a member can be defined as the total loss of its stiffness.

2.7. Examples

Example 1. A concrete block is cylindrical in shape with a diameter of 100 mm and a height of 300 mm. A compression load of 70 kN causes the cylinder to contract by 0.2 mm. Calculate the stress, the strain, the modulus of elasticity, the stiffness and the flexibility of the block.
Answer:

The area of the cross section $= \pi \times 100 \times 100/4$

$$A = 7860 \text{ mm}^2$$

Compressive stress $\sigma_c = p/A = 70/7860 = 8.9 \times 10^{-3}$ kN/mm²
Compressive strain $\epsilon_c = \delta L/L = 0.2/300 = 0.67 \times 10^{-3}$
Modulus of elasticity $E = \sigma_c/\epsilon_c = (8.9 \times 10^{-3})/(0.67 \times 10^{-3})$

$$E = 13.3 \text{ kN/mm}^2$$

Stiffness $k = EA/L = 13.3 \times 7860/300 = 348$ kN/mm
Flexibility $f = 1/k = 1/348 = 2.88 \times 10^{-3}$ mm/kN

Example 2. The structure shown in *Figure 1.10*, section 1.6, is to be made from mild steel members with modulus of elasticity 200 kN/mm². Calculate the area of each member so that the stress developed is limited to 0.15 kN/mm² in tension and 0.1 kN/mm² in compression.| What is the axial deformation in each member?

Answer: From equations (1.5), $\mathbf{P} = \mathbf{B}\ \mathbf{W}$, with $W = 10$ kN, is found to be

$$\mathbf{P} = \{p_1 \quad p_2 \quad p_3 \quad p_4 \quad p_5 \quad p_6 \quad p_7 \quad p_8\} =$$
$$\{-2 \quad \sqrt{2} \quad 1 \quad -1 \quad -1 \quad \sqrt{2} \quad 0 \quad 0\} [10] =$$
$$\{-20 \quad 14.14 \quad 10 \quad -10 \quad -10 \quad 14.14 \quad 0 \quad 0\} \text{ kN}$$

The areas of the members are given by equations (2.5), i.e. $\mathbf{A} = \mathbf{S_w}\ \mathbf{P}$. There are no forces in members 7 and 8 and they are disregarded, thus

$$\mathbf{A} = \begin{bmatrix} A_1 \\ A_2 \\ A_3 \\ A_4 \\ A_5 \\ A_6 \end{bmatrix} = \begin{bmatrix} 1/-0.1 & & & & & \\ & 1/0.15 & & & 0 & \\ & & 1/0.15 & & & \\ & & & 1/-0.1 & & \\ & 0 & & & 1/-0.1 & \\ & & & & & 1/0.15 \end{bmatrix} \begin{bmatrix} -20 \\ 14.14 \\ 10 \\ -10 \\ -10 \\ 14.14 \end{bmatrix}$$

$\therefore \quad A_1 = (-1/0.1) \times (-20) = 200 \text{ mm}^2, \; A_2 = 14.14/0.15 = 94.3 \text{ mm}^2$
and $A_3 = 10/0.15 = 66.7 \text{ mm}^2$. Similarly $A_4 = 100 \text{ mm}^2, \; A_5 = 100 \text{ mm}^2$
$A_6 = 94.3 \text{ mm}^2$.

The axial deformation in each member is calculated using the flexibility equations (2.18), i.e. $\delta = \mathbf{f}\,\mathbf{P}$. The member lengths are shown in *Figure 1.10*, thus

$$\begin{bmatrix} \delta L_1 \\ \delta L_2 \\ \delta L_3 \\ \delta L_4 \\ \delta L_5 \\ \delta L_6 \end{bmatrix} = \frac{1}{200} \begin{bmatrix} 10\,000/200 & & & & & \\ & 14\,140/94.3 & & & 0 & \\ & & 10\,000/66.7 & & & \\ & & & 10\,000/100 & & \\ & 0 & & & 10\,000/100 & \\ & & & & & 14\,140/94.3 \end{bmatrix} \begin{bmatrix} -20 \\ 14.14 \\ 10 \\ -10 \\ -10 \\ 14.14 \end{bmatrix}$$

$\delta L_1 = (1/200) \times (10\,000/200) \times -20 = -5$ mm i.e. contraction
$\delta L_2 = (1/200) \times (14\,140/94.3) \times 14.14 = 10.6$ mm i.e. extension
Similarly $\delta L_3 = 7.5$ mm, $\delta L_4 = -5$ mm, $\delta L_5 = -5$ mm and $\delta L_6 = 10.6$ mm

Example 3. Calculate the areas of the members of the structure ADB shown in *Figure 2.4* when the load W is 10 kN. What are the horizontal and the vertical deflections at joint D? The stress in a tensile member is 0.15 kN/mm² and in a compressive member is -0.1 kN/mm². Take E as 200 kN/mm².
Answer: Resolving the forces at D horizontally and then vertically we obtain

$$P_{AD} = p_1 = -\sqrt{3} \times W = -10\sqrt{3} \text{ kN (compression)}$$
$$P_{BD} = p_2 = 2W = 20 \text{ kN (tension)}$$
$$\text{area of member 1} = A_1 = (-10\sqrt{3})/-0.1 = 173 \text{ mm}^2$$
$$\text{area of member 2} = A_2 = 20/0.15 = 133 \text{ mm}^2$$

$$\text{Contraction of member 1} = \delta L_1 = \frac{L_1}{EA_1} p_1 = \frac{-10\sqrt{3} \times 10 \times 10^3}{200 \times 173} = -5 \text{ mm}$$

$$\text{Extension of member 2} = \delta L_2 = \frac{20 \times 10^3 \times 20}{200 \times 133 \times \sqrt{3}} = 8.66 \text{ mm}.$$

30   *Structural Properties*

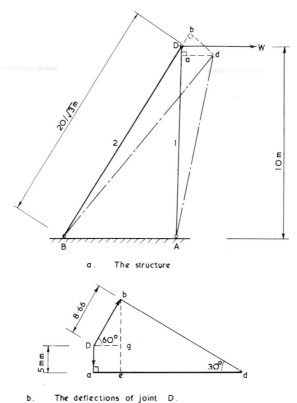

a.   The structure

b.   The deflections of joint D.

Figure 2.4

Had member AD been unrestrained by member BD at D, the compressive force would have shortened it by −5 mm vertically. On the other hand had member BD been free, the tensile force would have elongated it by 8.66 mm. The end of AD would have moved to point a in *Figure 2.4* and the end of BD would have moved to point b. In the figure Db and Da are drawn to an exaggerated scale for clarity. As the members are connected together at D, AD tends to rotate BD and b moves to d on the circumference of a circle with centre at B. Similarly BD tends to rotate AD and a moves to d on the circumference of a circle with centre at A. In this manner the ends of the members remain attached to one another. Point D moves to d and the structure takes the deformed shape BdA. This means the compatibility of joint D is preserved and the structure remains in one piece. Because the lengths of the curves bd and ad are very small compared to the lengths of the members, these are shown as straight lines perpendicular to BD and AD respectively.

The deflection diagram Dbda is reproduced in *Figure 2.4b*. From this diagram it is clear that the vertical deflection y of joint D is equal to Da = 5 mm downwards. The horizontal deflection x of joint D is equal to ad.

Now $ae = 8.66 \cos 60$

     $bg = 8.66 \sin 60$

∴ $be = bg + Da = 8.66 \sin 60 + 5$

     $ed = be \cot 30$

$$\therefore \quad x = ed + ae = 8.66 \cos 60 + be \cot 30$$
$$x = 8.66 \cos 60 + (5 + 8.66 \sin 60) \cot 30 = 25.98 \text{ mm}$$

Alternatively, using matrix algebra, consider a horizontal load W_1 and a vertical load W_2 are acting at D, remembering that in the actual example $W_2 = 0$. Resolving the forces at D horizontally and vertically, we obtain

$$p_1 = -\sqrt{3} \times W_1 + W_2 \qquad p_2 = 2W_1 + 0 \times W_2$$

$\therefore$ The equations $\mathbf{P} = \mathbf{B}\,\mathbf{W}$ become

$$\begin{bmatrix} p_1 \\ p_2 \end{bmatrix} = \begin{bmatrix} -\sqrt{3} & 1 \\ 2 & 0 \end{bmatrix} \begin{bmatrix} W_1 \\ W_2 \end{bmatrix} = \begin{bmatrix} -\sqrt{3} \times W_1 + W_2 \\ 2W_1 + 0 \times W_2 \end{bmatrix} \quad (2.20)$$

and
$$\mathbf{B} = \begin{bmatrix} -\sqrt{3} & 1 \\ 2 & 0 \end{bmatrix}$$

The transpose of $\mathbf{B}$ is $\mathbf{B}^T = \begin{bmatrix} -\sqrt{3} & 2 \\ 1 & 0 \end{bmatrix}$

i.e. the rows of $\mathbf{B}$ become the columns of $\mathbf{B}^T$.

With $A_1 = 173$ mm^2 and $A_2 = 133$ mm^2, equations (2.18), using the results obtained in equation (2.20), give

$$\begin{bmatrix} \delta L_1 \\ \delta L_2 \end{bmatrix} = \frac{1}{200} \begin{bmatrix} \dfrac{10 \times 10^3}{173} & 0 \\ 0 & \dfrac{20 \times 10^3}{\sqrt{3} \times 133} \end{bmatrix} \begin{bmatrix} -\sqrt{3} \times W_1 + W_2 \\ 2W_1 + 0 \times W_2 \end{bmatrix}$$

Since $W_2 = 0$ and $W_1 = 10$ kN

$\therefore \quad \delta L_1 = (1/200) \times (10\,000/173) \times -10\sqrt{3} = -5$ mm

and $\quad \delta L_2 = (1/200) \times (20\,000/133\sqrt{3}) \times 2 \times 10 = 8.66$ mm

i.e.
$$\begin{bmatrix} \delta L_1 \\ \delta L_2 \end{bmatrix} = \begin{bmatrix} -5 \\ 8.66 \end{bmatrix}$$

Now premultiply this vector by the matrix $\mathbf{B}^T$

$$\mathbf{X} = \mathbf{B}^T \boldsymbol{\delta} = \begin{bmatrix} -\sqrt{3} & 2 \\ 1 & 0 \end{bmatrix} \begin{bmatrix} -5 \\ 8.66 \end{bmatrix} = \begin{bmatrix} -\sqrt{3} \times -5 + 2 \times 8.66 \\ 1 \times -5 + 0 \times 8.66 \end{bmatrix} = \begin{bmatrix} 25.98 \\ -5 \end{bmatrix}$$

$$\mathbf{X} = \begin{bmatrix} 25.98 \\ -5 \end{bmatrix} = \begin{bmatrix} x \\ y \end{bmatrix}$$

It is noticed that the first element of vector $\mathbf{X}$ is 25.98 which is in fact the horizontal deflection of joint D. The second element of $\mathbf{X}$ is -5 which is the vertical deflection of D. It will be proved, in Chapter 9, that generally the deflection **X** of all the joints in a structure is in fact given by

$$\mathbf{X} = \mathbf{B}^T \boldsymbol{\delta} \tag{2.21}$$

Example 4. With $A_1 = 173$ mm², $A_2 = 133$ mm² and $E = 200$ kN/mm² what external load applied at D to the structure shown in *Figure 2.4* would cause joint D to deflect horizontally by 259.8 mm and vertically by −50 mm?
Answer: Intuition suggests that the load is 100 kN acting horizontally.

Deflections, like forces, are vectors. They have resultants and components. In *Figure 2.5* the vectors x and y are the horizontal and the vertical deflection components of joint D. Members AD and BD remain connected to joint D as it deflects.

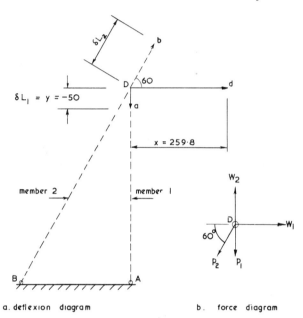

Figure 2.5

The deformation of member AD is equal to the vertical deflection of joint D and therefore

$$\delta L_1 = y = -50 \text{ mm} \tag{2.22}$$

The component of x in the direction BD is equal to $x \cos 60$, where $x = 259.8$ mm, while the component of y in the same direction is $y \sin 60$, where $y = -50$ mm. The net deflection of joint D in the direction BD is thus equal to $x \cos 60 + y \sin 60$. The deformation δL_2 of member BD has to be equal to the deflection of D in the direction BD, so that member BD remains connected to joint D, thus

$$\delta L_2 = x \cos 60 + y \sin 60 \tag{2.23}$$

i.e. $\qquad \delta L_2 = 259.8 \times 0.5 - 50 \times 0.5\sqrt{3} = 86.6$ mm

Equations (2.22) and (2.23) can be written in matrix form as

$$\begin{bmatrix} \delta L_1 \\ \delta L_2 \end{bmatrix} = \begin{bmatrix} 0 & 1 \\ \cos 60 & \sin 60 \end{bmatrix} \begin{bmatrix} x \\ y \end{bmatrix} \tag{2.24a}$$

or simply $\quad\quad\quad \boldsymbol{\delta} = \mathbf{D}\mathbf{X}$ \hfill (2.24b)

Matrix **D** contains sines and cosines of the angles of inclinations for members AD and BD and it is known as the displacement transformation matrix. It relates the deformations $\boldsymbol{\delta} = \{\delta L_1 \ \delta L_2\}$ to the deflections $\mathbf{X} = \{x \ y\}$ of joint D.

Once the member deformations are known, the member forces can be calculated from the stiffness equations (2.14) or (2.15), thus

$$\begin{bmatrix} p_1 \\ p_2 \end{bmatrix} = 200 \begin{bmatrix} 173/10\,000 & 0 \\ 0 & 133\sqrt{3}/20\,000 \end{bmatrix} \begin{bmatrix} -50 \\ 86.6 \end{bmatrix}$$

$\therefore \quad\quad\quad p_1 = 200 \times (173/10\,000) \times -50 = -173 \text{ kN}$

and $\quad\quad\quad p_2 = 200 \times (133\sqrt{3}/20\,000) \times 86.6 = 200 \text{ kN}$

So far we do not know the magnitude or the direction of the external force at D. Let the horizontal and vertical component of this force be W_1 and W_2 as shown in *Figure 2.5b*. Resolving the forces at D horizontally

$\quad\quad\quad W_1 - p_2 \cos 60 = 0$

$\therefore \quad\quad\quad W_1 = p_2 \cos 60 = 200 \times 1/2 = 100 \text{ kN}$

Resolving the forces at D vertically

$\quad\quad\quad W_2 - p_1 - p_2 \sin 60 = 0$

$\therefore \quad\quad\quad W_2 = p_1 + p_2 \sin 60 = -173 + 200 \times \sqrt{3}/2 = 0$

Thus the external force acting at D is $W = W_1 = 100$ kN.

Alternatively, notice that the transpose of matrix **D**, equation (2.24), is

$$\mathbf{D}^{\mathrm{T}} = \begin{bmatrix} 0 & \cos 60 \\ 1 & \sin 60 \end{bmatrix}$$

and post multiplying this matrix by the vector of member forces $\mathbf{P} = \{p_1 \ p_2\}$ gives

$$\begin{bmatrix} 0 & \cos 60 \\ 1 & \sin 60 \end{bmatrix} \begin{bmatrix} -173 \\ 200 \end{bmatrix} = \begin{bmatrix} 200 \cos 60 \\ -173 + 200 \sin 60 \end{bmatrix} = \begin{bmatrix} 100 \\ 0 \end{bmatrix} = \begin{bmatrix} W_1 \\ W_2 \end{bmatrix} \quad\quad (2.25)$$

It will be proved, in Chapter 12, that the results obtained in equations (2.24) and (2.25) are generally true, that is

and $\quad\quad\quad \left. \begin{array}{l} \boldsymbol{\delta} = \mathbf{D}\mathbf{X} \\ \mathbf{W} = \mathbf{D}^{\mathrm{T}}\mathbf{P} \end{array} \right\}$ \hfill (2.26)

2.8. Temperature stresses

When the temperature of a pin ended member is increased or decreased, the member extends or contracts. If this extension or contraction is prevented, temperature stresses develop in the member. Consider a pin ended member of initial length L and of linear coefficient of thermal expansion θ per degree centigrade. Let the temperature of the member be increased by $t°$ centigrade. If the member is free to expand, its length increases by $\theta L t$ to become $L + \theta L t$. If the expansion of the member is prevented, it is as if a member of length $L + \theta L t$ were compressed to a length L. Thus the compressive temperature strain $\epsilon°$ is given by

$$\epsilon° = -\theta L t / L (1 + \theta t) \simeq -\theta t \qquad (2.27)$$

Here it is assumed that $\theta L t$ is small compared to L and θt is small compared to unity. The compressive stress $\sigma°$ is obtained from

$$E = \sigma/\epsilon$$

i.e. $\qquad\qquad \sigma = E\epsilon,$

thus $\qquad\qquad \sigma° = -E\theta t \qquad (2.28)$

Similarly the tensile temperature stress in a member prevented from contracting, after a fall of $t°$ in its temperature, is $E\theta t$. The linear coefficient of thermal expansion for steel is 4×10^{-6} per degree centigrade and the product $E\theta$, with $E = 200$ kN/mm², is 0.8×10^{-3} kN/mm². As an example if the temperature of a pin ended member is changed by $50°C$ while the member is restrained, the strain $\epsilon°$ of the member will be $4 \times 10^{-6} \times 50 = 0.2 \times 10^{-3}$ and its stress $\sigma°$ will be $0.8 \times 10^{-3} \times 50 = 40 \times 10^{-3}$ kN/mm². If the area of the member is 1000 mm², then the force developed in it is $p = A\sigma° = 40$ kN.

Although the linear coefficient of thermal expansion is a property of the structural material, isostatic structures are at an advantage because thermal stresses do not develop in them. These stresses only become important in hyperstatic structures.

2.9. Example

Example 1. The pin ended member shown in *Figure 2.6* is 10 m long and has an area of 173 mm². The member is 2.89 mm too short. What force should be applied at B before it can be pinned into its support? Once it is pinned, what is the safe temperature to which it can be heated before the stress in the member reaches 0.1 kN/mm²? Take $E = 200$ kN/mm² and $\theta = 4 \times 10^{-6}$ mm/°C.

Answer:

$$p = EA\delta/L$$
$$p = 200 \times 173 \times 2.89/10\,000 = 10 \text{ kN}$$

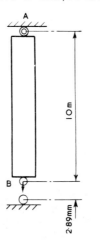

Figure 2.6

The tensile stress in the member $= \sigma_t = p/A = 10/173 = 0.0578 \text{ kN/mm}^2$
If the member is now heated by t_1°

$$\sigma = E\theta t_1$$

i.e. $0.0578 = 200 \times 4 \times 10^{-6} t_1$

Thus $t_1 = 0.0578/(200 \times 4 \times 10^{-6}) = 72.25^\circ C$

then all the tensile stress in the member will be relieved. The member can then be heated further by t_2°

$$t_2 = \sigma^\circ/E\theta = 0.1 \times 10^6/(200 \times 4) = 125^\circ C$$

before the compressive stress of 0.1 kN/mm is developed.
The total safe temperature is therefore

$$t = 72.25 + 125 = 197.25^\circ C$$

2.10. The principle of superposition

In examples 3 and 4 of section 2.7, the structure shown in *Figure 2.4* was subjected to two different horizontal loads of 10 kN and 100 kN. Under the first load, the resulting deflections were 25.98 mm horizontally and −5 mm vertically. Under the second load these deflections increased by ten times to 259.8 mm and −50 mm. If the structure is analysed with an external horizontal load of W = 50 kN, the resulting

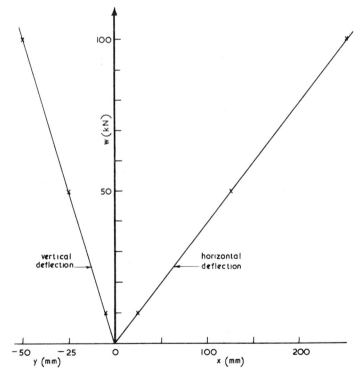

Figure 2.7

horizontal and vertical deflection will be 129.9 mm and −25 mm respectively. The results of these three analyses are plotted in *Figure 2.7* and it is noticed that the graph for each deflection against the applied loads is a straight line. This indicates that the deflection of a structure is linearly related to the external applied loads. This is known as the principle of superposition. If the applied load is doubled, the deflections are doubled. The load of 100 kN applied to the structure can be treated as two loads. The first is say 10 kN which gives the first set of deflections of 25.98 mm and −5 mm. The second is 90 kN which produces deflections of 233.82 mm and −45 mm, and the two loads acting together produced deflections of 259.8 mm and −50 mm.

Similarly the deflections at any point such as A, B, C of the structure shown in *Figure 2.8* are obtained as the sum of the deflections at that point due to the separate effects of the loads W_1 and W_2.

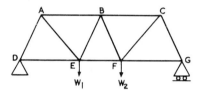

Figure 2.8

There are two exceptions to the principle of superposition. It does not apply to structures manufactured from material whose stress-strain relationship is not linear elastic or if the limit of proportionality of the stress-strain curve is exceeded. In example 4 of section 2.7, the structure is subject to a horizontal load of 100 kN at D. This produces a tensile force of 200 kN in member BD. With an area of 133 mm² the stress in this member is 1.50 kN/mm² which is five times the stress at the limit of proportionality for steel. Thus the resulting stresses are by far beyond the linear portion of the stress-strain curve for this material. In fact when the acting horizontal load W exceeds 20 kN, the material of the structure begins to yield and for higher loads plasticity develops in the material. For such high loads the stress-strain curve becomes non-linear and the principle of superposition becomes inapplicable.

The second exception is when the geometry of the structure is basically altered by the effect of the external loads. Here again the principle of superposition does

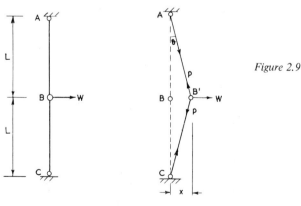

Figure 2.9

a. Two vertical members   b. The deflected shape

not apply. Consider, for instance, a pin jointed structure with two vertical members as shown in *Figure 2.9a*. As soon as the force W is applied to B, the members extend and the pin at B moves to the right to B' by an amount x. The structure takes the new shape shown in *Figure 2.9b*, in which

$$x = \theta L \tag{2.29}$$

Resolving the forces at B' horizontally

$$W - 2p \sin \theta = 0$$

i.e. $\quad p = 0.5W/\sin \theta$

Since θ is small, $\sin \theta \simeq \theta$ and

$$p = 0.5W/\theta \tag{2.30}$$

The strain in each member is given by

$$\epsilon = \text{change in length/original length}$$

$$\epsilon = \frac{\sqrt{(L^2 + x^2)} - L}{L} \simeq 0.5x^2/L^2 \tag{2.31}$$

On the other hand, the material is assumed to obey Hooke's law which gives the strain as

$$\epsilon = \sigma/E = p/EA \tag{2.32}$$

where A is the cross sectional area and E is the modulus of elasticity. Equating this strain to that given by equation (2.31), we obtain, using equations (2.29) and (2.30),

$$0.5x^2/L^2 = p/AE = W/(2\theta AE) = WL/(2xAE)$$

Hence

$$x = L(W/AE)^{1/3} \tag{2.32}$$

Thus the deflection x is not linearly related to the load W. This is in spite of the fact that the material obeys Hooke's law and the deflection x is small. A graph of x against W has the form shown in *Figure 2.10*.

With members subject to axial forces and bending moments (see next chapter) the principle of superposition becomes also inapplicable, either when the axial forces are too large or when the deflections are too large or both.

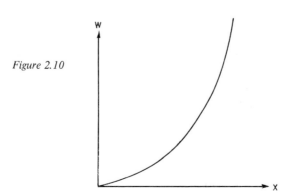

Figure 2.10

2.11. Strain energy

Under the action of gradually increasing external loads, the joints of a structure deflect and the members deform. The applied loads produce work at the joints to which they are applied and this work is stored in the structure in the form of energy which is known as strain energy. If the material of the structure is elastic, whether linear or non-linear, then gradual unloading of the structure relieves all the stresses and the strain energy is recovered.

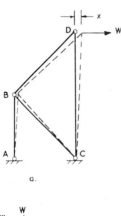

Figure 2.11

a.

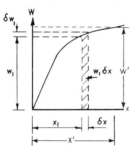

b. Work done by W.

c. Work done by force P in a member.

Consider an elastic structure subject to a single gradually increasing external load W as shown in *Figure 2.11*. The graph of W against the deflection x at D is of the shape shown in *Figure 2.11b*. If at a given instant the value of the load is W_1 and increases by an amount δW_1, then the work done is $W_1 \delta x$ as shown by the shaded area under the graph. The total work done by the load, during its gradual increase from zero to its final value W' is the sum of all such elemental areas. Hence the total work w is given by

$$w = \int_0^{x'} W dx \quad \text{kN mm} \tag{2.33}$$

where x' is the total deflection at D when $W = W'$.

In spite of these exceptions, the principle of superposition remains important because engineers often design their structures so that the limit of proportionality is not exceeded, the deflections are small and shapes are not grossly changed by the applied loads.

Figure 2.11c shows the force-deformation graph of one of the members of the structure. The strain energy stored in the member is, by a similar argument to the above, given by

$$U = \int_0^{\Delta'} p\, d\Delta \qquad (2.34)$$

where Δ' is the total deformation (extension or contraction) of the member. The total strain energy stored in all the members of the structure is U_t, where

$$U_t = \sum_1^N U = \sum_1^N \int_0^{\Delta'} p\, d\Delta \qquad (2.35)$$

and N is the total number of members. As the member force p occurs under the action of the external load W and the member deformation Δ corresponds to joint deflection x, then by the principle of the conservation of energy, $U_t = w$ and thus

$$w = U_t = \sum_1^N \int_0^{\Delta'} p\, d\Delta \qquad \text{kN mm} \qquad (2.36)$$

Consider now a single member of the structure. If the limit of proportionality of the stress-strain diagram is not exceeded for this member then the force deformation diagram will be linear and the area under the graph (*Figure 2.11c*) is that of a triangle, thus

$$U = \int_0^{\Delta'} p\, d\Delta = 0.5 p \Delta' \qquad (2.37)$$

where Δ' is the total deformation of the member and p is the total force in it.

But for linear elastic material which obeys Hooke's law, we have

$$\Delta' = pL/AE$$

It follows from equation (2.37) that

$$U = p^2 L/(2AE) \qquad (2.38)$$

Now p/A is the stress σ in the member, thus

$$U = 0.5 \sigma^2 AL/E \qquad \text{kN mm} \qquad (2.39)$$

Furthermore, since AL is the volume of the member, it follows that the strain energy per unit volume of the member u, is given by

$$u = 0.5 \sigma^2 / E \qquad \text{kN/mm}^2 \qquad (2.40)$$

For instance, for a steel member with $E = 200$ kN/mm and with the tensile stress at the limit of proportionality $\sigma_p = 0.3$ kN/mm, it is found from equation (2.40) that the amount of strain energy per mm³ which can be stored within the elastic range is $u = 0.225$ N/mm².

2.12. The work equation

Consider a pin jointed linear elastic structure with N members and subject to a total of M external loads $\mathbf{W} = \{W_1\ W_2\ \dots\ W_j\ \dots\ W_M\}$. Let the deformation in the members be $\boldsymbol{\delta} = \{\delta_1\ \delta_2\ \dots\ \delta_i\ \dots\ \delta_N\}$ and let the deflection under the external loads be $X = \{x_1\ x_2\ \ldots x_j \ldots x_M\}$. The work done by the external loads is thus

$$w = (W_1 x_1 + W_2 x_2 + \ldots + W_j x_j + \ldots + W_M x_M)/2$$

and the work done by the member forces is

$$U = (p_1 \delta_1 + p_2 \delta_2 + \ldots + p_i \delta_i + \ldots + p_N \delta_N)/2$$

Here U is also the total strain energy stored in the members and since $U = w$, it follows that

$$W_1 x_1 + W_2 x_2 + \ldots + W_j x_j + \ldots + W_M x_M = p_1 \delta_1 + p_2 \delta_2 + \ldots + p_i \delta_i + \ldots + p_N \delta_N \tag{2.41}$$

This is known as the work equation. In matrix form this equation is written as

$$\begin{matrix}[W_1 & W_2 & \ldots & W_j & \ldots & W_M] & \{x_1 & x_2 & \ldots & x_j & \ldots & x_M\} = \\ [p_1 & p_2 & \ldots & p_i & \ldots & p_N] & \{\delta_1 & \delta_2 & \ldots & \delta_i & \ldots & \delta_N\}\end{matrix} \tag{2.42a}$$

or simply $\qquad \mathbf{W}^T \mathbf{X} = \mathbf{P}^T \boldsymbol{\delta} \tag{2.42b}$

where $\mathbf{W}^T = [W_1\ W_2 \ldots W_j \ldots W_M]$ and $\mathbf{P}^T = [p_1\ p_2 \ldots p_i \ldots p_N]$ are row vectors. They are the transpose of the column vectors $\mathbf{W}$ and $\mathbf{P}$ respectively. The vectors $\mathbf{X}$ and $\boldsymbol{\delta}$ are column vectors, as usual.

Equations (2.41) or (2.42) are extremely useful for calculating the joint deflections in a structure.

2.13. Relationships among stress, strain and stiffness

Consider a gradually increasing load W applied to a pin ended linear elastic member as shown in *Figure 2.12*. Let the value of W increase to W' when the total deflection is $\Delta' = 1$ unit.

The total work done by the applied load is

$$w = \int_0^{\Delta'} 0.5 W \delta \Delta = 0.5 W' \Delta'$$

but $\qquad \Delta' = 1$ unit

$\therefore \qquad w = 0.5 W'$

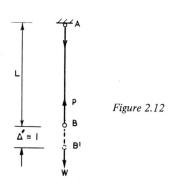

Figure 2.12

Structural Properties

When $W = W'$, the force in the member is p' and from equilibrium

$$p' - W' = 0$$
$$\therefore \quad p' = W'$$

Thus the total work done by the external load is

$$w = 0.5p'$$

The total extension of the member is δ' which is equal to $\Delta' = 1$ and since

$$p' = (EA/L)\delta' = k\delta'$$
$$\therefore \quad p' = k\delta' = k \times 1 = k$$

The total work done is thus

$$w = 0.5p' = 0.5k \tag{2.43}$$

Now the strain energy per unit volume, u, is from equation (2.40) equal to $0.5\sigma^2/E$ and the total strain energy is

$$U = \int_0^v (\sigma^2/2E)dv$$

where v is the total volume of the member.

Furthermore, from Hooke's law $\sigma = \epsilon E$, thus

$$U = \int_0^v (\sigma/2E)\epsilon E\, dv = 0.5 \int_0^v \sigma \epsilon\, dv \tag{2.44}$$

Equating the total strain energy in the member to the total work done by the external load, we obtain using equations (2.43) and (2.44)

$$k = \int_0^v \sigma \epsilon\, dv \tag{2.45}$$

Thus the stiffness of a member is the integral of the product of stress and strain over the volume for a unit deformation. This important formula is extensively used by engineers in the powerful 'finite element' method to derive the stiffness matrices of members or other elements, such as plates or solids. In matrix form equation (2.45) is written as

$$k = \int_0^v \boldsymbol{\sigma}^T \boldsymbol{\epsilon}\, dv \tag{2.46}$$

where $\boldsymbol{\sigma}^T$ is the transpose of the stress vector $\{\boldsymbol{\sigma}\}$.

2.14. Principles of structural analysis

With the aid of single members and pin jointed structures, the last two chapters have summarised the entire theory of structures using both classical and matrix methods. It should have been noticed that there are three basic principles involved. These are:

1. Make use of the stress-strain relationships of the material of the structure.
2. Make use of the equilibrium of the forces applied to the structure, those developed in the members and those developed at the supports.
3. Make use of the compatibility conditions in the structure, i.e. ensure that after applying the loads, the members remain connected to their supports and to the joints to which they were connected before loading.

Compatibility conditions are obtained by equating the deformation of the members to the deflections of the joints to which they are connected. It was stated that deflections and deformations are also vectors. Thus by equating these deformations and deflections we are in fact deriving equations of equilibrium for deflection vectors. Thus compatibility equations are merely statements of equilibrium of deflection. Looked at in this way, there are only two factors involved in the analysis of structures. These are stress-strain relationships and equilibrium of forces and deflections. Using these two factors every structure, linear or non-linear, isostatic or hyperstatic can be analysed. Nothing more is required.

2.15. Examples

Example 1. In the pin jointed frame shown in *Figure 2.13*, the area of each member is 100 mm^2 and $E = 200$ kN/mm^2. Calculate the strain energy per unit volume and the stiffness in each member. Using the work equation, what is the vertical deflection y at point D?

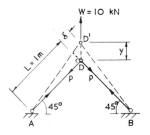

Figure 2.13

Answer: Resolving the forces at D vertically

$$W - 2p/\sqrt{2} = 0$$

∴ $p = 0.5W\sqrt{2} = 10/\sqrt{2}$ kN

The stress $= \sigma = p/A = 10/(100\sqrt{2}) = 0.071$ kN/mm^2

The strain energy per unit volume $= u = 0.5\sigma^2/E$

$$u = 0.5 \times 0.071 \times 0.071 \times 1/200 = 12.5 \times 10^{-6} \text{ kN/mm}^2$$

Stiffness $= EA/L = \dfrac{200 \times 100}{1000} = 20$ kN/mm

The strain energy in each member $= U = uv$

∴ $U = 12.5 \times 10^{-6} \times 100 \times 1000 = 1.25$ kN mm

Total strain energy in both members $= 2 \times 1.25 = 2.5$ kN mm.

The work done by the external load $= 0.5Wy$

∴ $0.5Wy = 2.5$

$$y = 2 \times 2.5/W = 5/10 = 0.5 \text{ mm}$$

Alternatively the extension in each member is δ, given by

$$p = k\delta$$

∴ $10/\sqrt{2} = 20\delta$

and $\delta = 1/(2\sqrt{2})$

From the work equation (2.41) or (2.42)

$$p_1 \delta_1 + p_2 \delta_2 = Wy$$
$$2p\delta = Wy$$
$$2 \times (10/\sqrt{2}) \times 1/(2\sqrt{2}) = 10y$$

giving $y = 0.5$ mm.

Example 2. Calculate the vertical deflection of joint D for the structure shown in Figure 2.14. All other data are given in the last example. What is the deflection at D when $W = 20$ kN?

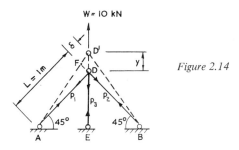

Figure 2.14

Answer: This structure is hyperstatic because there is not a sufficient number of equilibrium equations to find the forces in each member. Resolving forces at D horizontally

$$p_1 = p_2 \qquad (2.47)$$

Resolving forces at D vertically, using (2.47),

$$W - p_3 - 2p_1/\sqrt{2} = 0$$
$$p_3 = W - \sqrt{2} \times p_1 \qquad (2.48)$$

There are two unknowns p_1 and p_3 in this equation. To solve the problem a further equation is required. From the triangle DFD' (*Figure 2.14*) it is noticed that the compatibility condition is satisfied by

$$\delta = y \cos 45 = y/\sqrt{2} \qquad (2.49)$$

Now for member AD, $p_1 = EA\delta/L$, thus

$$\delta = p_1 L/EA$$

Similarly for member ED, $p_3 = EAy/(L/\sqrt{2})$

∴ $$y = p_3 L/(\sqrt{2} \times EA) \qquad (2.50)$$

Using these values for δ and y, equation (2.49) becomes

$$p_3 = 2p_1$$

substituting for p_3 in equation (2.48)

$$2p_1 = W - \sqrt{2} \times p_1$$

∴ $$p_1 = W/(2 + \sqrt{2}) = 10/3.141$$

and $$p_3 = 20/3.141$$

44   *Structural Properties*

Hence equation (2.50) gives

$$y = 20L/(3.141\sqrt{2} \times EA)$$
$$y = 20 \times 1000/(3.141\sqrt{2} \times 200 \times 100) = 0.225 \text{ mm}.$$

From the principle of superposition when $W = 20$ kN then $y = 0.45$ mm. Check that the stresses are below the limit of proportionality.

Example 3. Calculate the forces in the members of the pin jointed frame, shown in Figure 2.15 (a) when a horizontal load $W_1 = 5$ kN is acting at E, (b) when a horizontal load $W_2 = .10$ kN is acting at D and (c) when both forces are acting together.

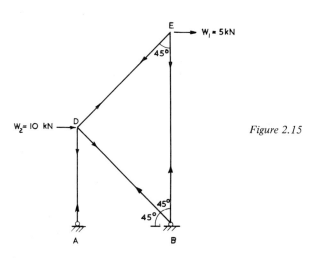

Figure 2.15

Answer: (a) With $W_1 = 5$ kN acting alone, resolving forces horizontally at E

$$W_1 - p_{ED}/\sqrt{2} = 0$$
$$\therefore \quad p_{ED} = \sqrt{2} \times W_1 = 5\sqrt{2} \text{ kN}$$

Resolving forces vertically at E $p_{EB} = -p_{ED}/\sqrt{2} = -5$ kN

At D resolving horizontally $p_{DE}/\sqrt{2} + p_{DB}/\sqrt{2} = 0$

$$\therefore \quad p_{DB} = -p_{DE} = -5\sqrt{2} \text{ kN}$$

Resolving vertically at D $-p_{DA} - p_{DB}/\sqrt{2} + p_{DE}/\sqrt{2} = 0$

$$\therefore \quad p_{DA} = -p_{DB}/\sqrt{2} + p_{DE}/\sqrt{2} = 10 \text{ kN}$$

(b) With $W_2 = 10$ kN acting alone, resolving the forces at E and D

$$p_{ED} = p_{EB} = 0, \quad p_{DB} = -10\sqrt{2} \text{ and } p_{DA} = 10 \text{ kN}$$

By the principle of superposition, the member forces, due to both W_1 and W_2 acting together are

$$\{p_{ED} \quad p_{EB} \quad p_{DA} \quad p_{DB}\} = \{5\sqrt{2} \quad -5 \quad 20 \quad -15\sqrt{2}\} \text{ kN}$$

2.16. Properties of symmetrical structures

In a symmetrical structure which is loaded symmetrically, such as that shown in Figure 2.16a, the forces p_i and p_j in two members i and j, which are arranged symmetrically on either side of the axis of symmetry, are equal. Thus to analyse such a structure it is sufficient to find the member forces in one half of the structure only. For each member, such as k, in that half there is a corresponding member n in the other half whose force p_n is equal to the known value of p_k. On the other hand, when a symmetrical structure is loaded antisymmetrically (*Figure 2.16b*) the forces p_i and p_j in the members i and j, which are arranged symmetrically on either side of the axis of symmetry, are equal in magnitude and opposite in sign. Thus, again it is sufficient to analyse only one half of the structure.

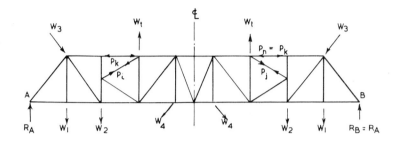

a. A symmetrical frame loaded symmetrically

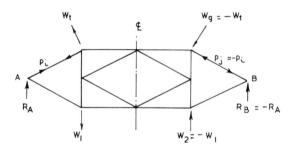

b. A symmetrical frame loaded antisymmetrically

Figure 2.16

When a symmetrical structure is loaded in a general unsymmetrical manner, it is possible to separate the actual loads into two systems, a symmetrical system and an antisymmetrical one. Half of the structure is then analysed once for each load system. The forces in the members due to the actual loads are then obtained from the results of the two analyses using the principle of superposition.

Consider the pin jointed frame shown in *Figure 2.17* which is symmetrical about the vertical line FG and subject to forces W_1 at D, W_2 at E and W_3 at C. The vertical reactions R_A and R_B are found by taking moments about B and A respectively, thus

$$W_1 \times 1.5L + W_2 \times 0.5L + W_3 \times L - R_A \times 2L = 0$$

∴ $$R_A = 0.25\,(3W_1 + W_2 + 2W_3) \tag{2.52}$$

and $$R_B \times 2L - W_1 \times 0.5L - W_2 \times 1.5L - W_3 \times L = 0$$

∴ $$R_B = 0.25\,(W_1 + 3W_2 + 2W_3) \tag{2.53}$$

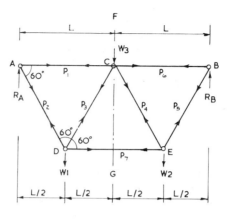

Figure 2.17

Notice that the coefficients of W_1 and W_2 in equations (2.52) and (2.53) have interchanged.

The member forces p_1, p_2, p_3 and p_7 are now found by the method of joints, thus

At A, resolving vertically

$$R_A - p_2 \cos 30 - 0$$

Using equation (2.52)

$$p_2 = 0.5(3W_1 + W_2 + 2W_3)/\sqrt{3}$$

At A, resolving horizontally

$$p_1 + p_2 \cos 60 = 0$$

$$\therefore \quad p_1 = -0.5p_2 = -0.25(3W_1 + W_2 + 2W_3)/\sqrt{3}$$

At D, resolving vertically

$$p_2 \cos 30 + p_3 \cos 30 - W_1 = 0$$

$$\therefore \quad p_3 = 0.5(W_1 - W_2 - 2W_3)/\sqrt{3}$$

At D, resolving horizontally

$$p_7 + p_3 \cos 60 - p_2 \cos 60 = 0$$

$$\therefore \quad p_7 = 0.5(W_1 + W_2 + 2W_3)/\sqrt{3}$$

The method of joints may be continued to calculate p_4, p_5 and p_6. However, it is evident that p_4, p_5 and p_6 are associated to W_1 and W_2 as p_3, p_2 and p_1 are associated to W_2 and W_1, thus interchanging the coefficients of W_1 and W_2

$$p_4 = 0.5(-W_1 + W_2 - 2W_3)/\sqrt{3}$$
$$p_5 = 0.5(W_1 + 3W_2 + 2W_3)/\sqrt{3}$$
$$p_6 = -0.25(W_1 + 3W_2 + 2W_3)/\sqrt{3}$$

Structural Properties

These member forces are written in matrix form $\mathbf{P} = \mathbf{B}\mathbf{W}$ as

$$\begin{bmatrix} p_1 \\ p_2 \\ p_3 \\ p_4 \\ p_5 \\ p_6 \\ p_7 \end{bmatrix} = 0.5/\sqrt{3} \begin{bmatrix} -1.5 & -0.5 & -1 \\ 3 & 1 & 2 \\ 1 & -1 & -2 \\ -1 & 1 & -2 \\ 1 & 3 & 2 \\ -0.5 & -1.5 & -1 \\ 1 & 1 & 2 \end{bmatrix} \begin{bmatrix} W_1 \\ W_2 \\ W_3 \end{bmatrix} \qquad (2.54)$$

Consider now the following three loading cases.

1. The unsymmetrical loading case where

$$W_1 = -W_2 \text{ and } W_3 \neq 0$$

That is, W_1 and W_3 are acting downwards while $W_2 = -W_1$ acting upwards as shown in *Figure 2.18a*. Equations (2.54) give

$$p_1 = 0.5(-1.5W_1 + 0.5W_1 - W_3)/\sqrt{3} = -0.5(W_1 + W_3)/\sqrt{3}$$

Similarly

$$\left.\begin{array}{l} p_2 = (W_1 + W_3)/\sqrt{3}, \quad p_3 = (W_1 - W_3)/\sqrt{3}, \quad p_4 = -(W_1 + W_3)/\sqrt{3} \\ p_5 = (-W_1 + W_3)/\sqrt{3}, \quad p_6 = 0.5(W_1 - W_3)/\sqrt{3} \text{ and } p_7 = W_3/\sqrt{3} \end{array}\right\} $$

$$(2.55)$$

2. The antisymmetrical loading case where

$$W_1 = -W_2 \text{ and } W_3 = 0$$

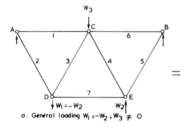

a. General loading $W_1 = -W_2$, $W_3 \neq 0$

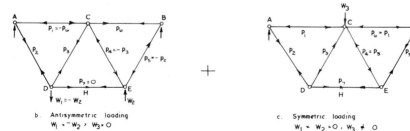

b. Antisymmetric loading $W_1 = -W_2$, $W_3 = 0$

c. Symmetric loading $W_1 = W_2 = 0$, $W_3 \neq 0$

Figure 2.18

That is W_1 is acting downwards while $W_2 = -W_1$ is acting upwards and W_3 is not acting as shown in *Figure 2.18b*. Equations (2.54) give

$$\left.\begin{aligned} p_1 &= -0.5W_1/\sqrt{3} & p_6 &= 0.5W_1/\sqrt{3} \\ p_2 &= W_1/\sqrt{3} & p_5 &= -W_1/\sqrt{3} \\ p_3 &= W_1/\sqrt{3} & p_4 &= -W_1/\sqrt{3} \\ p_7 &= 0 & & \end{aligned}\right\} \quad (2.56)$$

It is noticed that when this symmetrical frame is subjected to an antisymmetrical loading system then $p_6 = -p_1$, $p_5 = -p_2$ and $p_4 = -p_3$. This indicates that once the force in a member to the left of the axis FG of symmetry is determined, the force in a corresponding member to the right of this axis can be found by changing the sign of the force in the former member. Notice that members 4, 5 and 6 are the mirror images of the members 3, 2 and 1 respectively. Member 7 crosses the axis of symmetry and it is symmetrical about this axis. The force in this member is zero.

3. The symmetrical loading case where

$$W_1 = W_2 = 0 \text{ and } W_3 \neq 0$$

That is only W_3 is acting as shown in *Figure 2.18c*. Equations (2.54) give

$$\left.\begin{aligned} p_1 &= -0.5W_3/\sqrt{3} & p_6 &= -0.5W_3/\sqrt{3} \\ p_2 &= W_3/\sqrt{3} & p_5 &= W_3/\sqrt{3} \\ p_3 &= -W_3/\sqrt{3} & p_4 &= -W_3/\sqrt{3} \\ p_7 &= W_3/\sqrt{3} & & \end{aligned}\right\} \quad (2.57)$$

The load W_3 is symmetrically placed on the frame and it is noticed that $p_6 = p_1$, $p_5 = p_2$ and $p_4 = p_3$. This indicates that the values of the member forces are symmetrically distributed about the centre line of the frame and once these are calculated for one half of the frame the forces in the second half become known. The forces in members crossing the axis of symmetry, p_7 in this case, have also to be calculated.

Using the principle of superposition and adding the loading cases 2 and 3 (*Figure 2.18b* and *c*) the general unsymmetrical loading case 1, shown in *Figure 2.18a*, is obtained. Similarly, adding the values of the member forces obtained for cases 2 and 3, given by equations (2.56) and (2.57), results in the forces given by equations (2.55) which is for the general case 1. It is therefore concluded that the analysis of a symmetrical frame subject to a general state of loading is obtained by adding the results of two analyses, one with the symmetrical part of the loads and the second with the antisymmetrical.

2.17. The equivalent half frame

Cases 2 and 3 show that, for the symmetrical and the antisymmetrical load systems, only half the frame needs to be analysed. Thus the original symmetrical frame can be replaced by two equivalent half frames. Under the symmetrical loads, points on the axis of symmetry move only along this axis. Thus in the equivalent frame these points are supported by rollers free to move along this axis only. Any external load that may be acting along the axis of symmetry is shared by the two halves of the frame, thus in the equivalent half frame such loads are halved. For the symmetrical loading case shown in *Figure 2.18c*, the equivalent half frame is shown in *Figure 2.19a* where it is noticed that joint C and point H are supported by rollers free to move vertically. A load $W_3/2$ is acting at C.

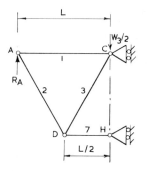
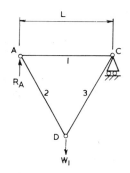

a. Equivalent half frame (symmetrical loading)

b. Equivalent half frame (antisymmetrical loading)

Figure 2.19

Under the antisymmetrical loads, points on the axis of symmetry move perpendicular to this axis. Thus in the equivalent frame these points are supported by rollers free to move perpendicular to the axis of symmetry. The force in a member which intersects the axis of symmetry perpendicularly is zero, thus in the equivalent frame such members are removed. For the antisymmetrical loading case shown in *Figure 2.18b*, the equivalent half frame is shown in *Figure 2.19b*, where joint C is supported by a roller free to move horizontally. Member 7 is removed.

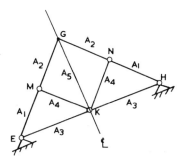

a. A symmetrical frame

Figure 2.20

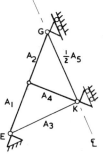

b. Equivalent half frame (symmetrical loading)

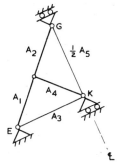

Equivalent half frame (antisymmetrical loading)

The calculation of the member forces in an isostatic pin jointed frame does not require the values of the cross sectional areas of the members. However, it was shown earlier that these areas are needed when the joint deflections of a frame are calculated. The symmetrical property of a frame can also be utilised in these deflection calculations. For this purpose, the area of any member that lies on the axis of symmetry is halved in the equivalent frame. The equivalent frames for that shown in *Figure 2.20a* are shown in *Figures 2.20b* and *2.20c*. It is noticed that the area A_5 of member GK is halved.

2.18. The equivalent loads

In *Figure 2.17* the joints to which external loads are applied are topologically symmetrical. Joint C is on the axis of symmetry while joints D and E symmetrically lie on either side of this axis. In a more general case, the frame may be symmetrical but the external loads may be applied unsymmetrically. Before making use of the symmetrical geometry of the frame, it is necessary to replace the unsymmetrically applied loads with equivalent systems which act at joints that are topologically symmetrical.

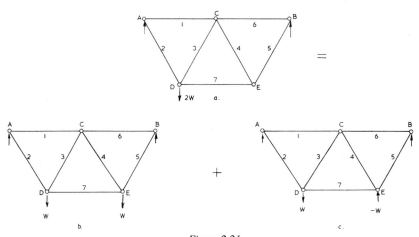

Figure 2.21

For the frame shown in *Figure 2.17* consider the following two further loading cases.

4. The unsymmetrical loading case where

$$W_1 = 2W, \quad W_2 = 0 \quad \text{and} \quad W_3 = 0$$

This is the case of a single vertical load $2W$ acting at D as shown in *Figure 2.21a*. The matrix equations (2.54) now give

$$\left.\begin{array}{ll} p_1 = -1.5W/\sqrt{3} & p_6 = -0.5W/\sqrt{3} \\ p_2 = 3W/\sqrt{3} & p_5 = W/\sqrt{3} \\ p_3 = W/\sqrt{3} & p_4 = -W/\sqrt{3} \\ p_7 = W/\sqrt{3} & \end{array}\right\} \quad (2.58)$$

5. The symmetrical loading case where

$$W_1 = W_2 = W \quad \text{and} \quad W_3 = 0$$

Structural Properties

Here both W_1 and W_2 are acting downwards while W_3 is removed as shown in *Figure 2.21b*. The matrix equations (2.54) give

$$\begin{aligned}
p_1 &= -W/\sqrt{3} & p_6 &= -W/\sqrt{3} \\
p_2 &= 2W/\sqrt{3} & p_5 &= 2W/\sqrt{3} \\
p_3 &= 0 & p_4 &= 0 \\
p_7 &= W/\sqrt{3}
\end{aligned} \qquad (2.59)$$

Thus the member forces are symmetrical as expected, Now adding the results of case 5, given by equations (2.59) to that of case 2, equations (2.55) with $W_1 = -W_2 = W$, we obtain the same results as those given by equations (2.58) for the unsymmetrical loading case with $2W$ acting at D. The loading case $W_1 = -W_2 = W$ is antisymmetrical and is shown in *Figure 2.21c*. It therefore follows that a single load, such as $2W$ acting at D, is equivalent to two loading systems. These are: A symmetrical system as shown in *Figure 2.21b* and an antisymmetrical system as shown in *Figure 2.21c*. In this manner any load can be replaced by its equivalent systems and applied to the equivalent frame. Some unsymmetrical loads and their equivalent frames and loads are shown in *Figure 2.22*.

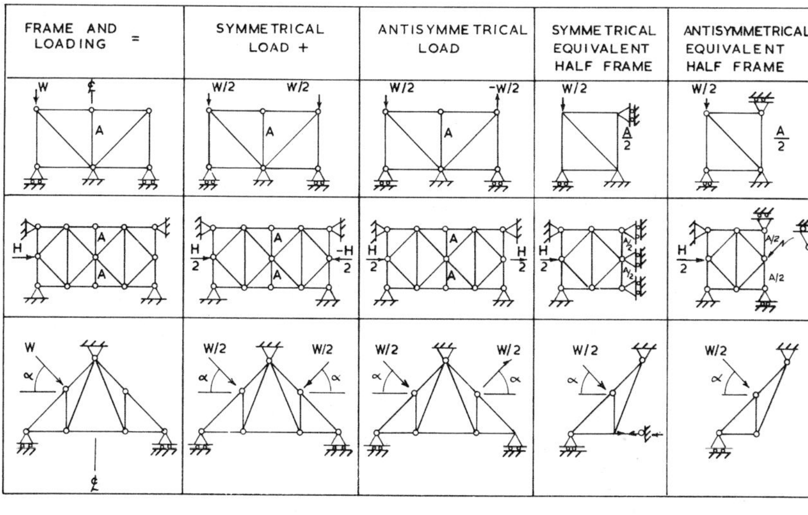

Figure 2.22

The analysis of the equivalent half frame requires the solution of half as many simultaneous equations as those for the whole frame. Even though two equivalent frames have to be analysed, the problem becomes simpler. When using a computer, there are two important advantages in making use of symmetry. These are:

1. A considerable amount of computer time is saved. It is estimated that the time required to solve a set of simultaneous equations increases cubically with the number of equations involved.
2. A considerable amount of computer storage is also saved because only half of the frame is given to the computer. In this manner the ability of a computer with a given storage to analyse large frames increases.

2.19. An advantage of matrix methods

In section 2.16 the member forces **P** were found by the method of joints and then expressed in the matrix form **P = B W** as shown by equations (2.54). Once this was done, the matrix equations were used repeatedly for five different loading cases. Thus the method of joints was used only once and the matrix equations made it possible to deal with the different load cases. Structures are often subjected to more than one loading case and an advantage of matrix equations is that a single analysis is sufficient to calculate the member forces for the various load cases.

2.20. The critical design load

As was stated earlier, a frame may be subjected to several individual load cases. To ensure that a member is safe it is necessary to calculate the forces in it for every individual load case and then design it for the most severe of these forces. The most severe external load case is not the same for all the members. This is especially true when it is realised that the limiting compressive stress in a member is numerically lower than the limiting tensile stress.

Consider the design of the pin jointed frame shown in *Figure 2.17*, section 2.16, to carry any one of the following three external load cases

(1) Load case 1: $W_1 = 10$ kN, $W_2 = -10$ kN and $W_3 = 0$
(2) Load case 2: $W_1 = W_2 = 0$, $W_3 = 10$ kN
(3) Load case 3: The combination of the first two
 i.e. $W_1 = 10$ kN, $W_2 = -10$ kN and $W_3 = 10$ kN

The limiting compressive stress is -0.18 kN/mm^2 and the limiting tensile stress is 0.27 kN/mm^2.

Once again the matrix equation (2.54) is used to express the member forces in terms of the external loads. The three separate external load cases are written as columns of the load matrix **W**, thus

$$\mathbf{W} = \begin{array}{c} \text{Case 1} \quad \text{Case 2} \quad \text{Case 3} \\ \begin{bmatrix} 10 & 0 & 10 \\ -10 & 0 & -10 \\ 0 & 10 & 10 \end{bmatrix} \end{array}$$

which is then used in equation (2.54) instead of the load vector $\mathbf{W} = \{W_1 \ W_2 \ W_3\}$. Equations (2.54) then become

$$\begin{bmatrix} p_1 \\ p_2 \\ p_3 \\ p_4 \\ p_5 \\ p_6 \\ p_7 \end{bmatrix} = 0.5/\sqrt{3} \begin{bmatrix} -1.5 & -0.5 & -1 \\ 3 & 1 & 2 \\ 1 & -1 & -2 \\ -1 & 1 & -2 \\ 1 & 3 & 2 \\ -0.5 & -1.5 & -1 \\ 1 & 1 & 2 \end{bmatrix} \begin{bmatrix} 10 & 0 & 10 \\ -10 & 0 & -10 \\ 0 & 10 & 10 \end{bmatrix} \quad (2.60)$$

or $\quad\quad\quad \mathbf{P = B W}$ (2.60a)

By matrix multiplication equations (2.60) give the member forces for each load case. For the first load case, p_1 is found by multiplying the first row of matrix **B** by

Structural Properties

the first column of matrix **W**, thus

Case 1: $p_1 = (0.5/\sqrt{3})(-1.5 \times 10 - 0.5 \times -10 - 1 \times 0) = -5/\sqrt{3}$

Similarly multiplying the first row of **B** by the second and third columns of **W** we obtain

Case 2: $p_1 = (0.5/\sqrt{3})(-1.5 \times 0 - 0.5 \times 0 - 1 \times 10) = -5/\sqrt{3}$
Case 3: $p_1 = (0.5/\sqrt{3})(-1.5 \times 10 - 0.5 \times -10 - 1 \times 10) = -10/\sqrt{3}$

It is noticed the most severe force in member 1 is caused by the load case 3. The forces in the other members are obtained from equation (2.60) similarly. These are given by equations (2.61) where the most severe force in each member is underlined.

$$\begin{bmatrix} p_{11} & p_{12} & p_{13} \\ p_{21} & p_{22} & p_{23} \\ p_{31} & p_{32} & p_{33} \\ p_{41} & p_{42} & p_{43} \\ p_{51} & p_{52} & p_{53} \\ p_{61} & p_{62} & p_{63} \\ p_{71} & p_{72} & p_{73} \end{bmatrix} = \frac{1}{\sqrt{3}} \begin{bmatrix} -5 & -5 & \underline{-10} \\ 10 & 10 & \underline{20} \\ \underline{10} & -10 & 0 \\ -10 & -10 & \underline{-20} \\ \underline{-10} & 10 & 0 \\ 5 & \underline{-5} & 0 \\ 0 & 10 & \underline{10} \end{bmatrix} \quad (2.61)$$

Since the limiting compressive stress is more critical than the limiting tensile stress, the first load case is the critical design load for member 5. For member 7 both the second and the third load cases are critical.

The area A_1 for member 1 is obtained by dividing the most severe force in that member by the maximum limiting stress, thus

$A_1 = -10/(\sqrt{3} \times -0.18) = 32 \text{ mm}^2$

Similarly

$A_2 = 20/(\sqrt{3} \times 0.27) = 43 \text{ mm}^2$; $\quad A_3 = -10/(\sqrt{3} \times -0.18) = 32 \text{ mm}^2$;
$A_4 = -20/(\sqrt{3} \times -0.18) = 64 \text{ mm}^2$; $\quad A_5 = -10/(\sqrt{3} \times -0.18) = 32 \text{ mm}^2$;
$A_6 = -5/(\sqrt{3} \times -0.18) = 16 \text{ mm}^2$; $\quad A_7 = 10/(\sqrt{3} \times 0.27) = 21.4 \text{ mm}^2$.

2.21. Example

Calculate the member forces in the pin jointed frame shown in *Figure 2.23*.

Answer: The equivalent half frames are shown in *Figures 2.23b* and *2.23c*. It is noticed that while this frame is hyperstatic, both its equivalent half frames are in fact isostatic.

For the symmetrical half frame (*Figure 2.23b*), resolving vertically at D we obtain $p_5 = 0$. Resolving vertically at C, $p_3 = 0$. Joint C is therefore unloaded.

Resolving the forces horizontally and vertically at B, we obtain

$p_2 = -\sqrt{3} \times W/4$ (compressive)
$p_1 = -W/4$ (compressive)

The only force acting at level ACJ is p_1 which is vertical and acting at the support A. Thus $H_A = H_J = p_4 = p_6 = 0$. Vertical equilibrium at A gives $R_A = W/4$ while horizontal equilibrium at D gives $H_D = -\sqrt{3} \times W/4$.

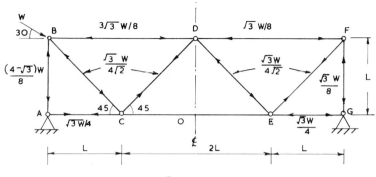

a. Frame and loading

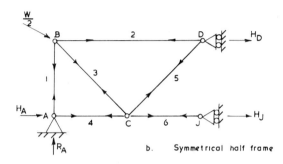

b. Symmetrical half frame

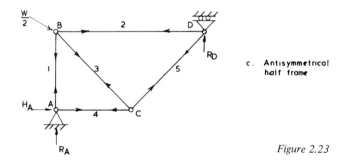

c. Antisymmetrical half frame

Figure 2.23

For the antisymmetrical half frame (*Figure 2.23c*), taking moments about A

$$L \times 0.5W \times \cos 30 - R_D \times 2L = 0$$

∴ $R_D = \sqrt{3} \times W/8.$

Resolving vertically at D

$$R_D - p_5 \cos 45 = 0$$

∴ $p_5 = \sqrt{6} \times W/8$ \qquad (tensile)

Resolving horizontally at D

$$p_2 + p_5 \cos 45 = 0$$

∴ $p_2 = -p_5/\sqrt{2} = -\sqrt{3} \times W/8$ \qquad (compressive)

Resolving vertically at C

$$p_5 \cos 45 + p_3 \cos 45 = 0$$

∴ $p_3 = -p_5 = -\sqrt{6} \times W/8$ (compressive)

Resolving horizontally at C

$$p_5 \cos 45 - p_3 \cos 45 - p_4 = 0$$

∴ $p_4 = \sqrt{3} \times W/4$ (tensile)

Resolving vertically at B

$$0.5W \sin 30 + p_1 + p_3 \cos 45 = 0$$

∴ $p_1 = -W(2 - \sqrt{3})/8$ (compressive)

The forces to the right of the axis of symmetry have the opposite sign to those calculated for the antisymmetrical case. In the symmetrical case the member forces are symmetrical with respect to this axis. The actual member forces are obtained by superposition of the two cases. These are shown in *Figure 2.23a*.

Exercises on Chapter 2

1. Calculate the elongation of a bar of length L and cross-sectional area A which hangs vertically under its weight W.

Ans. $\delta = WL/2AE$.

2. A bar rotates horizontally about a point midway between its ends at an angular velocity of π rad/s. It is 10 m long, and has a self weight of 10 kN. The cross sectional area of the bar is 100 mm^2 and its modulus of elasticity is 200 kN/mm^2. With the gravitational acceleration as 9.81 m/s^2 calculate its elongation.

Ans. 42 mm.

3. The cross sectional area of portion AB of the vertical member shown in *Figure 2.24* is 200 mm^2, that of BC is 100 mm^2 and the modulus of elasticity is 200 kN/mm^2. Calculate the value of the force W that makes the vertical displacement of point C equal to zero.

Ans. 4 kN.

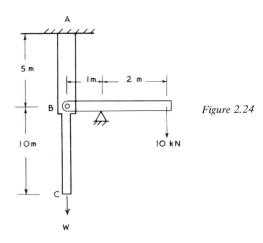

Figure 2.24

4. A 100 kN steam hammer falls from a height of 2 m on the free end of a 10 m long vertical pile. The other end of the pile is fixed and its area is 25×10^4 mm^2. Using the work equation calculate the stress and the contraction of the pile after impact. Take E as 24 kN/mm^2.

Ans. $\sigma = 0.0624$ kN/mm^2, $\delta = 26$ mm

5. The members shown in *Figure 2.25* have circular cross sections and each one carries a vertical load W of 10 kN as shown. Calculate the strain energy in each. Take $E = 200$ kN/mm^2.

Ans. $U_1 = 31.8$ kN mm, $U_2 = 13.9$ kN mm, $U_3 = 6.36$ kN mm.

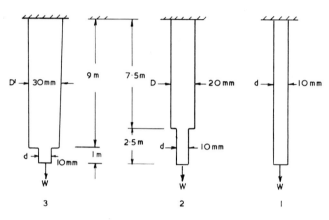

Figure 2.25

6. The cross sectional areas of the members of the pin jointed frame shown in *Figure 2.26* are all 100 mm^2 and $E = 200$ kN/mm^2. Calculate the horizontal deflection of joint E under the loading shown.

Ans. 2.91 mm.

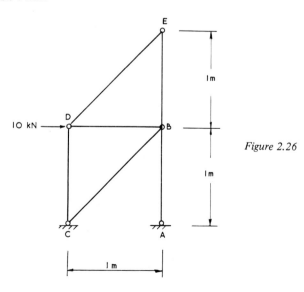

Figure 2.26

7. The members of the frame shown in *Figure 2.26* are made out of steel. Calculate the total deflection of joint E when members DE and BE are heated by 100°C.

Ans. 0.57 mm at 45° to the vertical.

8. Using symmetry, calculate the member forces in the frame shown in *Figure 2.27a*.
Ans. See *Figure 2.27b*.

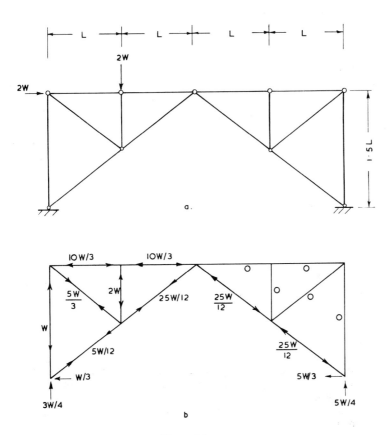

Figure 2.27

3

Isostatic beams

3.1. Introduction

The last two chapters examined the equilibrium of frames supporting loads applied at their joints. In many cases the members of a structure are subject to lateral forces along their lengths. Such members are known as beams. For instance, in *Figure 3.1* a beam is shown simply supported at its ends A and B, carrying a load W at point C. The load causes the beam to deform in two different manners. Firstly, the load bends the beam, as shown in *Figure 3.1b*. This causes tensile stresses to develop in the

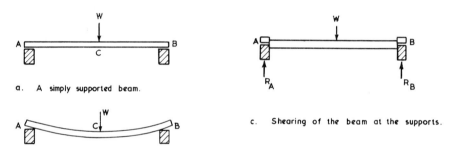

a. A simply supported beam.

b. Bending of the beam.

c. Shearing of the beam at the supports.

Figure 3.1

lower portion of the beam and compressive stresses to develop in the upper portion. Secondly, the load tends to push the beam downwards while the support reactions R_A and R_B prevent this action of the load with the result that the part of the beam between the supports may shear off as shown in *Figure 3.1c*. Thus shear stresses are also developed in the beam.

There are various types of beam and some of these are shown in *Figure 3.2*. The beam in *Figure 3.2a* is called a cantilever which is fixed to its support at A and free, i.e. unsupported, at B. That in *Figure 3.2b* is a simply supported beam. The supports may be idealised as knife edges or frictionless hinges. The beam in *Figure 3.2c* is called a propped cantilever. This beam is fixed at A and simply supported at B by a knife edge. *Figure 3.2d* shows an encastre beam which is built in, i.e. fixed, to the supports. The beam in *Figure 3.2e* is simply supported by knife edges at A, B, E and F. The parts AC, CD and DF are connected at C and D by frictionless hinges. This is a compound beam often used as a bridge. *Figure 3.2f* shows a beam continuous over the simple support at B. The beams shown in *Figure 3.2a, b* and *e* are isostatic because the equations of static equilibrium can, on their own, be used to calculate the support reactions and the internal forces. On the other hand, the beams shown in *Figure 3.2c, d* and *f* are hyperstatic. This chapter is only concerned with isostatic beams.

Various types of loads may be applied to a beam. These may be vertical point loads as in *Figure 3.2a, b, e* and *f*. They may be inclined point loads as in *Figure 3.2c*. The beam in *Figure 3.2g* is subject to a uniform load w per unit length along its span. The cantilever in *Figure 3.2h* is subject to an external moment M at its free end D, a vertical point load at C and a uniformly distributed load between B and D.

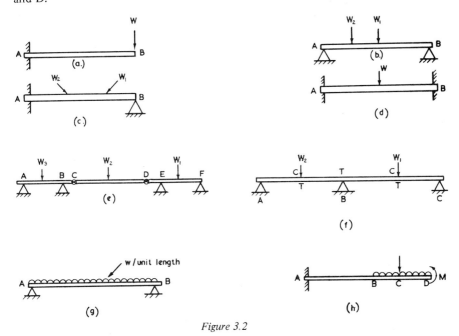

Figure 3.2

3.2. Shearing force and bending moments in beams

3.2.1. Definitions

The shearing force at a point in a beam is the algebraic sum of all the forces acting to one side of the point. The bending moment at a point is the algebraic sum of the moments of all the forces, acting to one side of the point, about that point. The calculation of the shearing forces and bending moments at all points along a beam produce graphs known as shearing force and bending moment diagrams for the beam. Such diagrams are constructed for various types of beams in the following sections.

3.2.2. A cantilever with an end load

As stated, a cantilever is a beam rigidly supported at one end and free at the other end. Consider first the simple case of a cantilever of span L supporting a vertical load W at its free end as shown in *Figure 3.3a*. Any portion of the beam such as AD is in vertical equilibrium because while the force W pushes AD downwards, the portion DB resists W by an internal upward force F as shown in *Figure 3.3b*. From the vertical equilibrium of AD

$$-W + F = 0$$
$$\therefore \quad F = W \tag{3.1}$$

Thus the upward force F is equal in magnitude to W and is called the shearing force at D. Vertical equilibrium at D gives the shearing effect of AD on DB as a downward force F acting at D.

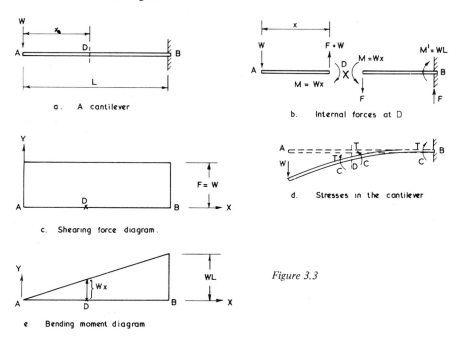

Figure 3.3

The force W also bends the beam in an anticlockwise direction as shown in *Figure 3.3d*. At D the portion DB resists the bending action of W by an internal moment M which is also shown in *Figure 3.3b*. The moment of W about point D is equal to Wx where the moment arm x is measured from the origin at A. For the rotational equilibrium of AD

$$Wx - M = 0$$
$$\therefore \quad M = Wx \tag{3.2}$$

The anticlockwise moment Wx of W about D is the bending moment at D. The convention is to consider this anticlockwise moment as positive. The resisting moment M acting at D on AD is clockwise and therefore negative. Point D, marked by a cross in *Figure 3.3b*, can be considered as a joint at which the portion AD and DB are rigidly connected. The rotational equilibrium at this joint shows that the portion DB is subjected to an anticlockwise positive moment $M = Wx$.

The portion DB of the beam is also in equilibrium. The vertical equilibrium of forces on DB gives the support reaction at B equal to F. The rotational equilibrium of DB is stated mathematically as

$$M + F(L - x) - M' = 0$$
$$\therefore \quad M' = M + F(L - x)$$

The moments M and $F(L - x)$ acting on DB are both anticlockwise and therefore positive while the clockwise resisting moment M' is negative. M' is the reactive moment of the support at B. Now since $M = Wx$ and $F(L - x) = W(L - x)$, it follows that

$$M' = Wx + W(L - x) = WL \tag{3.3}$$

Alternatively the reactive moment at B is calculated by considering the equilibrium of the whole beam AB, *Figure 3.3a*. Here the moment of W about B is directly given as WL.

In the case of this cantilever, it is evident that the value of the shearing force F, at any point, is independent of the position of that point. Thus a graph of the shearing force, for points along the cantilever, is a straight horizontal line as shown in *Figure 3.3c*. This graph is the shearing force diagram for the cantilever. Its ordinate at any point gives the value of the shearing force in the beam at that point.

On the other hand, the value of the bending moment at any point is equal to Wx which is dependent on the distance x of the point from the origin at A. A graph of the bending moment, for points along the beam, is thus a straight line as shown in *Figure 3.3e*. This is the bending moment diagram for the cantilever. The ordinate of this graph at A is zero and at B is WL.

3.2.3. Sign conventions

(a) For the co-ordinate axes: The X-Y axes are called the 'global' axes and refer to a whole structure of one or more members. The X axis is positive to the right and the Y axis is positive upwards. Each member of a structure has its own 'local' axes. These are the longitudinal and the lateral axes of the member and in the case of an inclined member these do not coincide with the global axes. The local axes of such an inclined member are specified in section 4.8. In *Figure 3.3a*, the positive longitudinal axis of the cantilever is from A to B which coincides with the positive X axis. The positive lateral axis is from A upwards and coincides with the positive Y axis.

(b) For moments and shear forces: By the right hand screw rule, anticlockwise moments and rotations are considered positive. Forces tending to rotate a portion of a member anticlockwise cause positive shear force in that portion. For instance, the force W and F in *Figure 3.3b* tend to rotate AD anticlockwise. The shearing force in AD is therefore positive.

(c) For shearing force and bending moment diagrams: The ordinate of a positive shearing force lies on the same side as the positive lateral axis of a member. In *Figure 3.3*, the shearing force is positive everywhere along the beam. The shearing force diagram therefore lies above the line AB as shown in *Figure 3.3c*.

Bending a member causes tension on one side and compression on the other side of the member. *Figure 3.3d* shows the deflected shape of the cantilever. The top portion of the beam has elongated and is in tension while the bottom portion is in compression. The letters T and C, shown in the figure, clarify this. The ordinates of the bending moment diagram, drawn from the longitudinal axis of a member, lie on the side of the member which is in tension. When these ordinates appear on the side of the positive lateral axis they are considered to be positive. In *Figure 3.3e*, the bending moment diagram, drawn on the tension side of the beam, lies above AB, i.e. on the side of the positive lateral axis. The ordinates of the diagram are therefore positive. In a different beam, tension may develop on the same side as the positive lateral axis only for parts of its length. The ordinates of the bending moment diagram will be positive only in these parts. Elsewhere, these ordinates will be negative. For instance, in *Figure 3.2f*, it is indicated that tension develops at the top of the beam at B and at the bottom of the beam under the loads W_1 and W_2. The bending moment diagram therefore lies above the longitudinal axis of the beam at B and below it under the loads. Considering this axis to be positive from left to right, the ordinate of the bending moment diagram will be upwards and positive at B and negative under the loads.

It is easy to identify the side of a member which is in tension. This is because the arrow head of the symbol (↻) for moments, in the deflected members, is always on the tension side. In *Figure 3.3d*, for instance, all the arrow heads at D and B are on the top side which is in tension.

3.2.4. A cantilever carrying an inclined force

Consider a cantilever carrying an inclined force W as shown in *Figure 3.4a*. This force is resolved into its horizontal and vertical components in *Figure 3.4b*. The horizontal component $P = W \cos \theta$ causes an axial tension equal to P to develop in the beam. This is shown in *Figure 3.4c*. The vertical component $W \sin \theta$ is shown applied to the end of the beam at A in *Figure 3.4d*. At any point, such as D, distance x from A, this component causes a shearing force $F = W \sin \theta$ and a bending moment $M = Wx \sin \theta$ to develop. Thus the shearing force and the bending moment diagrams for this cantilever are similar to those for the cantilever shown in *Figure 3.3*, except that the ordinates of the graphs are scaled by $\sin \theta$. These diagrams are shown in *Figures 3.4e* and *3.4f* respectively.

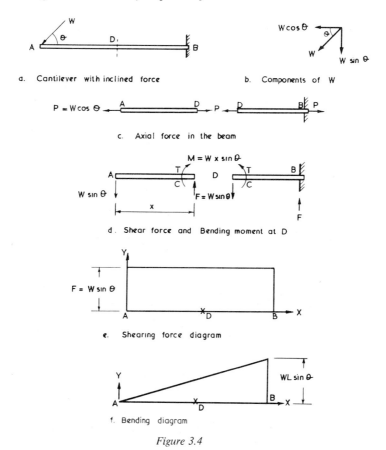

Figure 3.4

3.2.5. A uniformly loaded cantilever

The cantilever shown in *Figure 3.5a* carries a uniformly distributed load of intensity w per unit length of the beam. The shearing force at any point D, a distance x from

A, is F, which is equal to the sum of the forces acting to the left of the point. Now each unit length of AD carries a load w and the total load carried by AD is wx, thus

$$F = wx \tag{3.4}$$

The shearing force is therefore directly proportional to x and the shearing force diagram is a straight line whose ordinate increases from zero at A to a maximum wL at B as shown in *Figure 3.5b*.

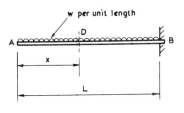

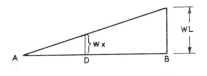

b.   Shearing force diagram

a.   A uniformly load cantilever

Figure 3.5

c.   Bending moment diagram

The centre of gravity of the load wx acting on AD is at a distance $x/2$ from D. The moment of this force about point D is $wx \times 0.5x$; thus the bending moment M at D is

$$M = 0.5wx^2 \tag{3.5}$$

Similarly the bending moment at B is

$$M_B = 0.5wL^2 \tag{3.6}$$

The bending moment at a point is thus proportional to x^2 and the bending moment diagram is parabolic as shown in *Figure 3.5c*.

3.2.6. A cantilever with several loads

Finally consider a cantilever subject to several point loads. In *Figure 3.6*, for instance, the cantilever is subject to three point loads. There is no load acting between A and C. Thus, by definition, the shearing force and the bending moment between these two points have zero value.

At any point between C and D the shearing force is equal to W_1. Between D and E the shearing force is $W_1 + W_2$. Finally between E and B the shearing force is $W_1 + W_2 + W_3$. The shearing force diagram is shown in *Figure 3.6b*.

At a point distance x to the right of C the bending moment is

$$M_x = W_1 x$$

and at D, where $x = L_2$, the bending moment is

$$M_D = W_1 L_2$$

The bending moment at a distance y to the right of point D is

$$M_y = W_1(L_2 + y) + W_2 y$$

and at E, where $y = L_3$, the bending moment is

$$M_E = W_1(L_2 + L_3) + W_2 L_3$$

Finally, the bending moment at a point, distance z from E, is

$$M_Z = W_1(L_2 + L_3 + z) + W_2(L_3 + z) + W_3 z$$

and at B where $z = L_4$, the bending moment is

$$M_B = W_1(L_2 + L_3 + L_4) + W_2(L_3 + L_4) + W_3 L_4$$

The bending moment diagram is shown in *Figure 3.6c*.

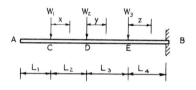

a. Cantilever with three loads

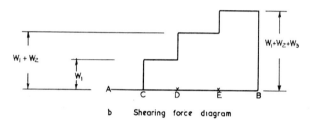

b. Shearing force diagram

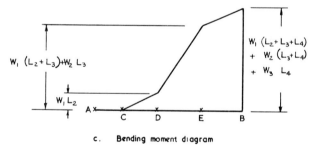

Figure 3.6

c. Bending moment diagram

3.2.7. Simply supported beams

A simply supported beam is constrained at its ends by knife edges or frictionless hinges. The supports are thus incapable of resisting bending moments. This means that the bending moment at each end of the beam is zero. Consider a simply supported beam of span L carrying a point load W at a distance a from the left hand support as shown in *Figure 3.7*.

The vertical reaction R_A at support A is calculated by taking moments about B, thus

$$Wb - R_A L = 0$$

$$\therefore \quad R_A = Wb/L \tag{3.7}$$

Notice that the moment Wb about B is anticlockwise and therefore positive while the moment $R_A L$ of R_A about B is clockwise and negative.

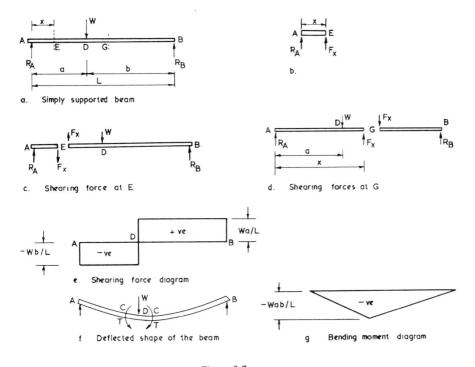

Figure 3.7

The reaction R_B, at support B, is calculated from the vertical equilibrium of the forces, thus

$$R_B + R_A - W = 0$$

Substituting for R_A from equation (3.7), we obtain

$$R_B = Wa/L \tag{3.8}$$

From the vertical equilibrium of a portion AE, with E being at a distance x from A, *Figure 3.7b*

$$R_A + F_X = 0$$

∴ $$F_X = -R_A = -Wb/L \tag{3.9}$$

This means F_X at E is downwards as shown in *Figure 3.7c*. The forces R_A and F_X tend to rotate AE in a clockwise manner. The shearing force diagram for AE therefore has negative ordinates. When $x = a$, at D, the shearing force, which is independent of x, remains at its constant value of $-Wb/L$.

When $x > a$, at a point such as G between D and B (*Figure 3.7d*), the vertical equilibrium of AG gives

$$R_A + F_X - W = 0$$

∴ $$F_X = W - R_A = W - Wb/L = Wa/L = R_B \tag{3.10}$$

This means that F_X at G acting on the portion AG of the beam is upwards and positive. Comparing equations (3.9) and (3.10), or *Figures 3.7c* and *3.7d*, it is noticed that the shearing force has changed its sign once $x > a$. The forces acting on

DG rotate it anticlockwise. The ordinate of the shearing force diagram also changes its sign and becomes positive. Once again, the shearing force remains constant at Wa/L up to point B. The shearing force diagram is shown in *Figure 3.7e*.

The deflected shape of the beam is shown in *Figure 3.7f* which indicates that the lower side of the beam is in tension throughout. The bending moment diagram, drawn on the tension side of the beam, therefore has negative ordinates throughout. The bending moment M_X at E is obtained by taking moments about E, of all the forces to the left of E. There is only one such force, R_A, thus

$$M_X = -R_A x = -Wbx/L$$

Notice the moment $R_A x$ is clockwise and therefore negative.

When $x = a$, at D, the bending moment becomes

$$M_D = -Wab/L \qquad (3.11)$$

For $x > a$, between D and B, the bending moment M_X is

$$M_X = -R_A x + W(x - a)$$

i.e. $\qquad M_X = -(Wbx/L) + W(x - a) \qquad (3.12)$

and when $x = L$, that is at B, equation (3.12) gives M_B as

$$M_B = -(WbL/L) + W(L - a) = 0$$

On the other hand, when $x = a$, that is at D, equation (3.12) gives $M_D = -Wab/L$ which is the same as we obtained in equation (3.11). The bending moment diagram is shown in *Figure 3.7g*.

It is noticed that at D the shearing force diagram changes its sign, crossing the horizontal axis of the beam and that the bending moment has its greatest magnitude at this point. Mathematically, however, because the bending moment diagram has a negative ordinate at D, its value is minimum at this point.

For this beam, when $a = b = L/2$, which is the case of a simply supported beam loaded at its midspan, the shearing force between B and D becomes $W/2$ and that between A and D becomes $-W_2$. The minimum bending moment at midspan D, becomes $-WL/4$.

Consider next a simply supported beam of span L, carrying a uniformly distributed load of intensity w per unit length, as shown in *Figure 3.8*. Taking moments of the applied load and the reaction R_A about B, we obtain

$$R_A = 0.5wL$$

and from the vertical equilibrium of the beam

$$R_B = 0.5wL$$

From the vertical equilibrium of the portion AD, it follows that the shearing force F_X at D is

$$F_X = -R_A + wx$$

i.e. $\qquad F_X = -0.5wL + wx \qquad (3.13)$

and when $x = L/2$, at midspan E, equation (3.13) gives $F_X = 0$. The shearing force at midspan is therefore zero. For values of $x > L/2$, equation (3.13) shows that the value of the shearing force F_X becomes positive and when $x = L$, at B,

$$F_B = -0.5wL + wL = +0.5wL \qquad (3.14)$$

The deflected shape of the beam indicates that its lower side is in tension throughout. The ordinates of the bending moment diagram are therefore negative for the whole span.

Isostatic beams

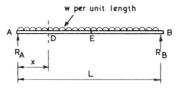

a. A simply supported beam with uniform loading.

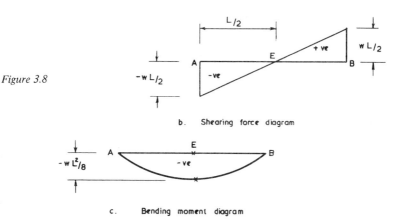

Figure 3.8

b. Shearing force diagram

c. Bending moment diagram

The bending moment M_X at D is

$$M_X = -R_A x + wx \times 0.5x$$

i.e. $$M_X = -0.5wLx + 0.5 wx^2 \tag{3.15}$$

and when $x = L/2$, the bending moment M_E is, from equation (3.15), given as

$$M_E = -0.125wL^2 \tag{3.16}$$

When $x = L$, at B, equation (3.15) gives

$$M_B = -0.5wL^2 + 0.5wL^2 = 0$$

Equation (3.15) also shows that the shape of the bending moment diagram is parabolic. At E the shear force is zero and the bending moment is mathematically a minimum.

3.2.8. Examples

Example 1. Calculate the shearing forces and the bending moments at A, D, E, G and B for the beam shown in *Figure 3.9* and tabulate the results in terms of W_1, W_2 and W_3 as a matrix equation $\mathbf{P} = \mathbf{B}\,\mathbf{W}$

Answer: Taking moments about B for the whole beam

$$-10R_A + W_1(10-2) + W_2(10-4) + W_3(10-6) = 0$$

$$\therefore \quad R_A = 0.8W_1 + 0.6W_2 + 0.4W_3$$

At point H, distance x from A, the shearing force F_X is

$$F_X = W_1 + W_2 + W_3 - R_A = R_B$$

$$F_X = W_1 + W_2 + W_3 - 0.8W_1 - 0.6W_2 - 0.4W_3$$
$$F_X = 0.2W_1 + 0.4W_2 + 0.6W_3$$

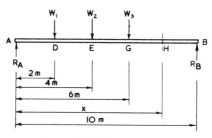

a. beam and loading

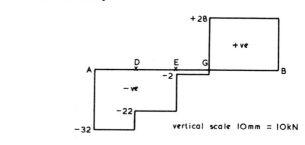

b. Shearing force diagram

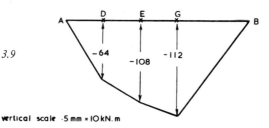

Figure 3.9

c. Bending moment diagram

This is the value of the shearing force between G and B. The shearing force at a point is the sum of all the forces to one side of that point. Thus the forces to the left of G are W_1, W_2 and R_A. The shearing force at G is therefore given by

$$F_G = W_1 + W_2 - R_A = 0.2W_1 + 0.4W_2 - 0.4W_3$$

Similarly at E, F_E becomes

$$F_E = W_1 - R_A = W_1 - 0.8W_1 - 0.6W_2 - 0.4W_3$$
$$F_E = 0.2W_1 - 0.6W_2 - 0.4W_3$$

and at D, F_D becomes

$$F_D = -R_A = -0.8W_1 - 0.6W_2 - 0.4W_3$$

The value of the shearing force at A is also equal to

$$-R_A = -0.8W_1 - 0.6W_2 - 0.4W_3$$

Isostatic beams

The values of the shearing forces for AD, DE, EG and GB are thus written in matrix form as

$$\begin{bmatrix} F_{AD} \\ F_{DE} \\ F_{EG} \\ F_{GB} \end{bmatrix} = \begin{bmatrix} -0.8 & -0.6 & -0.4 \\ 0.2 & -0.6 & -0.4 \\ 0.2 & 0.4 & -0.4 \\ 0.2 & 0.4 & 0.6 \end{bmatrix} \begin{bmatrix} W_1 \\ W_2 \\ W_3 \end{bmatrix} \quad (3.17)$$

At point H, distance x from A, the bending moment M_x is

$$M_x = -R_A x + W_1 \{x-2\} + W_2 \{x-4\} + W_3 \{x-6\}$$
$$M_x = -x(0.8W_1 + 0.6W_2 + 0.4W_3) + W_1\{x-2\} + W_2\{x-4\} + W_3\{x-6\} \quad (3.18)$$

At B, where $x = 10$, equation (3.18) gives

$$M_B = 0$$

The bending moment at a point is the sum of all the moments of the loads acting to one side of that point. The forces to the left of G are W_1, W_2 and R_A. At this point, the term $W_3\{x-6\}$ in equation (3.18) is therefore disregarded. Notice that when $x = 6$, at G, $\{x-6\} = 0$ and the bending moment M_G at G becomes

$$M_G = -6(0.8W_1 + 0.6W_2 + 0.4W_3) + W_1\{6-2\} + W_2\{6-4\}$$
$$M_G = -0.8W_1 - 1.6W_2 - 2.4W_3$$

For points between E and G, $\{x-6\}$ in equation (3.18) becomes negative and the term $W_3\{x-6\}$ is again disregarded. Physically this means that W_3 is to the right of the point being considered. With this in mind and with $x = 4$, equation (3.18) gives the bending moment M_E at E as

$$M_E = -4(0.8W_1 + 0.6W_2 + 0.4W_3) + W_1\{4-2\} + W_2\{4-4\}$$

i.e.
$$M_E = -1.2W_1 - 2.4W_2 - 1.6W_3$$

For points between D and E, both $\{x-6\}$ and $\{x-4\}$ in equation (3.18) become negative. Disregarding therefore terms with negative values in the curly brackets, the bending moment M_D at D, with $x = 2$, is given by equation (3.18) as

$$M_D = -2(0.8W_1 + 0.6W_2 + 0.4W_3) + W_1\{2-2\}$$
$$M_D = -1.6W_1 - 1.2W_2 - 0.8W_3$$

Between A and D all the terms in the curly brackets become negative and therefore disregarded and with $x = 0$, equation (3.18) gives the bending moment $M_A = 0$.
In matrix form, the bending moments at A, D, E, G and B, are tabulated as

$$\begin{bmatrix} M_A \\ M_D \\ M_E \\ M_G \\ M_B \end{bmatrix} = \begin{bmatrix} 0 & 0 & 0 \\ -1.6 & -1.2 & -0.8 \\ -1.2 & -2.4 & -1.6 \\ -0.8 & -1.6 & -2.4 \\ 0 & 0 & 0 \end{bmatrix} \begin{bmatrix} W_1 \\ W_2 \\ W_3 \end{bmatrix} \quad (3.19)$$

Example 2. For the beam in *Figure 3.9*, draw the shearing force and the bending moment diagrams for the case when

$$W_1 = 10 \text{ kN}, \quad W_2 = 20 \text{ kN} \text{ and } W_3 = 30 \text{ kN}.$$

Answer: From the matrix equation (3.17), with $W_1 = 10$, $W_2 = 20$ and $W_3 = 30$

70   Isostatic beams

$$F_{AD} = -0.8 \times 10 - 0.6 \times 20 - 0.4 \times 30 = -32 \text{ kN}$$
$$F_{DE} = 0.2 \times 10 - 0.6 \times 20 - 0.4 \times 30 = -22 \text{ kN}$$
$$F_{EG} = 0.2 \times 10 + 0.4 \times 20 - 0.4 \times 30 = -2 \text{ kN}$$

and
$$F_{GB} = 0.2 \times 10 + 0.4 \times 20 + 0.6 \times 30 = +28 \text{ kN}$$

The shearing force diagram is shown in *Figure 3.9b*.

From the matrix equation (3.19), with $W_1 = 10$, $W_2 = 20$ and $W_3 = 30$

$$M_A = 0 \times 10 + 0 \times 20 + 0 \times 30 = 0 \text{ kN m}$$
$$M_D = -1.6 \times 10 - 1.2 \times 20 - 0.8 \times 30 = -64 \text{ kN m}$$
$$M_E = -1.2 \times 10 - 2.4 \times 20 - 1.6 \times 30 = -108 \text{ kN m}$$
$$M_G = -0.8 \times 10 - 1.6 \times 20 - 2.4 \times 30 = -112 \text{ kN m}$$

and
$$M_B = 0 \times 10 + 0 \times 20 + 0 \times 30 = 0 \text{ kN m}$$

The bending moment diagram is shown in *Figure 3.9c*. Notice that at G the shearing force diagram crosses the X axis of the beam. At this point, the bending moment is at its minimum at -112 kN m.

3.3. Relationships for the load, the shearing force and the bending moment

Consider an element ABED of a beam under a general system of loading as shown in *Figure 3.10*. The length of the element is δx and it is subject to a load of intensity

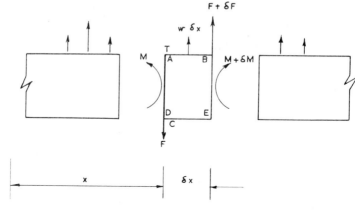

Figure 3.10

w which is assumed to be uniform over the small distance δx. The internal shearing force is assumed to rotate the element in an anticlockwise direction and therefore the shearing force diagram is considered to have positive ordinates over the length of the element. The magnitude of the shearing force at the vertical section AD is F and at the vertical section BE is $F + \delta F$. The bending moments at these sections are M and $M + \delta M$ respectively. These develop tension in the top portion of the element and thus the ordinate of the bending moment diagram is positive.

The vertical equilibrium of the element gives

$$w\delta x + (F + \delta F) - F = 0$$

and if δx is infinitesimally small,

Isostatic beams

$$\frac{dF}{dx} = -w \tag{3.20}$$

Thus the rate of change of F with respect to x is equal to $-w$.

Integrating between two points with distances x_1 and x_2 from the origin at the left end of the beam

$$\int_{x_1}^{x_2} dF = -\int_{x_1}^{x_2} w\,dx$$

If F_1 and F_2 are the shearing forces at x_1 and x_2 respectively, then

$$F_2 - F_1 = -\int_{x_1}^{x_2} w\,dx$$

or
$$F_1 - F_2 = \int_{x_1}^{x_2} w\,dx \tag{3.21}$$

Thus the decrease in the shearing force between x_1 and x_2 is equal to the area below the load distribution curve over this length of the beam.

Taking moments about point D for the element

$$(F + \delta F)\,\delta x + w\delta x \times 0.5\delta x + M - (M + \delta M) = 0$$

Neglecting small quantities with higher order $\delta F \delta x$, δx^2, it follows that

$$\delta M = F\delta x$$

and for infinitesimal δx

$$dM/dx = F \tag{3.22}$$

Thus the rate of change of M with respect to x is equal to the shearing force. Equation 3.22 also indicates that the bending moment M is stationary when the shearing force F is zero. Furthermore equations (3.22) and (3.20) give

$$d^2M/dx^2 = dF/dx = -w \tag{3.23}$$

Thus when the shearing force is zero, the bending moment is mathematically a maximum since d^2M/dx^2 is negative. This is the case when upward forces are considered positive. Consequently for a beam subject to forces acting downwards the bending moment is minimum when the shear force is zero.

Integrating equation (3.22) between the limits x_1 and x_2

$$\int_{x_1}^{x_2} dM = \int_{x_1}^{x_2} F\,dx$$

∴
$$M_2 - M_1 = \int_{x_1}^{x_2} F\,dx \tag{3.24}$$

Thus the increase in bending moment from M_1 at x_1 to M_2 at x_2 is given by the area below the graph for shearing force between the two points x_1 and x_2. Equations (3.21) and (3.24) are used to calculate the shearing force and the bending moment for beams subject to irregular loads.

From equation (3.21), the shearing force at a distance x from the first, i.e. left, end of the beam is given by

$$F = F_1 - \int_{x_1}^{x} w \, dx \tag{3.25}$$

Substituting this into equation (3.24), it follows that

$$M_2 = M_1 + F_1(x_2 - x_1) - \int_{x_1}^{x_2} \left[\int_{x_1}^{x} w \, dx \right] dx \tag{3.26}$$

3.4. Examples

Example 1. Draw the shearing force and the bending moment diagrams for the beam shown in *Figure 3.11*. The mass of the beam is 1.02 kg/mm and it is subjected to external moments of 100 kN m at A and −50 kN m at B.

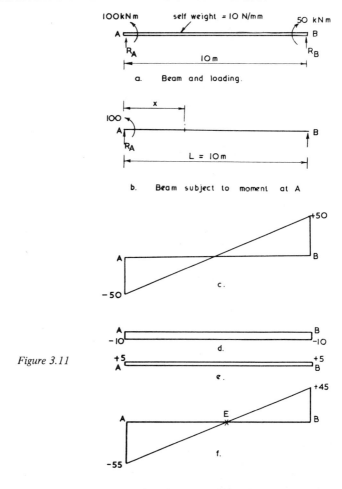

Figure 3.11

Answer: Weight of beam = mg = 1.02 × 9.81 = 10 N/m. Due to the self weight of the beam the shear force at A is $-wL/2 = -10 \times 10 \times 10^3/2 = -50$ kN
That at B is + 50 kN.
Due to the self weight of the beam, the bending moment is parabolic with a minimum $-wL^2/8 = -125$ kN m.

Consider the beam with a moment of 100 kN m acting alone at A (*Figure 3.11b*). Taking moments about B

$$-R_A \times 10 + 100 = 0$$
$$\therefore \quad R_A = 10 \text{ kN}$$

From vertical equilibrium of the beam:

$$R_B = -10 \text{ kN}$$

∴ Shear force at A = –10 kN and it is constant throughout the beam.
At a distance x from A

$$M_x = -R_A x + 100 = -10x + 100$$

At A, $x = 0$, $M_A = 100$ while at B, $x = 10$, $M_B = 0$.

Similarly for the beam subject to a moment of –50 kN m acting alone at B, we obtain $R_B = 5$ and $R_A = -5$.
The shear force at A = 5 kN and remains constant throughout.
At distance x from A

$$M_x = +5x$$

At A, $x = 0$ and $M_A = 0$ while at B, $x = 10$ and $M_B = +50$ kN m.

The shearing force diagrams due to the three cases, i.e. self weight, moment of 100 at A and moment of –50 at B, are shown in *Figure 3.11c, d* and *e* respectively. By the principle of superposition the shearing force at A for the actual loading is –50 –10 + 5 = –55 kN and at B is +50 –10 +5 = +45 kN. The shearing force diagram is shown in *Figure 3.11e*.

The bending moment diagrams for the three separate loading cases are shown in *Figure 3.12a, b* and *c* respectively. By the principle of superposition the bending moment at any point is obtained by the algebraic sum of the ordinates at the point for *Figure 3.12a, b* and *c*. Thus at A the bending moment is 100 + 0 + 0 = 100 kN m, at B it is +50 kN m and at the midspan it is –125 + 50 + 25 = –50 kN m.

From *Figure 3.11f*, the shearing force is zero at E, and the similar triangles in the figure give

$$55/AE = 45/(10 - AE)$$
$$\therefore \quad AE = 5.5 \text{ m.}$$

The minimum bending moment occurs where the shear force is zero, i.e. at E.

$$M_E = [\text{Moment at E due to self weight}] + [\text{Due to } M \text{ at A}] + [\text{Due to } -M \text{ at B}]$$
$$M_E = [-10 \times 10 \times 5.5 \times 0.5 + 10 \times (5.5)^2 \times 0.5] + [-10 \times 5.5 + 100]$$
$$+ [5 \times 5.5]$$
$$M_E = -275 + 151.25 - 55 + 100 + 27.5 = -51.25 \text{ kN m}$$

which is mathematically the minimum.

Example 2. Draw the shear force and the bending moment diagram for the beam shown in *Figure 3.13*.

Answer: Moments about B:

$$10 - R_A \times 10 = 0$$
$$\therefore \quad R_A = 1 \text{ kN}$$
and $\quad R_B = -1$ kN downwards.

Shear force at A = −1 kN and remains constant throughout.
Between A and D at distance x from A

$$M_X = -R_A x = -x$$
$$\text{At } x = 7.5, \; M_D = -7.5 \text{ kN m.}$$

a. Bending moment diagram due to self weight

b. Bending moment diagram due to 100 kNm at A

c. Bending moment diagram due to 50 kNm at B

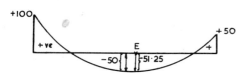

d. Final bending moment diagram

Figure 3.12

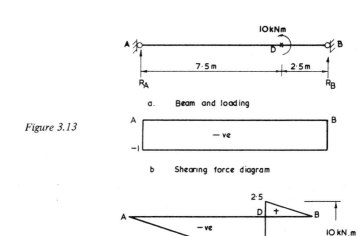

Figure 3.13

Between D and B, at distance x from A

$$M_X = -R_A x + 10 = -x + 10$$

At $x = 7.5$, $M_D = -7.5 + 10 = 2.5$ kN m.
At $x = 10$, $M_B = 0$

The shearing force and bending moment diagrams are shown in the figure.

Example 3. The beam AB has a weight of 1 kN per metre run and is resting on the ground. It is required to be moved upright by a force W so that it rotates about point A as shown in *Figure 3.14*. Where should the load W be applied so that the beam suffers from the least bending effects?

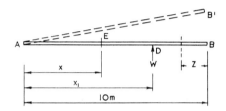
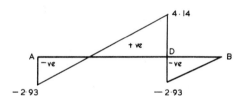
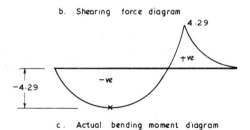

Figure 3.14

a. Beam and loading

b. Shearing force diagram

c. Actual bending moment diagram

Answer: Let the force W act at a distance x_1 from A. Taking moments about A

$$Wx_1 - 1 \times 10 \times 5 = 0$$
$$\therefore W = 50/x_1$$

From the vertical equilibrium of the beam

$$R_A + W - 1 \times 10 = 0$$
$$\therefore R_A = 10 - 50/x_1$$

76   Isostatic beams

The shearing force at $A = F_A = -R_A = -(10 - 50/x_1)$ kN. At E which is at a distance x from A, the shearing force F_X is

$$F_X = -(10 - 50/x_1) + 1 \times x = -10 + x + 50/x_1$$

When the shearing force is zero the bending moment is a maximum or a minimum, thus for $F_X = 0$

$$-10 + x + 50/x_1 = 0$$
$$x = 10 - 50/x_1$$

The largest numerical value of the bending moment M_E is

$$\begin{aligned} M_E &= -R_A x + 1 \times x^2/2 \\ &= -(10 - 50/x_1)(10 - 50/x_1) + 0.5(10 - 50/x_1)^2 \\ &= -0.5(10 - 50/x_1)^2 \end{aligned} \qquad (3.27)$$

The bending moment at a distance Z from B is

$$M_Z = 1 \times Z^2/2$$

When $Z = 10 - x_1$ at D

$$M_D = (10 - x_1)^2/2 \qquad (3.28)$$

As point D moves towards A, away from B, the numerical value of M_D given by equation (3.28) increases while the numerical value of $|M_E| = 0.5(10 - 50/x_1)^2$, see equation (3.27), decreases.

Thus keeping point D nearer B causes a numerically large bending moment at E which affects the beam excessively. On the other hand, moving point D nearer A causes a large bending moment at D which also affects the beam excessively. These two effects are balanced if $|M_E| = M_D$, i.e.

$$0.5(10 - 50/x_1)^2 = 0.5(10 - x_1)^2 \qquad (3.29)$$

Multiplying both sides of this equation by 2 and taking square roots

$$10 - 50/x_1 = 10 - x_1$$

which gives $x_1 = \sqrt{50} = 7.071$ m
and $x = 10 - 50/x_1 = 10 - 50/7.071 = 2.93$ m

This is where M_E has its largest numerical value. Equation (3.27) gives

$$M_E = -0.5(10 - 50/7.071)^2 = -4.29 \text{ kN m}$$

Equation (3.28) gives

$$M_D = (10 - 7.071)^2/2 = 4.29 \text{ kN m}$$

The value of the force W itself is

$$W = 50/x_1 = 7.071 \text{ kN}$$

The shear force at A is

$$F_A = -(10 - 50/x_1) = -2.93 \text{ kN}$$

The shearing force at D, in AD is

$$F_{DA} = -2.93 + 7.071 \times 1 = 4.14 \text{ kN}$$

∴ The shearing force at D in DB is

$$F_{DB} = 4.14 - W = 4.14 - 7.071 = -2.93 \text{ kN}$$

The actual shearing force and bending moment diagrams are shown in *Figure 3.14b* and *c* respectively.

Example 4. Draw the shearing force and the bending moment diagram for the beam shown in *Figure 3.15*.

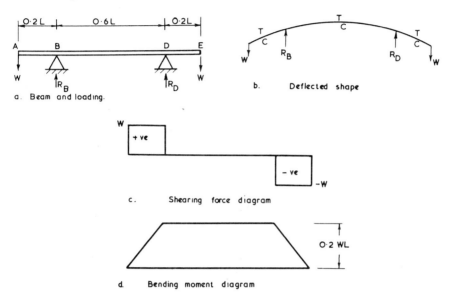

Figure 3.15

Answer: From symmetry and equilibrium

$$R_B = R_D = W$$

Between A and B, the shearing force $F_{AB} = W$
Between B and D the shearing force $F_{BD} = W - R_B = 0$
Between D and E the shearing force $F_{DE} = W - R_B - R_D = -W$

The forces W deflect the two ends down and the deflected shape of the beam is shown in *Figure 3.15b* with tension developing at the top surface of the beam throughout. Between A and B at distance x from A

$$M_X = Wx$$

When $x = 0.2L$ at B

$$M_B = W \times 0.2L = 0.2WL$$

Between B and D at distance x from A

$$M_X = Wx - R_B(x - 0.2L) = Wx - W(x - 0.2L) = 0.2WL$$

Between D and E at distance x from A

$$M_X = Wx - R_B(x - 0.2L) - R_D(x - 0.8L)$$
$$= Wx - W(x - 0.2L) - W(x - 0.8L) = WL - Wx$$

When $x = 0.8L$ at D

$$M_D = WL - 0.8WL = 0.2WL$$

When $x = L$ at E

$$M_E = WL - WL = 0$$

The shearing force and the bending moment diagrams are shown in *Figure 3.15c* and *d*.

3.5. Bending stresses in beams

The middle portion BD of the beam shown in *Figure 3.15* was noticed to be free from shearing force and had uniform bending moment of $0.2WL$. This condition of bending with zero shear is called pure bending. Assuming that the cross section of the beam is symmetrical about the X-Y plane and the loads are acting in this plane it follows that the bending of the beam only takes place in the plane. Consider that the material of the beam obeys Hooke's law and the modulus of elasticity in tension and compression is the same. In this case the deformation of the beam between B and D is uniform and takes the form of an arc of a circle as shown in *Figure 3.16*. Assume that each cross section, originally plane, remains

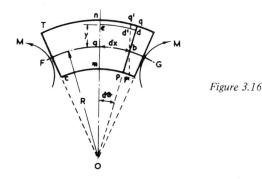

Figure 3.16

plane and normal to the longitudinal fibres of the beam. Fibres on the tension side of the beam are extended and those on the compression side are contracted. Somewhere in between the top and the bottom there is a layer of fibres which remain unchanged in length. This is called the neutral plane. The intersection of this plane with the longitudinal plane of symmetry of the beam is called the neutral axis of the beam. The intersection of the neutral plane with a cross section is called the neutral axis of the section. After bending takes place, the planes of two adjacent cross sections *mn* and *pq* intersect at the centre of the circle O. Let the angle between these planes be $d\theta$. From *Figure 3.16* it is noticed that

$$d\theta = dx/R \tag{3.30}$$

where R is the radius of curvature. Through b, on the neutral axis, the line $p'q'$ is drawn parallel to *mn*. This indicates the original position of the cross section *pq* before bending took place.

It is noticed that the segment ed' of a fibre, distance y from the neutral axis, extends by the amount dd', where

$$dd' = y d\theta \tag{3.31}$$

The original length of the fibre was $ed' = dx$, it follows that the strain in it is

$$\epsilon = y d\theta/dx = y/R \tag{3.32}$$

and from Hooke's law

$$\sigma = \epsilon E = Ey/R \tag{3.33}$$

For point G between E and B where $x = x_G$ equations (3.45) and (3.46) gives the shearing force F_G and the bending moment M_G.

For point J between C and E, $x = x_J$ and $\{x_J - b\}$ becomes negative. The quantities $W_2\{x - b\}^0$ and $W_2\{x - b\}$ are therefore disregarded. Equations (3.45) and (3.46) thus give

$$F_J = -R_A + W_1$$
and $$M_J = -R_A x_J + W_1\{x_J - a\}$$

For point D between A and C, $x = x_D$ and both $\{x_D - a\}$ and $\{x_D - b\}$ become negative. Equations (3.45) and (3.46) then give

$$F_D = -R_A \quad \text{and} \quad M_D = -R_A x_D$$

3.7.2. A beam with uniform loads

For a beam carrying a uniform load w per unit length covering the portion from C to B as shown in *Figure 3.20a*, the shearing force F_X and the bending moment M_X at a point H, which is at a variable distance x from the origin A, are

$$\left. \begin{array}{l} F_X = -R_A + w\{x - a\} \\ M_X = -R_A x + \{w\,x - a\}^2/2 \end{array} \right\} \quad (3.47)$$

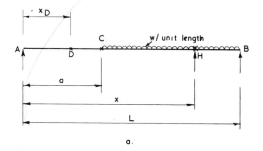

a.

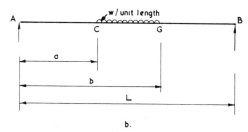

b.

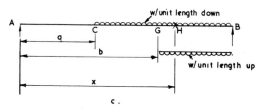

c.

Fig. 3.20

84   *Isostatic beams*

when $x < a$ between A and C, $\{x - a\}$ becomes negative and the quantities $w\{x - a\}$ and $w\{x - a\}^2/2$ are disgregarded. For a point such as D, therefore, $x_D < a$ and equations (3.47) give

$$F_D = -R_A$$
and $$M_D = -R_A x_D$$

This is provided that the uniform load extends from C to the end of the beam at B. In a beam carrying a uniform load which does not extend to the end, such as that shown in *Figure 3.20b*, the procedure is to extend this load to the end of the beam and balance it by an upward force of the same intensity (i.e. w/unit length) as shown in *Figure 3.20c*. It is noticed that the loads covering the length GB cancel each other and thus the net external load acting on the beam in this figure is the same as the original load acting on the beam shown in *Figure 3.20b*.

Referring to *Figure 3.20c*, the shearing force F_X and the bending moment M_X at H are

$$F_X = -R_A + w\{x - a\} - w\{x - b\}$$
and $$M_X = -R_A x + 0.5w\{x - a\}^2 - 0.5w\{x - b\}^2 \quad (3.48)$$

Between C and G the quantities $w\{x - b\}$ and $w\{x - b\}^2/2$ are disregarded while between A and C the quantities $w\{x - a\}$, $w\{x - b\}$, $w\{x - a\}^2/2$ and $w\{x - b\}^2/2$ are all disregarded.

3.7.3. A beam with external moments

When a beam is subjected to an external moment M at a point C along its span as shown in *Figure 3.21a*, the shearing force F_X and the bending moment M_X at H are

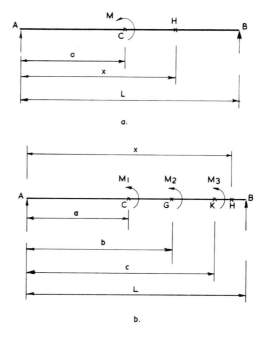

Figure 3.21

$$F_X = -R_A + M\{x - a\}^{-\infty}$$
and
$$M_X = -R_A x + M\{x - a\}^0 \tag{3.49}$$

Here $\{x - a\}^{-\infty} = 0$ and $\{x - a\}^0 = 1$. The quantity $M\{x - a\}^{-\infty}$ is thus equal to zero and left out of the expression for F_X while the quantity $M\{x - a\}^0 = M$ is disregarded when $\{x - a\}$ becomes negative between A and C.

For several external moments M_1, M_2, M_3 etc. (*Figure 3.21b*)

$$F_X = -R_A + M_1\{x - a\}^{-\infty} + M_2\{x - b\}^{-\infty} + M_3\{x - c\}^{-\infty} = -R_A$$
and $M_X = -R_A x + M_1\{x - a\}^0 + M_2\{x - b\}^0 + M_3\{x - c\}^0 \tag{3.50}$

The quantity associated with each curly bracket is disregarded when the value inside the bracket becomes negative.

3.8. Example

The beam ABK shown in *Figure 3.22a* is pin ended at A and simply supported at B. Draw the shearing force and the bending moment diagram for the loading shown in the figure.

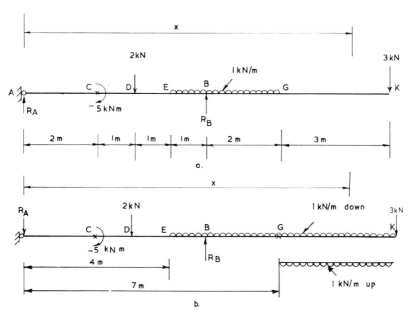

Figure 3.22

Answer: Taking moments about the support at B for the whole beam

$$-R_A \times 5 - 5 + 2 \times 2 + 1 \times 0.5 - 2 \times 1 - 3 \times 5 = 0$$
∴ $R_A = -3.5$ kN (down)

For the vertical equilibrium of the beam:

$$R_A + R_B - 2 - 3 - 3 = 0$$
∴ $R_B = 8 - R_A = 8 + 3.5 = 11.5$ kN (up)

Extending the uniform downward load from G to the end K and applying an upward uniform load between G and K as shown in *Figure 3.22b*, the shearing force F_X at distance x from A becomes

$$F_X = 3.5 - 5\{x - 2\}^{-\infty} + 2\{x - 3\}^0 + 1\{x - 4\} - 11.5\{x - 5\}^0 - \{1\ x - 7\}$$

At K $x = 10$

$$F_K = 3.5 + 2 + 1 \times 6 - 11.5 - 1 \times 3 = -3\text{ kN}$$

Just to the right of G $x_G = 7$

$$F_{GK} = 3.5 + 2 + 1 \times 3 - 11.5 = -3\text{ kN} = F_{GB}$$

Just to the right of B $x = 5$

$$F_{BG} = 3.5 + 2 + 1 - 11.5 = -5\text{ kN}$$

Just to the left of B $x = 5$

$$F_{BE} = 3.5 + 2 + 1 = 6.5\text{ kN}$$

At E $x = 4$

$$F_E = 3.5 + 2 = 5.5\text{ kN}$$

Just to the right of D $x = 3$

$$F_{DE} = 3.5 + 2 = 5.5\text{ kN}$$

Just to the left of D $x = 3$

$$F_{DC} = 3.5\text{ kN} = F_C = F_A$$

The shearing force diagram is shown in *Figure 3.23a*.
The bending moment M_X at distance x from A is distance x from A is

$$M_X = 3.5x - 5\{x - 2\}^0 + 2\{x - 3\} + \tfrac{1}{2}\{x - 4\}^2 - 11.5\{x - 5\} - \tfrac{1}{2}\{x - 7\}^2$$

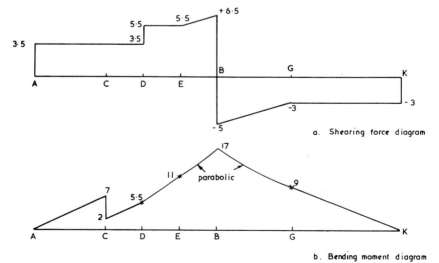

Figure 3.23

At K $x = 10; M_K = 0$
At G $x = 7$
        $M_G = 3.5 \times 7 - 5 + 2 \times 4 + \frac{1}{2} \times 9 - 11.5 \times 2 - \frac{1}{2} \times 0 = 9 \text{ kN m}$
At B $x = 5$
        $M_B = 3.5 \times 5 - 5 + 2 \times 2 + \frac{1}{2} \times 1 - 11.5 \times 0 = 17 \text{ kN m}$
At E $x = 4$
        $M_E = 3.5 \times 4 - 5 + 2 \times 1 + \frac{1}{2} \times 0 = 11 \text{ kN m}$
At D $x = 3$
        $M_D = 3.5 \times 3 - 5 + 2 \times 0 = 5.5 \text{ kN m}$
Just to the right of C $x = 2$
        $M_{CD} = 3.5 \times 2 - 5 = 2 \text{ kN m}$
Just to the left of C $x = 2$
        $M_{CA} = 3.5 \times 2 = 7 \text{ kN m}$
At A $x = 0, M_A = 0$

The bending moment diagram is shown in *Figure 3.23b* where it is noticed that the portions between EB and BG are parabolic and concave upwards. At B the shearing force changes sign and the bending moment is maximum.

Exercises on Chapter 3

1. Draw the shearing force and the bending moment diagrams for the beams shown in *Figure 3.24*. Calculate the maximum bending moment in each case.

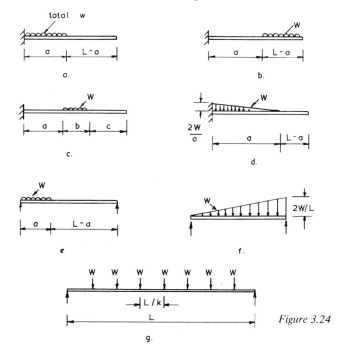

Figure 3.24

Ans. (a) $0.5Wa$
(b) $W[a + 0.5(L - a)]$
(c) $W(a + 0.5b)$
(d) $Wa/3$
(e) $0.5Wa(1 - 0.5a/L)^2$
(f) $0.128WL$
(g) $(k^2 - 1)WL/8k$, when k is odd and $kWL/8$ when k is even.

2. Draw the bending moment diagram for the beam shown in *Figure 3.25*. Find the value of x which makes the numerically largest bending moment a minimum.

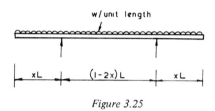

Figure 3.25

Ans. $x = 0.207$.

3. Draw the shearing force and the bending moment diagrams for the beam shown in *Figure 3.26*. What is the value of the maximum bending moment?

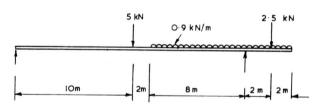

Figure 3.26

Ans. 33.3 kN m

4. For $R_A = 0$, calculate the shearing forces and the bending moments at points 1 to 10 for the beam shown in *Figure 3.27* and express the results in the matrix form

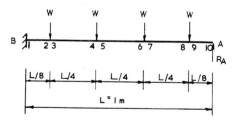

Figure 3.27

$P = BW$. For $W = 0$, calculate the shearing forces and the bending moments at the same points in terms of R_A and tabulate the results in matrix form. For $W = 10$ kN and $R_A = 95W/64$, what is the bending moment at points 1 and 8?

Ans. $M_1 = 33 \times 10^4/64$ kN mm, $M_8 = 157 \times 10^4/512$ kN mm.

5. Two loads $W_1 = 1$ kN and $W_2 = 2$ kN are 5 metres apart and act on a 20 metre span simply supported beam AB. The load W_1 is x metres from A and W_2 is $x + 5$ metres from the same support. If x is variable find its value so that the bending moment under W_2 is maximum.

Ans. 5.83 metres.

4

Bent members and structures

4.1. Introduction

In the last chapter we discussed straight beams and showed that these are subject to shearing forces and bending moments. Many structures are manufactured from curved or bent members which are subject to shearing forces and/or bending moments as well as axial loads or torques or both. In calculating the stresses all these factors are taken into consideration.

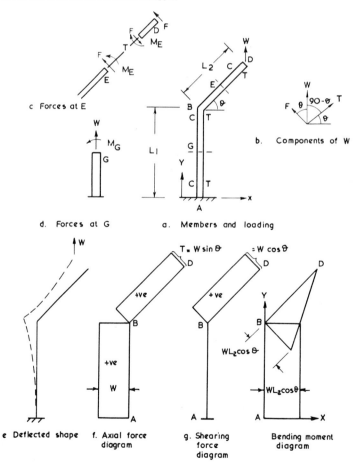

Figure 4.1

4.2. Plane bent members

Consider first the case of a member bent at some point such as that shown in *Figure 4.1*. The portion AB is vertical with length L_1 and fixed at A. The portion BD is inclined at an angle θ, its length is L_2 and it is rigidly connected to AB at B. A load W is acting at D which can be resolved into its components T and F as shown in *Figure 4.1b*. These components are

$$\left.\begin{array}{l} T = W \sin \theta \\ F = W \cos \theta \end{array}\right\} \quad (4.1)$$

Thus at any point, such as E, between B and D a tensile force T and a shear force F are developed. Both these forces remain constant throughout BD. The shear force F is positive because it tends to rotate DE anticlockwise.

Apart from these two forces, the load W causes a bending moment M_E at E. Measuring X and Y from the origin at the fixed support at A, the magnitude of M_E is

$$M_E = W(L_2 \cos \theta - x) \quad (4.2)$$

As the load W acts on the member it bends the member as shown in *Figure 4.1e*. Tension therefore develops on the surface on the side of the positive X axis as indicated with letters T in *Figure 4.1a*. The bending moment diagram for BD therefore lies on this side. When $x = 0$ at B, equation (4.2) gives

$$M_B = WL_2 \cos \theta \quad (4.3)$$

while when $x = L_2 \cos \theta$ at D, equation (4.2) gives $M_D = 0$.

At any point, such as G, between A and B, vertical equilibrium gives the tensile force at G as W, which has no horizontal component. There is therefore no shearing force at G or anywhere along AB.

The magnitude of the bending moment at G is $M_G = WL_2 \cos \theta$ which remains constant along AB. The axial load, shearing force and bending moment diagrams are shown in *Figure 4.1f, g* and *h* respectively.

4.3. Plane curved members

Next consider the case of the curved member shown in *Figure 4.2*. It is fixed at A and carries a vertical load W at B. The member has the shape of a quarter of a circle with radius R. At any section, such as E, the force W can be resolved into two components, T_E normal to the section and F_E across the section. These are

and $$\left.\begin{array}{l} T_E = W \sin \phi \\ F_E = W \cos \phi \end{array}\right\} \quad (4.4)$$

These forces are the tensile force and the shearing force at E.

Measuring the coordinates x and y from the origin at the support A, the magnitude of the bending moment at E (see *Figure 4.2b*) is

$$M_E = W(R - x) \quad (4.5)$$

But $$\sin \phi = (R - x)/R$$

$\therefore$ $$M_E = WR \sin \phi \quad (4.6)$$

The bending effect of W causes tension at the inner surface of the member and compression at the outer surface as shown by letters T and C in the figure.

Noting that $y/R = \cos \phi$, equations (4.4), (4.5) and (4.6) are rewritten in terms of ϕ or x and y as

$$T_E = W \sin \phi = W(1 - x/R)$$
$$F_E = W \cos \phi = Wy/R \qquad (4.7)$$
and
$$M_E = WR \sin \phi = W(R - x)$$

When $x = R$ and $\phi = 0$ at B
$$T_B = 0, \quad F_B = W \text{ and } M_B = 0 \qquad (4.8)$$
and when $x = 0$ and $\phi = 90°$ at A
$$T_A = W, \quad F_A = 0 \text{ and } |M_A| = WR \qquad (4.9)$$

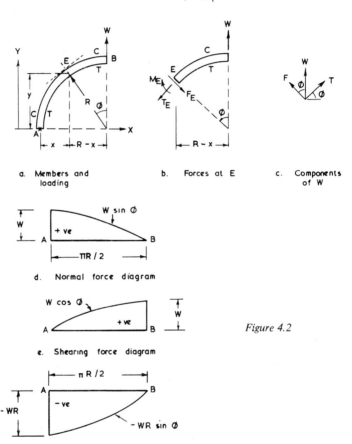

a. Members and loading

b. Forces at E

c. Components of W

d. Normal force diagram

e. Shearing force diagram

f. Bending moment diagram

Figure 4.2

The normal force, shearing force and the bending moment diagrams are shown in *Figure 4.2d, e* and *f* respectively. The member is straightened for this purpose.

4.4. Examples

Example 1. Calculate the axial force, the shearing force and the bending moments at points 1 to 6 for the bent member shown in *Figure 4.3*. Tabulate the results in the matrix form **P = B W**. The member is subject to external loads H_E and V_E and an external moment M_E at its end.

Bent members and structures

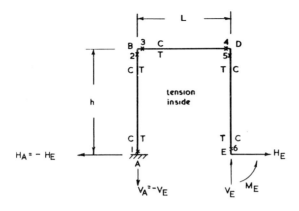

a. Member and loading.

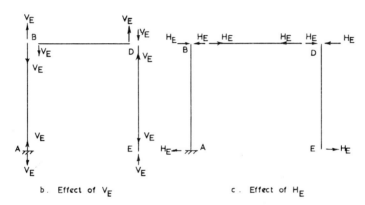

b. Effect of V_E c. Effect of H_E

Figure 4.3

Answer: From vertical equilibrium of the whole member

$$V_A + V_E = 0$$

∴ $V_A = -V_E$, i.e. acting down.

This causes tension in member AB.

From the equilibrium of AB, BD and DE and also of joints B and D, the effect of V_E in producing forces in the members and at the joints are obtained. These are shown in *Figure 4.3b*.

∴ $p_{AB} = V_E,\ F_{BD} = V_E\ \text{and}\ p_{DE} = -V_E$ (4.10)

From the horizontal equilibrium of the whole member

$$H_A + H_E = 0$$

∴ $H_A = -H_E$

From the above reasoning it follows

$$F_{AB} = -H_E,\ p_{BD} = H_E\ \text{and}\ F_{DE} = H_E \quad (4.11)$$

In matrix form equations (4.10) and (4.11) become

$$\begin{bmatrix} p_{1,2} \\ p_{3,4} \\ p_{5,6} \end{bmatrix} = \begin{bmatrix} 0 & 1 \\ 1 & 0 \\ 0 & -1 \end{bmatrix} \begin{bmatrix} H_E \\ V_E \end{bmatrix} ; \begin{bmatrix} F_{1,2} \\ F_{3,4} \\ F_{5,6} \end{bmatrix} = \begin{bmatrix} -1 & 0 \\ 0 & 1 \\ 1 & 0 \end{bmatrix} \begin{bmatrix} H_E \\ V_E \end{bmatrix}$$

The forces V_E and H_E and the moment M_E open up the bent member and thus cause tension in the inside surfaces of the member. The bending moment diagram therefore lies inside the bent member.

The magnitudes of the bending moments are

$$M_1 = H_E \times 0 + V_E \times L + M_E$$
$$M_2 = H_E h + V_E L + M_E$$
$$M_3 = H_E h + V_E L + M_E$$
$$M_4 = H_E h + V_E \times 0 + M_E$$
$$M_5 = H_E h + V_E \times 0 + M_E$$
$$M_6 = H_E \times 0 + V_E \times 0 + M_E$$

In matrix form these become

$$\begin{bmatrix} M_1 \\ M_2 \\ M_3 \\ M_4 \\ M_5 \\ M_6 \end{bmatrix} = \begin{bmatrix} 0 & L & 1 \\ h & L & 1 \\ h & L & 1 \\ h & 0 & 1 \\ h & 0 & 1 \\ 0 & 0 & 1 \end{bmatrix} \begin{bmatrix} H_E \\ V_E \\ M_E \end{bmatrix}$$

Example 2. Calculate the normal force, the shearing force and the bending moments at points 1 to 5 in the semicircle shown in *Figure 4.4a*.

Answer: At any section such as E (*Figure 4.4c*) the force V_B is resolved into

$$C = V_B \cos \phi = 1 \times \cos \phi = \cos \phi$$

causing compression in BE and

$$F_1 = V_B \sin \phi = \sin \phi$$

directed towards the centre.

Similarly the force H_B is resolved into

$$T = H_B \sin \phi = 10 \sin \phi$$

causing tension in BE and

$$F_2 = H_B \cos \phi = 10 \cos \phi$$

directed towards the centre.

The total normal force at E is

$$T = 10 \sin \phi - \cos \phi \tag{4.11}$$

The total shearing force at E is

$$F_E = \sin \phi + 10 \cos \phi \tag{4.12}$$

The bending moment due to H_B, V_B and M_B causes tension inside the semicircle and the bending moment diagram lies inside the semicircle. Tha magnitude of the bending moment at E is M_E

Bent members and structures

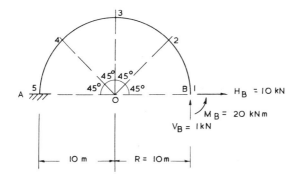

a. Member and loading

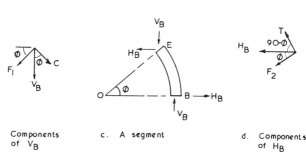

Figure 4.4

b. Components of V_B c. A segment d. Components of H_B

$$M_E = H_B R \sin\phi + V_B R(1 - \cos\phi) + M_B$$
$$M_E = 10 \times 10 \times \sin\phi + 1 \times 10(1 - \cos\phi) + 20$$
$$M_E = 100 \sin\phi - 10 \cos\phi + 30 \qquad (4.13)$$

using equations (4.11), (4.12) and (4.13) for points 1 to 5

$T_1 = 0 - 1 = -1$ kN (compression)
$F_1 = 0 + 10 = 10$ kN
$M_1 = 0 - 10 + 30 = 20$ kN m

$T_2 = (10 \times 1/\sqrt{2}) - (1/\sqrt{2}) = 9/\sqrt{2}$ kN (tension)
$F_2 = (1/\sqrt{2}) + (10/\sqrt{2}) = 11/\sqrt{2}$ kN
$M_2 = (100/\sqrt{2}) - (10/\sqrt{2}) + 30 = 132.42/\sqrt{2}$ kN m

$T_3 = 10 - 0 = 10$ kN
$F_3 = 1 + 0 = 1$ kN
$M_3 = 100 - 0 + 30 = 130$ kN m

$T_4 = (10/\sqrt{2}) - (-1/\sqrt{2}) = (10/\sqrt{2}) + (1/\sqrt{2}) = 11/\sqrt{2}$ kN
$F_4 = (1/\sqrt{2}) - (10/\sqrt{2}) = -9/\sqrt{2}$ kN
$M_4 = (100/\sqrt{2}) + (10/\sqrt{2}) + 30 = 152.42/\sqrt{2}$ kN m

$T_5 = 0 - (-1) = 1$ kN
$F_5 = 0 - 10 = -10$ kN
$M_5 = 0 + 10 + 30 = 40$ kN m

4.5. Bent member loaded out of its plane

The member ABD in *Figure 4.5a* is bent in the X-Y plane and a force W acts horizontally in the direction of +Z axis.

The portion BD is under a shearing force W and a bending moment $W(L_2 - x)$ where x is measured from B. At B the bending moment becomes WL_2. Tension will develop in the vertical surface on the $-Z$ side. The vertical column AB is subject to a shear force W, a constant torque of WL_2 (units of force × units of length) and a varying bending moment. At B on the column the bending moment is zero. At a distance y from A, the bending moment is $W(L_1 - y)$ which is linearly related to y and reaches a numerical maximum of WL_1 at A. Tension is also developed on the surface which is on the $-Z$ side. The shear force, bending moment and torque in each member are shown in *Figure 4.5b*.

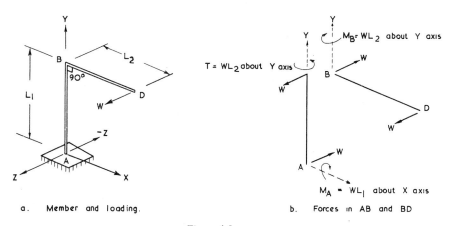

a. Member and loading.   b. Forces in AB and BD

Figure 4.5

4.6. Curved member loaded out of its plane

Consider next the horizontal quarter circular arc of radius R loaded vertically at its free end A (*Figure 4.6*).

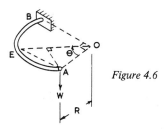

Figure 4.6

The circular cantilever is subject to a constant shearing force W as well as bending and torsion throughout. At any cross section E, defined by the angle θ, the bending moment is

$$M_E = WR \sin \theta \tag{4.14}$$

The twisting moment (torque) at the same section is

$$T_E = WR(1 - \cos \theta) \tag{4.15}$$

At A, where $\theta = 0$, $M_A = T_A = 0$ while at B, where $\theta = 90°$

and
$$\left.\begin{array}{r} M_B = WR \\ T_B = WR \end{array}\right\} \tag{4.16}$$

4.7. Three pinned arches

These are isostatic frames made from bent or curved members. A three pinned arch is supported at its ends by two pins and a third pin joins the parts of the frame somewhere between the supports as shown in *Figure 4.7*. Any cross section of the arch is subject to a normal force, a shearing force and a bending moment. It has been stated that a frictionless pin does not transmit a bending moment, and therefore the bending moment at each of the three pins is equal to zero. This fact is used to calculate the support reactions and the forces and bending moments everywhere in the arch. However it should be stressed that these hinges do transmit shear forces.

In *Figure 4.7*, the four support reaction components are shown as positive. These are calculated first as follows:
For the whole arch, taking moments about point B

$$W(L - x) + H_A h_2 - R_A L = 0 \tag{4.17}$$

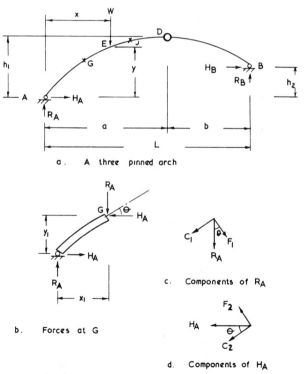

a. A three pinned arch

b. Forces at G

c. Components of R_A

d. Components of H_A

Figure 4.7

For the whole arch, taking moments about point A
$$R_B L - H_B h_2 - Wx = 0 \tag{4.18}$$
For the portion DB, taking moments about the pin at D
$$R_B b + H_B(h_1 - h_2) = 0 \tag{4.19}$$
and from the horizontal equilibrium of the arch
$$H_A + H_B = 0$$
i.e.
$$H_A = -H_B \tag{4.20}$$
Solving these equations we obtain
$$H_A = Wxb/(Lh_1 - ah_2) = -H_B \tag{4.21}$$
Thus H_B acts from right to left
$$R_A = W\left[1 - x(h_1 - h_2)/(Lh_1 - ah_2)\right] \tag{4.22}$$
$$R_B = Wx(h_1 - h_2)/(Lh_1 - ah_2) \tag{4.23}$$
Note that $R_A + R_B = W$ which satisfies the condition of vertical equilibrium.
Since the bending moment at D is zero and member BD is not loaded, the resultant of H_B and R_B passes through point D.
The vertical shear force at the pin D is
$$V_D = -R_A + W = R_B \tag{4.24}$$
At any point G between A and E, the normal force is (see *Figure 4.7c and d*)
$$P_G = C_1 + C_2$$
$$P_G = -R_A \sin\theta - H_A \cos\theta \tag{4.25}$$
where θ is the angle of inclination of the tangent at G.
The shearing force at G is
$$F_G = F_2 - F_1$$
$$F_G = H_A \sin\theta - R_A \cos\theta \tag{4.26}$$
The bending moment at G is
$$M_G = -R_A x_1 + H_A y_1 \tag{4.27}$$
where x_1 and y_1 are the coordinates of G.
At any point, such as J between E and D, the normal force p_J is
$$p_J = -(R_A - W)\sin\phi - H_A \cos\phi \tag{4.25b}$$
where ϕ is the angle of inclination of the tangent at J.
The shearing force at J is
$$F_J = H_A \sin\phi - (R_A - W)\cos\phi \tag{4.26b}$$

4.8. The local axes of a member

When a structure consists of several members, each with different inclination, the drawing of the shearing force and the bending moment diagrams may cause confusion as to which side of a member the diagrams should be. This confusion can be resolved by defining its local coordinates. A member is defined by an arrow the head of which points to the second end of the member. In *Figure 4.8a*, the arrow on the member AB is pointing to B. Thus A is the first end and B is the second end as indicated by numbers 1 and 2. The engineer may choose the direction of the arrow at will but once this is done the first and second ends of the member are defined.

Bent members and structures

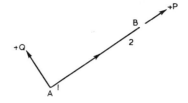

a. The ends of a member

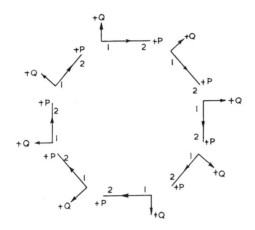

b. The position P and Q axis for members in any direction.

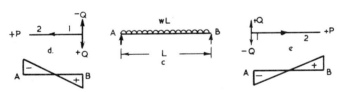

Figure 4.8

End one of a member is considered as its origin. The direction of the arrow defines the positive longitudinal P axis of the member. This is the axis along the member. The positive lateral Q axis of the member is defined by the right hand screw rule. The positive Q axis for member AB is shown in *Figure 4.8a*. In *Figure 4.8b* various members are shown together with their positive P and Q axes. The figure covers all inclinations. Use is made of the local P-Q axes of a member in drawing the shear force diagram. If the shearing forces acting at end 1 and 2 of a member tend to rotate the member in a positive anticlockwise manner then the shear force is, as stated earlier, positive. The diagram for a positive shear force lies on the side of the positive Q axis of the member. The diagram for a negative, clockwise, shearing force lies on the side of the negative Q axis.

Thus for member AB, loaded uniformly with wL, if an arrow is placed on the member to point from B to A (*Figure 4.8d*) then the positive Q axis is downwards. The shearing force at A is $-0.5wL$ and this lies on the side of $-Q$ axis, i.e. above the line AB in this case. The shearing force at B is positive and therefore lies on the side of $+Q$ as shown in *Figure 4.8d*.

The arrow in *Figure 4.8e* is towards B. Thus +Q is upwards and positive shearing force at B lies on +Q side.

The bending moment diagram is always drawn on the side where tension is developed. If the diagram also lies on +Q side then its sign is positive.

4.9. Examples

Example 1. Draw the shearing force and the bending moment diagrams for the three pinned arch shown in *Figure 4.9*.

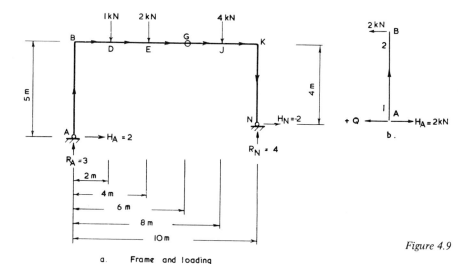

Figure 4.9

Answer: Taking moments about N for the whole frame
$$R_A \times 10 - H_A \times 1 - 1 \times 8 - 2 \times 6 - 4 \times 2 = 0$$
$$10 R_A - H_A = 28$$
From the vertical equilibrium
$$R_A + R_N = 7$$
From the horizontal equilibrium
$$H_A + H_N = 0$$
For the bent portion GKN taking moments about the pin at G
$$4 \times 2 - R_N \times 4 - H_N \times 4 = 0$$
$$\therefore \quad R_N + H_N = 2$$
From these equations we obtain
$$\{R_A \; H_A \; R_N \; H_N\} = \{3 \; 2 \; 4 \; -2\}$$
The shearing force throughout AB is
$$F_{AB} = H_A = 2 \text{ kN}$$

In *Figure 4.9b* member AB is shown with an arrow pointing to B. Thus A is the first end of the member, B is the second end and +Q axis is pointing to the left. Since the shearing force in AB is positive, the shearing force diagram for AB lies to

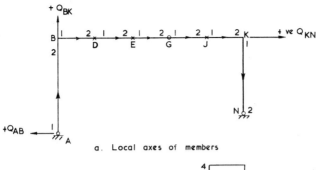
a. Local axes of members

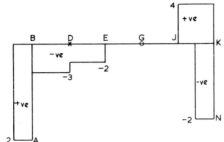
b. Shearing force diagram

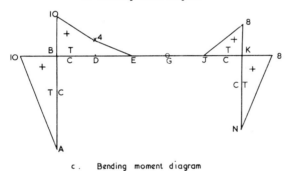
c. Bending moment diagram

Figure 4.10

the left of AB. *Figure 4.10a* defines the ends and the positive P and Q axes for all the members.

The shearing force in BD is

$$F_{BD} = -R_A = -3 \text{ kN}$$

This is because since R_A is acting upwards it tends to rotate BD in a clockwise manner.

Similarly $\quad F_{DE} = -3 + 1 = -2 \text{ kN}$

$$F_{EG} = -2 + 2 = 0$$

Since H_N is -2 kN, i.e. acting to the left, it tends to cause a clockwise, negative, rotation in KN

$\therefore \quad\quad\quad F_{KN} = H_N = -2 \text{ kN}$

Similarly $\quad\quad F_{JK} = R_N = 4 \text{ kN}$

The shearing force diagram is shown in *Figure 4.10b*.

The force R_A causes no bending moment in AB. The force $H_A = 2$ kN causes a bending moment of $2y$ kN m at any point in AB, distance y from A. This moment causes tension on the left side of the member as indicated by letters T and C in *Figure 4.10c*. The magnitude of the bending moment at B, in AB, is

$$M_B = 2 \times 5 = 10 \text{ kN m}$$

Similarly M_B in BD is 10 kN m.
At D

$$M_D = H_A \times 5 - R_A \times 2 = 2 \times 5 - 3 \times 2 = 4 \text{ kN m}$$

At E

$$M_E = H_A \times 5 - R_A \times 4 + 1 \times 2 = 10 - 12 + 2 = 0$$

The force $H_N = -2$ kN bends member KN and causes tension on the right. The magnitude of M at a point distance y from N is $2y$ and at K

$$M_K = 2 \times 4 = 8 \text{ kN m}$$

Similarly M_K in JK is 8 kN m with tension developing at the top fibres.
Finally

$$M_J = R_N \times 2 + H_N \times 4$$
$$M_J = 4 \times 2 - 2 \times 4 = 0$$

The shearing force and the bending moment along E G J turn out to be zero. The bending moment diagram is shown in *Figure 4.10c*. For each member the bending moment diagram lies on its +Q side. Thus the bending moment is positive throughout as shown in *Figure 4.10c*. It is helpful to show the forces acting on each member throughout a structure. This checks that equilibrium is satisfied for the structure, in each member and in each joint. It also facilitates the calculation of the shearing force, axial force and bending moments anywhere in the structures.

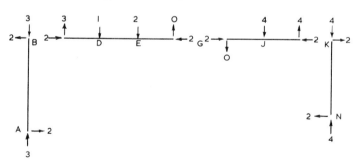

Forces in each member of the frame shown in figure 4.9

Figure 4.11

Figure 4.11 shows the forces acting on each member of the arch. It is advantageous to start this diagram by indicating the forces acting on the members at each support, i.e. deal with A in AB and N in KN first. Then deal with end B of member AB and end K of member NK. Continue the process to point G moving towards it from B and K.

Example 2. Draw the vertical shearing force and the bending moment diagram for the parabolic arch shown in *Figure 4.12*. What is the bending moment, shearing force and the normal thrust at a point J, 50m horizontally to the right of A?

Bent members and structures

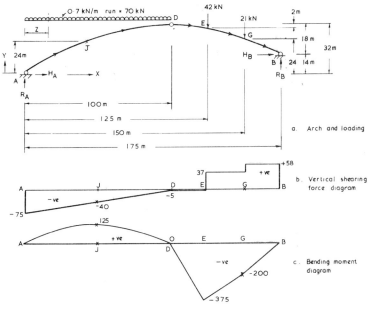

Figure 4.12

Answer: For the whole frame, taking moments about B

$$H_A \times 14 - R_A \times 175 + 70 \times 125 + 42 \times 50 + 21 \times 25 = 0$$

From the vertical equilibrium of the frame

$$R_A + R_B - 70 - 42 - 21 = 0$$

From the horizontal equilibrium of the whole frame

$$H_A + H_B = 0$$

For the portion DB, taking moments about the pin at D

$$R_B \times 75 + H_B \times 18 - 42 \times 25 - 21 \times 50 = 0$$

Solving these equations we obtain

$$\{H_A \; R_A \; H_B \; R_B\} = \{125 \; 75 \; -125 \; 58\} \text{ kN}$$

Vertical shearing forces F_X:
Between A and D, at a distance x from A, ther vertical shearing force is F_X given by

$$F_X = -R_A + 0.7x$$

When $x = 0$, $\quad F_A = -R_A = -75$ kN

When $x = 50$ m at J

$$F_J = -75 + 0.7 \times 50 = -40 \text{ kN}$$

When $x = 100$ m at D

$$F_D = -75 + 0.7 \times 100 = -5 \text{ kN}$$

Between D and E, the shearing force is constant and equal to $F_D = -5$ kN and $F_{DE} = -5$ kN.

At a point to the right of point E between E and G
$$F_{EG} = -75 + 70 + 42 = 37 \text{ kN}$$
Finally
$$F_{GB} = -75 + 70 + 42 + 21 = 58 \text{ kN}$$
The vertical shearing force diagram is shown in *Figure 4.12b*.

Bending moments:
At a distance x from A between D and B
$$M_X = -R_A x + H_A y + 0.7 \times 100(x - 50) + 42 \{x - 125\} + 21 \{x - 150\}$$
A term involving a negative value inside a curly bracket is disregarded.
At G, $x = 150, y = 24$
$$M_G = -75 \times 150 + 125 \times 24 + 70 \times 100 + 42 \times 25 + 21 \times 0$$
$$= -200 \text{ kN m}$$
At E, $x = 125, y = 30$
$$M_E = -75 \times 125 + 125 \times 30 + 70 \times 75 + 42 \times 0 = -375 \text{ kN m}$$
At D, $x = 100, y = 32$
$$M_D = -75 \times 100 + 125 \times 32 + 70 \times 50 = 0$$
At distance x from A, between A and D
$$M_X = -R_A x + H_A y + 0.7 x^2 / 2$$
At D, $x = 100, y = 32$
$$M_D = -75 \times 100 + 125 \times 32 + 0.7 \times (100)^2 / 2 = 0$$
At J, $x = 50, y = 24$
$$M_J = -75 \times 50 + 125 \times 24 + 0.7(50)^2 / 2 = +125 \text{ kN m}$$
At A, $x = 0, y = 0$
$$M_A = 0$$
With origin at A let the equation of the parabola be
$$y = ax^2 + bx + c$$
At $x = 0, y = 0$. At hinge D when $x = 100, y = 32$ and at B when $x = 175, y = 14$.
$$\therefore \quad a = -0.0032, \ b = 0.64 \text{ and } c = 0$$
The equation of the parabola is therefore given by
$$y = -0.0032 x^2 + 0.64 x$$
$$dy/dx = -0.0064 x + 0.64$$
At J when $x = 50$,
$$dy/dx = 0.32$$
i.e. the slope of the tangent at J = 0.32. The tangent makes an angle θ with the horizontal axis (see *Figure 4.13*) and thus
$$\tan \theta = 0.32$$
$$\sin \theta = 0.3048$$
$$\cos \theta = 0.9524$$

Bent members and structures

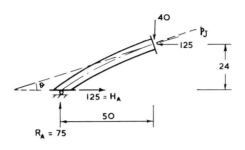

section at J

Figure 4.13

The normal force at J (see Figure 4.13) is
$$p_J = -R_A \sin\theta + 0.7 \times 50 \sin\theta - H_A \cos\theta$$
$$p_J = 0.3048(-75 + 35) - 125 \times 0.9524 = -131.2 \text{ kN}$$

The shearing force
$$F_J = -R_A \cos\theta + H_A \sin\theta + 0.7 \times 50 \cos\theta$$
$$F_J = 0.9524(-75 + 35) + 125 \times 0.3048 = 0$$

Thus at J the shearing force is zero and therefore the positive bending moment (see Figure 4.12c) is a maximum with $M_J = 125$ kN m.

Example 3. Calculate the bending moments, the axial forces and the shearing forces at the supports A and G of the structure shown in Figure 4.14. The supports are fixed and the portion BDE is a semicircle of radius $R = 10$m.

Answer: The structure consists of three parts, the column AB, the arch BDE and the column EG as shown in Figure 4.14b, c and d.
Consider the arch first. For vertical and horizontal equilibrium
$$R_B + R_E = 10$$
$$H_B + H_E = 0$$

Taking moments about E
$$20R_B = 10 \times 5$$
$$R_B = 2.5 \text{ kN}$$
and
$$R_E = 10 - 2.5 = 7.5 \text{ kN}$$

For the portion BD, taking moments about the pin at D
$$10R_B - 10H_B = 0$$
$$H_B = R_B = 2.5 \text{ kN}$$
and
$$H_E = -2.5 \text{ kN (acting from right to left)}$$

At B the reactions of the column AB on the arch are $R_B = 2.5$ kN upwards and

$H_B = 2.5$ kN to the right. Thus the actions of the arch on the column are, from the equilibrium of joint B, a vertical downward force of $R_B = 2.5$ kN and a horizontal force $H_B = -2.5$ kN acting to the left as shown in *Figure 4.14b*. For the horizontal equilibrium of the column AB, the horizontal reaction at A will be $H_B = 2.5$ acting to the right. The shearing force in AB is therefore +2.5 kN. The axial compressive force in AB = -2.5 kN. Finally the bending moment at A = $2.5 \times 10 = 25$ kN m, causing tension on the right side of the column.

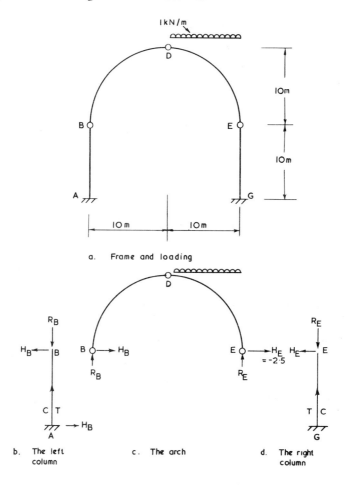

Figure 4.14

Similarly the horizontal force at E in the column GE is = $-H_E$ = +2.5 kN which is acting to the right. The shearing force in GE is thus -2.5 kN. The axial compressive force is $-R_E = -7.5$ kN and the bending moment at G is $2.5 \times 10 = 25$ kN m, causing tension on the left side of GE; because H_E is acting to the right.

Example 4. The hyperstatic frame in *Figure 4.15a* becomes an isostatic three pinned arch if a third pin is inserted at B as shown in *Figure 4.15b*. Calculate the bending moments at points 1 to 6 in this arch.

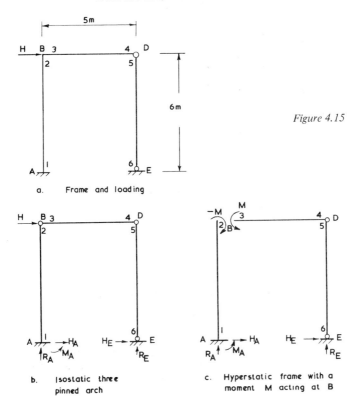

Figure 4.15

The value of the unknown redundant bending moment at B is M as shown in Figure 4.15c. Calculate the bending moments at points 1 to 6 in Figure 4.15c. Tabulate the results in the matrix form **P = B W**.

For $H = 10$ kN and $M = 5$ kN m, calculate the bending moments at points 1 to 6 in the original hyperstatic frame shown in Figure 4.15a.

Answer: In the three pinned arch shown in Figure 4.15b, because there are frictionless pins at B, D and E, with zero bending moments, it follows

$$M_2 = M_3 = M_4 = M_5 = M_6 = 0$$

For member DE, taking moments about D: $H_E \times 6 = 0$

∴ $H_E = 0$

The horizontal equilibrium of the frame then gives

$$H + H_A + H_E = 0$$
$$H_A = -H + 0 = -H$$

The negative sign indicates that H_A acts from right to left.
For member AB taking moments about B

$$6H_A + M_A = 0$$

∴ $M_A = -6H_A = -6 \times -H = 6H$

These moments are written in matrix form as

$$\begin{bmatrix} M_1 \\ M_2 \\ M_3 \\ M_4 \\ M_5 \\ M_6 \end{bmatrix} = \begin{bmatrix} 6 \\ 0 \\ 0 \\ 0 \\ 0 \\ 0 \end{bmatrix} [H] \quad (4.27)$$

In *Figure 4.15c*: Because of the pins at D and E

$$M_4 = M_5 = M_6 = 0$$

For member DE, taking moments about D: $H_E \times 6 = 0$

$$H_E = 0$$

From the horizontal equilibrium of the whole system shown in *Figure 4.15c*

$$H_A + H_E = 0$$

$$\therefore \quad H_A = 0$$

For member AB (*Figure 4.15c*) taking moments about B

$$M_A + H_A \times 6 - M = 0$$

$$\therefore \quad M_A = +M = M_1$$

The bending moment at point 2 is equal to the applied moment M and the bending moment at point 3 is equal to the applied moment M. Both these have the same sign because they cause tension at B on the outside surface of the frame. These values are written in matrix form as

$$\begin{bmatrix} M_1 \\ M_2 \\ M_3 \\ M_4 \\ M_5 \\ M_6 \end{bmatrix} = \begin{bmatrix} +1 \\ 1 \\ 1 \\ 0 \\ 0 \\ 0 \end{bmatrix} [M] \quad (4.28)$$

When $H = 10$ kN and $M = 5$ kN m, using the principle of superposition (i.e. the bending moment at a point due to the combined effect of H and M is the sum of the bending moments at the point due to the separate effects of H and M), equations (4.27) and (4.28) give

$$\begin{bmatrix} M_1 \\ M_2 \\ M_3 \\ M_4 \\ M_5 \\ M_6 \end{bmatrix} = \begin{bmatrix} 6 \\ 0 \\ 0 \\ 0 \\ 0 \\ 0 \end{bmatrix} [10] + \begin{bmatrix} +1 \\ 1 \\ 1 \\ 0 \\ 0 \\ 0 \end{bmatrix} [5] = \begin{bmatrix} 60+5 \\ 5 \\ 5 \\ 0 \\ 0 \\ 0 \end{bmatrix} = \begin{bmatrix} 65 \\ 5 \\ 5 \\ 0 \\ 0 \\ 0 \end{bmatrix} \quad (4.29)$$

These are the bending moments in the original hyperstatic frame provided that the value of the redundant moment at B is 5 kN m. This is not necessarily the value of

M_B. However, once the actual value of M_B is found, by methods given later in Chapter 9, it is substituted in equation (4.29) instead of 5 and the bending moments throughout the frame will then be given correctly.

Exercises on Chapter 4

1. The column CD in *Figure 4.16* is pinned to the beam AB at C. The strut EF is pinned at E to the column and at F to the beam. A horizontal load of 2 kN is applied at D. Calculate the axial forces in CE and EF. Draw the bending moment diagrams for AB and CD and obtain the maximum values of the bending moments in both.

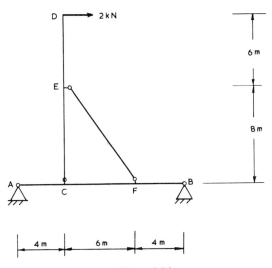

Figure 4.16

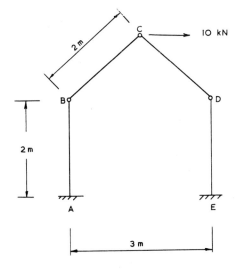

Figure 4.17

Ans. $P_{CE} = 4.67$ kN, $P_{EF} = 5.83$ kN, $M_{max\ CD} = 12$ kN m,
$M_{max\ AB} = 8$ kN m.

2. The columns of the pitched roof frame shown in *Figure 4.17* have an area of 100 mm² and a section modulus of 1 000 mm³. Calculate the maximum tensile and compressive stresses at A and E.

Ans. $\sigma_{A\ max} = 10.04$ kN/mm²   (tensile)
$\sigma_{E\ max} = 30.04$ kN/mm²   (compressive)

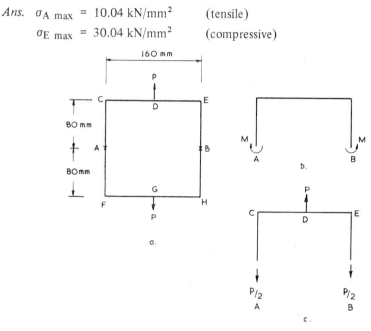

Figure 4.18

3. The square frame shown in *Figure 4.18* is acted upon by two equal and opposite forces P at D and G. The bending moments at A and B are M kN mm as shown in *Figure 4.18b*. For M = 0, calculate the bending moments at C, D and E in terms of P. For P = 0, calculate the bending moments at C, D and E in terms of M. With P = 10 kN, M = 100 kN mm, what is the bending moment at D?

Ans. 300 kN mm.

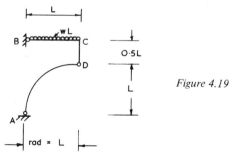

Figure 4.19

4. *Figure 4.19* is a diagramatic representation of the entrance to a building. Calculate the bending moment at C.

Ans. $wL^2/6$.

5

Slopes and deflections

5.1. Introduction

The design of a structure requires that the stresses anywhere in the structure should be satisfactory. It also requires that the deflections of the deformed structure should not exceed certain limits. The stresses can be calculated when the axial forces, shearing forces and the bending moments are known. Once this is done, it is necessary to calculate the deflections at certain critical points, or sometimes throughout the structure. In this chapter we begin with the calculation of deflections of simple beams. Before doing this, however, it is necessary to obtain expressions which give the deflections of a general member. Once these expressions are available, they may be applied to calculate the deflection of a member with any specific support or loading conditions.

5.2. Elastic bending of beams

In Chapter 3, it was shown that under pure bending, the shape of a straight beam changes to a circular arc of radius R given by

$$\frac{1}{R} = \frac{M}{EI} \tag{5.1}$$

where I is the second moment of area and M is the bending moment in the part which is under pure bending.

When a beam is not in pure bending, i.e. when it is also subjected to a shearing force, which is often the case, the shape of the deformed beam is no longer circular. However, it is assumed that equation (5.1) is applicable and gives the radius of curvature of the beam at any point where the bending moment is M. This means that as the bending moment varies along the beam, so does the radius of the curvature.

Consider a short length AB along the unloaded axis of a beam as shown in *Figure 5.1a*. When the beam is subsequently loaded, points A and B deflect. Let A move to A$'$ by a vertical amount y and B move to B$'$. The radius of curvature R at any section is then given by

$$1/R = \pm \frac{d^2 y/dx^2}{[1 + (dy/dx)^2]^{1 \cdot 5}} \tag{5.2}$$

If the deflection y is small, which is the case in most practical structures, then dy/dx will be small and $(dy/dx)^2$ can be neglected. Equation (5.2) then becomes

$$1/R = \pm d^2 y/dx^2 \tag{5.3}$$

and equation (5.1) changes to

$$\pm EI(d^2 y/dx^2) = M \tag{5.4}$$

In *Figure 5.1b* an element of a member is shown subject to moments which cause tension at the top fibres. The ordinate of the bending moment diagram for this

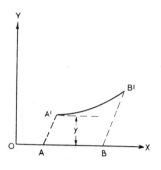

a. Element AB of a beam deflected to A' B'

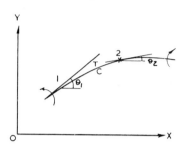

b. Bending of a member

Figure 5.1

element will be positive. Moving along the deformed beam, it is noticed that its slope is reducing from θ_1 at point 1 to θ_2 at point 2. For small values of θ; $\tan \theta = \theta = dy/dx$ and it follows that as the value of x increases dy/dx reduces. Thus a bending moment diagram with positive ordinates gives rise to a negative curvature, thus

$$\left. \begin{array}{r} M = -EI\,(d^2y/dx^2) \\ EI\,(d^2y/dx^2) = -M \end{array} \right\} \quad (5.5)$$

or

In Chapter 3, it was shown that

$$d^2M/dx^2 = dF/dx = -w \quad (5.6)$$

where F is the shearing force and w is the intensity of loading. Substituting for M from equation (5.5) into equation (5.6), we obtain

$$(d^2/dx^2)\,[-EI\,(d^2y/dx^2)] = dF/dx = -w \quad (5.7)$$

This equation is valid when EI changes along a beam with a variable cross section. If, on the other hand, EI is constant, then equation (5.7) becomes

$$-EI\,(d^4y/dx^4) = dF/dx = -w \quad (5.8)$$

which states that the product of $-EI$ and the fourth derivative of the deflection of an element of a beam gives the intensity of loading.

5.3. The deflection function

The deflection or displacement function relates the deflection y at a point along a beam to its x coordinate. This function is obtained by integrating equation (5.8) four times.

Consider a general member AB, whose first end A is at the origin of the coordinate axes and whose length is L, as shown in *Figure 5.2*. The member is subject, at its first end, to a bending moment M_{AB} and a shearing force F_{AB}. It is subject at its second end to a moment M_{BA} and a shearing force F_{BA}. Both M_{AB} and M_{BA} are positive because they are anticlockwise. The shearing forces F_{AB} and F_{BA} tend to rotate the member in an anticlockwise manner and therefore they are also positive. These forces and moments deform the member AB to the shape shown by AB' in the figure. The slope of the beam at the first end A is the anticlockwise rotation $\theta_{AB} = (dy/dx)_A$. That at the second end is $\theta_{BA} = (dy/dx)_B$. The deflection of the second end with respect to the first end is v as shown in the figure. The

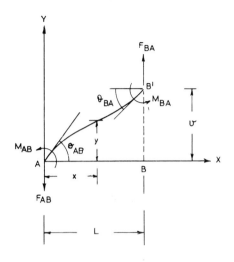

Figure 5.2

deflection at a distance x from A is y. It is noticed that the member is subject to loads only at its ends and not along its length. Thus $w = 0$ and equation (5.8) for this member becomes:

$$d^4y/dx^4 = 0 \tag{5.9}$$

Integrating this equation four times we obtain

$$y = ax^3 + bx^2 + cx + d \tag{5.10}$$

This equation is called the deflection function of the beam. It is cubic in x and it gives the deflection y of the beam at any given x along it. The displacement function in its form given by equation (5.10) is used extensively in the finite element method.

5.4. The slope-deflection equation

In the deflection function (5.10), the quantities a, b, c and d are arbitrary constants which can be found from the end (boundary) conditions. These are

(i) At point A in *Figure 5.2*, when $x = 0$, $y = 0$.
(ii) At $x = 0$, $dy/dx = \theta_{AB}$
(iii) At $x = 0$, $EI(d^2y/dx^2) = -M_{AB}$, cf. equation (5.5).
(iv) At $x = L$, $EI(d^2y/dx^2) = -M_{BA}$

Using the deflection function (5.10) and its first and second derivatives, the above four conditions yield

$$d = 0$$
$$c = \theta_{AB}$$
$$b = -M_{AB}/(2EI)$$

and $\quad a = (M_{AB} - M_{BA})/(6EIL)$

The deflection function (5.10) therefore becomes

Slopes and Deflections

$$y = (M_{AB} - M_{BA})x^3/(6EIL) - 0.5M_{AB}x^2/(EI) + x\theta_{AB} \quad (5.11)$$

Now when $x = L$, the deflection of the member at B (*Figure 5.2*) is v, which is the sway in the member and equation (5.11) gives this sway as

$$v = (M_{AB} - M_{BA})L^2/(6EI) - M_{AB}L^2/(2EI) + \theta_{AB}L \quad (5.12)$$

Also when $x = L$, $dy/dx = \theta_{BA}$ which is the rotation of end B. Equation (5.11) therefore gives

$$\left(\frac{dy}{dx}\right)_B = \theta_{BA} = \frac{3(M_{AB} - M_{BA})L}{6EI} - \frac{M_{AB}L}{EI} + \theta_{AB} \quad (5.13)$$

Substituting from equation (5.13), for $(M_{AB} - M_{BA})/6EI$, into equation (5.12) and simplifying, we obtain

$$M_{AB} = -\frac{6EI}{L^2}v + \frac{4EI}{L}\theta_{AB} + \frac{2EI}{L}\theta_{BA}$$

Similarly
$$\left. \begin{array}{c} \\ \\ \end{array} \right\} \quad (5.14)$$

$$M_{BA} = -\frac{6EI}{L^2}v + \frac{2EI}{L}\theta_{AB} + \frac{4EI}{L}\theta_{BA}$$

These are the well known and exceptionally useful 'slope deflection equations.' Their derivation in one form or another was carried out long ago but, since the advent of computers and matrix methods, they have become perhaps the most important set of equations in the analysis of rigidly jointed frames. Their use in calculating the bending moments and deflections of isostatic beams is not common, but in this chapter, this will be fully demonstrated. It will be shown that the equations, here too, are powerful to the extent that every other method, such as the moment-area method, may be disregarded. Furthermore, calculating deflections by means of the slope deflection equation improves the engineer's understanding of the boundary conditions of the structure in question.

5.5. Deflection and slope of a cantilever

The cantilever shown in *Figure 5.3* has length L and carries a point load W at end 2. It is noticed that at A, $\theta_{AB} = 0$ because the cantilever is fixed there and that $M_{AB} = +WL$ because it is an anticlockwise moment. Denoting $k = EI/L$, the second of the slope deflection equations (5.14) gives

$$M_{BA} = 0 = -6kv/L + 4k\theta_{BA}$$

∴ $\quad 2k\theta_{BA} = 3kv/L \quad (5.15)$

and the first gives $M_{AB} = WL = -6kv/L + 0 + 2k\theta_{BA} =$

$$M_{AB} = WL = -6kv/L + 3kv/L = -3kv/L \quad (5.16)$$

Thus the deflection at B is

$$v = -WL^2/(3k) = -WL^3/(3EI) \quad (5.17a)$$

The slope θ_{BA} is, from equation (5.15) $\theta_{BA} = \frac{3}{L}v \times \frac{1}{2k} = -\frac{3}{2} \cdot \frac{WL^3}{3EI} \cdot \frac{1}{L}$

$$\theta_{BA} = -WL^2/(2EI) \quad (5.18)$$

It is noticed that the deflection v is negative which means it is downwards in the $-y$ direction. Similarly θ_{BA} is negative which means it is clockwise.

To calculate the deflection anywhere else along the cantilever, say v' at D, half-way along the beam, the slope deflection equations are simply used again, this time for the portion AD or DB. Now for AD = $L/2$, denoting $2EI/L = k'$, the slope deflection equations (5.14) gives

$$M_{AD} = 2k'\theta_{DA} - 12k'v'/L = WL$$

and

$$M_{DA} = 4k'\theta_{DA} - 12k'v'/L = -0.5\,WL$$

It is noticed, from *Figure 5.3b*, that M_{DA} is clockwise and therefore negative.

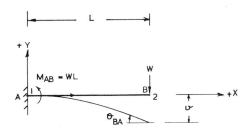

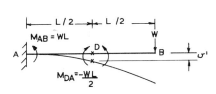

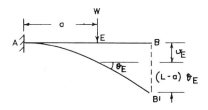

Figure 5.3

a. Deflection and rotation at end B

b. Deflection at D

c. Cantilever loaded along its span

Subtracting M_{DA} from M_{AD}, we obtain

$$2k'\theta_{DA} = -1.5\,WL$$

∴ $$M_{AD} = -1.5\,WL - 12k'v'/L = WL$$

which gives

$$v' = -5WL^2/24k' = -5WL^3/48EI$$

When the cantilever is loaded at some point such as E, distance a from the support at A, the beam between E and B carries no bending moment and therefore remains straight. The deflection at E is v_E, which is calculated from equation (5.17) with $L = a$ as

$$v_E = -Wa^3/(3EI)$$

The rotation, i.e. the slope, at E is θ_E which is given by equation (5.18), with $L = a$ as

$$\theta_E = -Wa^2/(2EI)$$

The deflection at the free end B is then calculated, using *Figure 5.3c*, as

$$v_B = v_E + (L - a)\theta_E$$

∴ $$v_B = -Wa^3/(3EI) - (L - a)Wa^2/(2EI)$$

$$v_B = -Wa^2(3L - a)/(6EI) \qquad (5.17b)$$

5.6. Deflection of a simply supported beam loaded at midspan

The simply supported beam shown in *Figure 5.4* has span L and carries a point load W at its midspan. The two parts of the beam are shown in *Figure 5.4b*, where

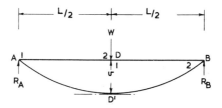

a.   Beam in its deflected form

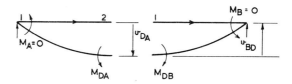

b.   The deflections of A D and DB

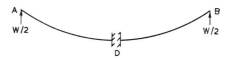

c.   The beam as two cantilevers

Figure 5.4

it is noticed that because A and B are simple supports

$$M_{AD} = M_{BD} = 0$$

Since DA is attached to DB and the rotation θ_D at point D is the same for both parts of the beam then

$$\theta_{DA} = \theta_{DB} = \theta_D$$

This simply states that the member is continuous at D.

Furthermore it is known that the bending moment at D in DA is $WL/4$ and that in DB = $-WL/4$.

In *Figure 5.4b* end A is specified as the first end of AD while end D is specified as the first end of DB, thus if the deflection of D relative to A is v_{DA} and the deflection of B relative to D is v_{BD}, then it is clear from the figure that

$$v_{BD} = -v_{DA} \tag{5.19}$$

because while v_{DA} is a downward movement, v_{BD} is an upward movement. It is also noticed that the net sway of a member is always given as the deflection of the second end relative to the first end. Placing an arrow to define the positive P axis of the member is therefore necessary. These facts fully describe the simple beam and can be used when applying the slope deflection equations. Denoting $k = EI/(0.5L) = 2EI/L$ for member AD

Slopes and Deflections 117

$$M_{AD} = 4k\theta_A + 2k\theta_D - 12kv_{DA}/L = 0$$
∴
$$2k\theta_A = -k\theta_D + 6kv_{DA}/L \tag{5.20}$$
$$M_{DA} = 2k\theta_A + 4k\theta_D - 12kv_{DA}/L = WL/4$$

This and equation (5.20) gives

$$3k\theta_D = 0.25\,WL + 6kv_{DA}/L \tag{5.21}$$

For BD

$$M_{BD} = 4k\theta_B + 2k\theta_D - 12kv_{BD}/L = 0$$

But $v_{BD} = -v_{DA}$, c.f. equation (5.19)
∴
$$M_{BD} = 4k\theta_B + 2k\theta_D + 12kv_{DA}/L = 0$$

which gives

$$2k\theta_B = -k\theta_D - 6kv_{DA}/L \tag{5.22}$$

and

$$M_{DB} = 3k\theta_D + 6kv_{DA}/L = -WL/4 \tag{5.23}$$

Hence

$$3k\theta_D = -0.25\,WL - 6kv_{DA}/L \tag{5.24}$$

Adding equations (5.21) and (5.24), we obtain

$$\theta_D = 0$$

On the other hand subtracting equation (5.24) from equation (5.21), we obtain

$$v = v_{DA} = -WL^3/(48EI) \tag{5.25}$$

Of course this result can be obtained simply by treating the beam as two cantilevers each with span $0.5L$ and carrying a point load $W/2$ at its end as shown in *Figure 5.4c*. Equation (5.17) then gives

$$v = -(W/2) \times (L/2)^3/3EI = -WL^3/(48EI)$$

Substituting $\theta_D = 0$ and $v_{DA} = -WL^3/(48EI)$ in equation (5.20), we obtain the slope of the simply supported beam at A as

$$\theta_A = -WL^2/(16EI) \tag{5.26}$$

5.7. A beam carrying a point load unsymmetrically

In the case of a simply supported beam carrying a point load W at D, distance a from the left hand support A with $a < L/2$, $k_1 = EI/a$ for AD, $k_2 = EI/(L-a)$ for DB and $M_{DA} = -M_{DB} = Wa(L-a)/L$, a procedure exactly similar to the one given above gives

$$v_D = -Wa^2(L-a)^2/(3EIL) \tag{5.27}$$

5.8. Beam with end moments and midspan load

Figure 5.5a shows a simply supported beam with a load W acting at midspan and two external moments M and $-M$ acting at A and B respectively. It is noticed that the beam is loaded symmetrically and thus $\theta_D = 0$.

The deflection of the beam due to the point load was found to be $-WL^3/48EI$, cf. equation (5.25). It is therefore sufficient to obtain only the deflection at D due to the end moments and then use the principle of superposition to calculate the

118   Slopes and Deflections

total deflection at D. The beam is shown in *Figure 5.5b* subject to two external moments M at A and $-M$ at B. For AD with $k = 2EI/L$ and $\theta_D = 0$

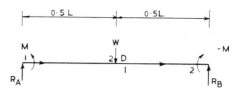

a. Beam and loading

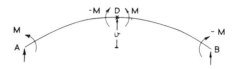

b. External moments M and −M - at A and B

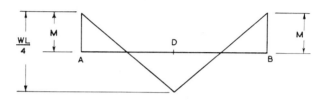

c. Bending moment diagram

Figure 5.5

$$M_{AD} = 4k\theta_A - 12kv/L = M$$
∴ $$2k\theta_A = 0.5M + 6kv/L$$
$$M_{DA} = 2k\theta_A - 12kv/L = -M \quad (5.28)$$
∴ $$M_{DA} = 0.5M + 6kv/L - 12kv/L = -M$$

Hence

$$6kv/L = 1.5 M$$

i.e.

$$v = ML^2/(8EI) \quad (5.29)$$

and equation (5.28) gives

$$\theta_A = 0.5 ML/EI \quad (5.30)$$

When the load W and the end moments are acting together, equations (5.25) and (5.29) give

$$v = [ML^2/(8EI)] - [WL^3/(48EI)] \quad (5.31a)$$

and equations (5.26) and (5.30) give

$$\theta_A = [ML/(2EI)] - [WL^2/(16EI)] \quad (5.31b)$$

The bending moment at D due to the end moments and the central point load is also obtained by superposition, thus

$$M_D = M - 0.25WL \qquad (5.31c)$$

The bending moment diagram is shown in *Figure 5.5c*

5.9. Fixed ended beam carrying a central load

When both ends of a beam are fixed, no rotation takes place at these ends. *Figure 5.6* shows such a beam and indicates that the beam resists the end rotations by its

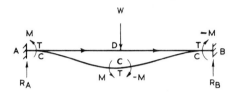

a. Fixed ended beam subject to a central point load w.

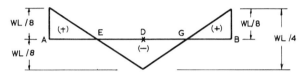

b. The bending moment diagram

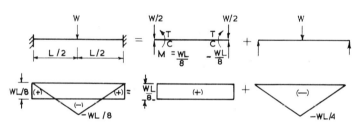

c. A fixed beam = a simple beam with end couples + a simple beam loaded along its span.

Figure 5.6

resisting moments M at A and $-M$ at B. The magnitude of M must be sufficiently large to reduce θ_A and θ_B to zero. Now in the last section, the value of θ_A was found to be given by equation (5.31b). Thus making $\theta_A = 0$ in this equation, we obtain

$$\theta_A = [ML/2(EI)] - [WL^2/(16EI)] = 0$$

which gives the value of the fixed end moment at A as

$$M = WL/8$$

Thus the clockwise end moment at B is $-M = -WL/8$

Equation (5.31a) then gives the central deflection of the fixed ended beam as

$$v_D = -WL^3/(192EI) \qquad (5.32)$$

Finally equation (5.31c) gives the central bending moment M_D as

$$M_D = (WL/8) - (WL/4) = -WL/8$$

In *Figure 5.6*, it is noticed that tensile stresses develop in the top fibres of the beam at A and B, while tensile stresses develop in the bottom fibres at midspan D. The letters T and C clarify this. The bending moment diagram is drawn in *Figure 5.6b* from which it is concluded that the top fibres of portions AE and GB are in tension while those of EG are in compression.

This fixed ended beam is hyperstatic and it was analysed by superimposing the results of two isostatic, simply supported, beams. Such a procedure of analysing a hyperstatic structure by the superposition of the results of two isostatic structures is always possible. Alternatively, the fixed ended beam can also be analysed by the direct application of the slope deflection equations. With $\theta_A = \theta_D = 0$, $M_{AD} = M_{DA} = M$, $k = 2EI/L$

$$M_{DA} = -12k\,v/L = M$$

i.e.
$$v = ML^2/(24EI) \tag{5.33}$$

Consider now the equilibrium of AD by taking moments about point D

$$-0.5\,R_A L + M_{AD} + M_{DA} = 0$$

$\therefore$
$$R_A = 4M/L$$

Similarly $R_B = 4M/L$. The vertical equilibrium of the beam gives

$$R_A + R_B = W$$

$\therefore$
$$8M/L = W$$

i.e.
$$M = WL/8$$

and from equation (5.33), the value of v is once again given as $-WL^3/(192EI)$.

The calculation of the moments and the deflections of this fixed ended beam showed that such a beam is indeed equivalent to two simply supported beams, the first carrying a central point load W and the second carrying external moments $\pm WL/8$ at its ends. Similarly, whenever a rigidly jointed member is loaded somewhere between its ends, this load is replaced by its equivalent end moments. The member is then analysed and the results are added to that obtained for a simply supported beam which is loaded at the same point as the original rigid member. This procedure is shown diagrammatically in *Figure 5.6c*.

5.10. Uniformly loaded simply supported beam

The slope deflection equations were obtained for a member loaded only at its ends. The same procedure, however, can be adopted for deriving similar equations starting with equation (5.8) instead of (5.9) which can then be used to analyse beams carrying uniform loads. Alternatively, the slope and the deflection at any point of any beam can be obtained using equation (5.5) which states

$$EI(d^2y/dx^2) = -M$$

In the case of the uniformly loaded simply supported beam shown in *Figure 5.7*, measuring x from A, the bending moment at a point D is given by

$$M_D = -0.5\,wLx + 0.5\,wx^2$$

Thus

$$EI(d^2y/dx^2) = -M_D = 0.5\,wLx - 0.5\,wx^2$$

and integrating this equation we obtain

$$EI(dy/dx) = 0.25\,wLx^2 - (wx^3/6) + A \tag{5.34}$$

Integrating again
$$EIy = (wLx^3/12) - (wx^4/24) + Ax + B \qquad (5.35)$$
Now when $x = 0$ at end A, the deflection is $y = 0$, thus $B = 0$. Similarly when $x = L$

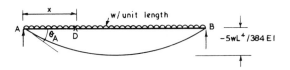

Figure 5.7

at end B, y is also zero, thus: $A = -wL^3/24$. Equation (5.35) then becomes
$$EIy = (wLx^3/12) - (wx^4/24) - (wxL^3/24)$$
i.e.
$$EIy = wx(2Lx^2 - x^3 - L^3)/24$$
and at the midspan when $x = 0.5 L$
$$y = -5wL^4/(384EI) \qquad (5.36)$$
Equation (5.34) gives the end rotation θ_A at $x = 0$ as
$$EI\theta_A = A = -wL^3/24$$
i.e.
$$\theta_A = -wL^3/(24EI) \qquad (5.37)$$
In section 5.16 the case of uniform loading will be once again dealt with in more detail.

5.11. A uniformly loaded fixed ended beam

Using the same procedure as in section 5.9, a fixed ended beam has no rotation at its fixed ends, thus equations (5.30) and (5.37) give
$$[ML/(2EI)] - [wL^3/(24\ EI)] = 0$$
$$\therefore \qquad M = wL^2/12 \qquad (5.38)$$
causing tension at the top fibres. M is the fixed end moment.

Equations (5.29) and (5.36) give the central deflection of the beam as
$$v = [ML^2/(8EI)] - [5\ wL^4/(384EI)]$$
and using equation (5.38)
$$v = -wL^4/(384EI) \qquad (5.39)$$
The bending moment at midspan, by the principle of superposition and making use of *Figure 5.6b*, is given by
$$M_D = (wL^2/12) - (wL^2/8) = -wL^2/24 \qquad (5.40)$$
which causes tension at the bottom fibres as shown in *Figure 5.8*.

Once again, it is clear that a uniformly loaded fixed ended beam is equivalent to two beams, one simply supported and loaded uniformly and the other loaded with end couples $wL^2/12$.

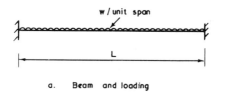

a. Beam and loading

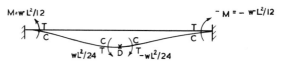

b. Deflected form of beam

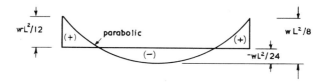

c. Bending moment diagram

Figure 5.8

5.12. Treatment of several loads

When a member is subjected to several point loads it is treated as if it consisted of several members connected to one another at joints under the point loads. For instance, Member AB in *Figure 5.9* carries loads W_1, W_2 and W_3 at points D, E and G respectively.

This member is treated as four members AD being member 1 joined to member 2, which is DE at D. An arrow is placed on each member to specify its first and second ends. The slope deflection equations are applied to each of these members realising that

$$\left.\begin{array}{l} M_A = M_B = 0 \\ \theta_{DA} = \theta_{DE} = \theta_D \\ \theta_{ED} = \theta_{EG} = \theta_E \\ \theta_{GE} = \theta_{GB} = \theta_G \end{array}\right\} \quad (5.41)$$

and that for member AD which is member 1, the sway v_1 is

also
$$\left.\begin{array}{l} v_1 = v_D - 0, \\ v_2 = v_E - v_D \\ v_3 = v_G - v_E \\ v_4 = 0 - v_G \end{array}\right\} \quad (5.42)$$
and

Slopes and Deflections

The first two of equations (5.42) yield positive results since $|v_E| > |v_D|$ and $|v_D| > 0$. Thus the term $-6k\,v/L$ in the slope deflection equations remain negative for members 1 and 2. On the other hand, because v_E is numerically larger than v_G and v_G is numerically larger than zero, both v_3 and v_4 become negative. Thus when using the slope deflection equations for members 3 and 4, the term $-6k\,v/L$ changes to $+6k\,v/L$.

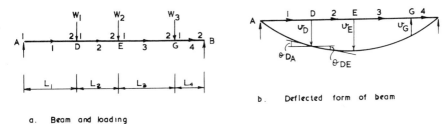

a. Beam and loading

b. Deflected form of beam

Figure 5.9

5.13. Examples

Example 1. The simply supported beam shown in *Figure 5.10* is subject to an external moment M at D which is at a distance a from end A, with $a < L/2$. Calculate

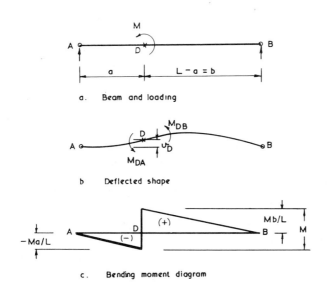

a. Beam and loading

b. Deflected shape

c. Bending moment diagram

Figure 5.10

the deflection at D. For $a = 4$m, $L = 10$ m and $E = 200$ kN/mm², $I = 10^6$ mm⁴, and $M = 10$ kN m, what is the deflection v_D?

Answer: Let $k_1 = EI/a$, $k_2 = EI/b = ak_1/b$
Applying the slope deflection equation to AD

$$M_{AD} = 0 = 4k_1\theta_A + 2k_1\theta_D - 6k_1 v_D/a$$

124　　　　　　　　　　　*Slopes and Deflections*

$$\therefore \quad 2k_1\theta_A = -k_1\theta_D + 3k_1v_D/a$$

The anticlockwise moment M_{DA} is, from *Figure 5.10c*, equal to Ma/L thus

$$M_{DA} = 2k_1\theta_A + 4k_1\theta_D - 6k_1v_D/a = Ma/L$$
$$\therefore \quad M_{DA} = 3k_1\theta_D - 3k_1v_D/a = Ma/L$$

i.e. $\quad 3k_1\theta_D = Ma/L + 3k_1v_D/a \quad$ (5.43a)

Applying the slope deflection equations to DB

$$M_{BD} = 4k_2\theta_B + 2k_2\theta_D + 6k_2v_D/b = 0$$
$$\therefore \quad 2k_2\theta_B = -k_2\theta_D - 3k_2v_D/b$$
and $\quad M_{DB} = 3k_2\theta_D + 3k_2v_D/L = Mb/L$

i.e. $\quad 3k_2\theta_D = Mb/L - 3k_2v_D/b \quad$ (5.43b)

Solving equations (5.43a) and (5.43b) for v_D and θ_D

$$v_D = Mab(b^2 - a^2)/(3EIL^2) \quad (5.44)$$
and
$$\theta_D = M(b^3 + a^3)/(3EIL^2) \quad (5.45)$$

With $a = 4000$ mm, $L = 10\,000$ mm, $M = 10\,000$ kN mm

$$v_D = \frac{10\,000 \times 4\,000 \times 6\,000}{3 \times 200 \times 10^6 \times 10^8}[(6\,000)^2 - (4\,000)^2] = 80\,\text{mm}$$

Example 2. The simply supported beam shown in *Figure 5.11* is loaded at its third points with loads W. Calculate the maximum deflection at its midspan.

Answer: The beam can be treated as two cantilevers fixed at G as shown in *Figure 5.11b*. The portion AG is subjected to an upward force $R_A = W$ as shown in *Figure 5.11c* and a downward force W at a distance of $a = L/6$ from G as in *Figure 5.11d*.

Using equation (5.17a), the upward deflection due to $R_A = W$ is

$$v_A = W(0.5L)^3/(3EI) = WL^3/(24EI)$$

Using equation (5.17b), with $a = L/6$, the downward deflection at A due to W acting at D is

$$v'_A = -W(L/6)^2 (1.5L - L/6)/6EI$$
$$v'_A = -WL^3/(162EI)$$

$\therefore \qquad$ The net upward deflection at $A = v_A - v'_A$

i.e. $\quad v_A - v'_A = [WL^3/(24EI)] - [WL^3/(162EI)] = 23WL^3/(648EI)$

Alternatively, treating the cantilever ADG, *Figure 5.11b* as two members; AD as member 1 with length of $L/3$ and DG as member 2 with length $L/6$, we obtain

$$k_1 = EI/(L/3) = 3EI/L; \; k_2 = EI/(L/6) = 6EI/L = 2k_1$$

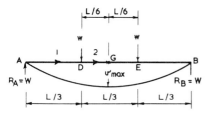
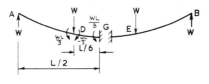

b. The beam as two cantilevers

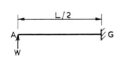
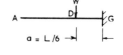

c. Reaction at A   d. Load at D acting alone

Figure 5.11

The slope deflection equations for member 2 are

$$M_{DG} = 4k_2\theta_D - 36k_2 v_{GD}/L = -WL/3$$
$$M_{GD} = 2k_2\theta_D - 36k_2 v_{GD}/L = WL/3$$

These give

$$v_{GD} = -WL^2/36k_2 = -WL^2/72k_1$$

and

$$\theta_D = -WL/3k_2 = -WL/6k_1$$

Similarly the slope deflection equations for member 1 are

$$M_{AD} = 4k_1\theta_A + 2k_1\theta_D - 18k_1 v_{DA}/L = 0$$
$$M_{DA} = 4k_1\theta_D + 2k_1\theta_A - 18k_1 v_{DA}/L = WL/3$$

and these together with $\theta_D = -WL/6k_1$ give

$$v_{DA} = -5WL^2/54k_1$$

The total deflection at G then becomes

$$v_G = v_{GD} + v_{DA} = -[WL^2/(72k_1)] - [5WL^2/(54k_1)] = -23WL^3/(648EI)$$

Example 3. The fixed ended beam shown in *Figure 5.12* is subjected to a load W at a distance a from A. Calculate the deflection and the rotation under the load. What are the bending moments at A, D and B? Draw the bending moment diagram.

Answer: Since the ends are fixed, $\theta_A = \theta_B = 0$.
Let $k_1 = EI/a$, $k_2 = EI/b = ak_1/b$, the slope deflection equations then become

$$\left.\begin{aligned}M_{AD} &= 2k_1\theta_D - 6k_1 v/a \\ M_{DA} &= 4k_1\theta_D - 6k_1 v/a \\ M_{BD} &= 2k_2\theta_D + 6k_2 v/b \\ M_{DB} &= 4k_2\theta_D + 6k_2 v/b\end{aligned}\right\} \quad (5.46)$$

For the equilibrium of joint D

$$M_{DA} + M_{DB} = 0$$

∴
$$4(k_1 + k_2)\theta_D - 6k_1 v/a + 6k_2 v/b = 0 \quad (5.47)$$

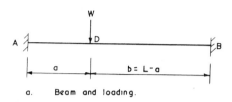

a. Beam and loading.

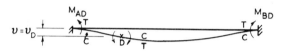

b. Deflected shape

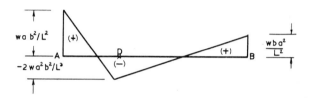

c. Bending moment diagram

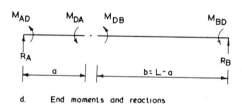

d. End moments and reactions

Figure 5.12

For member AD, taking moments about D (*Figure 5.12d*)

$$-R_A a + M_{AD} + M_{DA} = 0$$

∴
$$R_A = (M_{DA} + M_{AD})/a$$

For member DB, taking moments about D

$$R_B b + M_{BD} + M_{DB} = 0$$

∴
$$R_B = -(M_{BD} + M_{DB})/b$$

For the vertical equilibrium of the whole beam

$$R_A + R_B = W$$

∴
$$(M_{AD} + M_{DA})/a - (M_{BD} + M_{DB})/b = W$$

Using equations (5.46) for M_{AD}, M_{DA} etc

$$(6k_1 \theta_D - 12k_1 v/a)/a - (6k_2 \theta_D + 12k_2 v/b)/b = W \quad (5.48)$$

Solving equations (5.47) and (5.48) for θ_D and v

$$\left. \begin{array}{l} \theta_D = Wa^2b^2(a-b)/(2EIL^3) \\ \\ v = -Wa^3b^3/(3EIL^3) \end{array} \right\} \quad (5.49)$$

and

Substituting these values in equations (5.46)

$$\left. \begin{array}{l} M_{AD} = Wab^2/L^2, M_{DA} = +2Wa^2b^2/L^3, M_{DB} = -2Wa^2b^2/L^3 \\ M_{BD} = -Wba^2/L^2 \end{array} \right\} \quad (5.50)$$

These moments with their correct signs are shown in *Figure 5.12b*. It is evident that at A and B tension develops at the top fibres while at D tension develops at the bottom fibres. The bending moment diagram, drawn as usual on the tension side, is shown in *Figure 5.12c*. Notice that when deriving equations (5.47) and (5.48) the moments M_{AD}, M_{DA}, M_{DB} and M_{BD} were considered positive throughout (see *Figure 5.12d*). Only after solving equations (5.48) and (5.47) and substituting for v and θ_D back in equations (5.46) did we find the correct sign of the moments.

5.14. Deflection and moment at an unloaded specific point

When the deflection or the bending moment is required at a specific point, this point is treated as an extra joint in the member. The slope deflection equation is then applied to each portion in the whole member as usual. In *Figure 5.13*, for

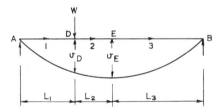

Figure 5.13

instance, the beam AB is loaded at D. The deflection as well as shearing force and the bending moment at E can be calculated by considering E as an extra joint. Thus the member AB is treated as three members. AD is the first one, DE is the second and EB is the third as shown numbered in the figure. When applying the slope deflection equation to member DE, the net sway in it is taken as $v_E - v_D$.

5.15. Examples

Example 1. In the beam shown in *Figure 5.12*, for a value of $W = 10$ kN, $L = 10$ m, $E = 200$ kN/mm², $I = 10^6$ mm⁴ and $a = 4$ m calculate the deflection at a point which is 4.545 m from the support at A.

Answer: From equations (5.50) and *Figure 5.12c*

$M_A = M_{AD} = Wab^2/L^2 = 10 \times 4 \times 36/100 = 14.4$ kN m
$M_B = M_{BD} = Wba^2/L^2 = 10 \times 6 \times 16/100 = 9.6$ kN m
$M_D = 2Wa^2b^2/L^3 = 2 \times 10 \times 16 \times 36/1000 = 11.52$ kN m

The bending moment diagram for this beam is shown in *Figure 5.14b*, where from the similar triangle GDK and NBK, it is found that DK = 3.273. From the similar triangles DGK and JEK

$$11.52/3.273 = M_E/2.728$$
$$\therefore \quad M_E = 9.602 \text{ kN m}$$

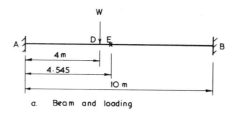

a. Beam and loading

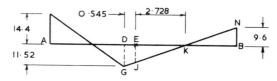

b. Bending moment diagram

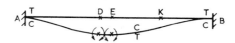

c. Deflected form

Figure 5.14

The ordinate EJ for M_E is negative, thus at E tension develops at the bottom fibres. This means that $M_{ED} = +9.602$ kN m, i.e. anticlockwise while M_{EK} is -9.602 kN m, i.e. clockwise, as shown in *Figure 5.14c* where the bending moments are shown with their correct signs.

The sway of E relative to D is $v = v_E - v_D$. With $k = EI/0.545$ the slope deflection equations for DE become

$$\begin{aligned} M_{DE} &= 4k\theta_D + 2k\theta_E - 6kv/0.545 = -11.52 \\ M_{ED} &= 2k\theta_D + 4k\theta_E - 6kv/0.545 = 9.602 \end{aligned} \quad (5.51)$$

The value of θ_D is calculated from the first of equations (5.49) as

$$\theta_D = 10 \times 16 \times 36(4-6) \times 10^{-3}/(2 \times 200) = -0.0288 \text{ rad}$$

Using $\theta_D = -0.0288$ in equations (5.51) and solving them for θ_E and v, we obtain

$$v = -8 \text{ mm} \text{ and } \theta_E = 0$$

Now from the second of equation (5.49)

$$v_D = -Wa^3 b^3/(3EIL^3)$$

$$v_D = -\frac{10 \times 64 \times 216 \times 10^{18}}{3 \times 200 \times 10^{18}} = -230 \text{ mm}$$

$\therefore$ the deflection at E = $-230 - 8 = -238$ mm, which is, incidentally, the maximum deflection in the beam because $\theta_E = dy/dx = 0$.

Example 2. Calculate the deflection at midspan point D of the beam shown in *Figure 5.15* when a moment $-M$ is acting at A. The span of the beam is $2L$.

Answer: The bending moment and deflected form diagrams for the beam are shown in *Figures 5.15b* and *c*. Treating point D as a joint connecting members 1 and 2

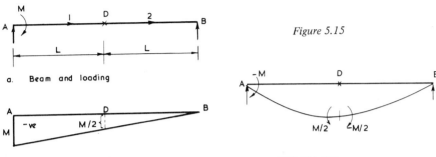

Figure 5.15

a. Beam and loading

b. Bending moment diagram

c. Deflected form

and applying the slope deflection equations to AD and DB with $k = EI/L$

$$M_{AD} = 4k\theta_A + 2k\theta_D - 6kv/L = -M$$
$$M_{DA} = 2k\theta_A + 4k\theta_D - 6kv/L = M/2$$
$$M_{BD} = 4k\theta_B + 2k\theta_D + 6kv/L = 0$$
$$M_{DB} = 4k\theta_D + 2k\theta_B + 6kv/L = -M/2$$

These equations give

$$v = v_D = -ML^2/(4EI) \tag{5.52}$$

5.16. Treatment of uniform loading

When a beam is subjected to a uniformly distributed load over the whole or a part of its span, that part is treated as a member. A joint is inserted at the point where the uniform load starts and another joint is placed where the uniform load terminates. In the beam shown in *Figure 5.16a*, for instance, the portion DE is subjected to a uniform load of w/unit length. This beam is treated as three members. AD is the first member which is connected to the second member DE at joint D. The third member is EB connected to member 2 at E.

In section 5.11 it was shown that a beam with a uniformly distributed load of w/unit length is equivalent to two simply supported beams one of which is loaded uniformly by w/unit length and the other is subject to two end couples of $wL^2/12$ and $-wL^2/12$. The actual beam ADEB (*Figure 5.16a*) is therefore equivalent to system 1 shown in *Figure 5.16b* plus system 2 shown in *Figure 5.16c*. The system in *Figure 5.16b* consists of an unloaded beam and a simply supported beam. The bending moment diagram for this system is shown in *Figure 5.16d*. The system in *Figure 5.16c* is a beam simply supported at A and B and subjected to two vertical loads of $wL_2/2$ at D and E together with external moments of $-wL_2^2/12$ at D and $wL_2^2/12$ at E. The bending moment diagram for this beam is shown in *Figure 5.16e*. Notice that when the two systems are added together, i.e. add *Figures 5.16b* and *c*, the vertical loads and the end couples cancel each other, leaving the uniformly distributed load as in the original beam shown in *Figure 5.16a*. Thus the beam in system 2, *Figure 5.16c* is analysed by slope deflection and to its results, i.e. those shown in *Figure 5.16e* are added the results of the simply supported system 1, i.e. those shown in *Figure 5.16d*. The final and the actual results thus obtained are shown on the bending moment diagram in *Figure 5.16f*. This is the bending moment diagram for the actual beam.

130   *Slopes and Deflections*

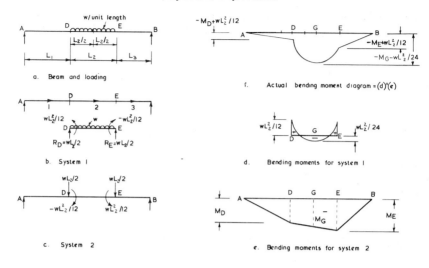

Figure 5.16

5.17. Example

Example 1. Calculate the deflection at the midspan D for the simply supported beam AB shown in *Figure 5.17a*. The span is $2L$ and half of the beam, AD, is subjected to a total load of W.

Answer: The beam and the loading is equivalent to the two systems shown in *Figures 5.17b* and *c*. System b is by the principle of superposition, equivalent to those shown in *Figures 5.17d, e* and *f*. The bending moment diagrams for *Figures 5.17c, d, e* and *f* are shown to the right of each beam. The algebraic sum of these figures gives the bending moment diagram shown to the right of the beam with the actual loading shown in *Figure 5.17a*.

At D, for instance

$$M_{DG} = -\frac{WL}{4} - \frac{WL}{12} \times \frac{1}{2} - \frac{WL}{24} + \frac{WL}{12} = -WL/4$$

$$M_{DB} = -\frac{WL}{4} - \frac{WL}{12} \times \frac{1}{2} + \frac{WL}{24} = -WL/4$$

and at G

$$M_G = -\frac{WL}{4} \times \frac{1}{2} - \frac{WL}{12} \times \frac{3}{4} - \frac{WL}{24} \times \frac{1}{2} - \frac{WL}{24} = -WL/4$$

To answer the actual question of finding the deflection at point D, equation (5.25) gives the deflection at D due to the loading shown in *Figure 5.17d* as

$$v_{D1} = -0.5W(2L)^3/48EI = -WL^3/12EI$$

Equation (5.44) gives the deflection at D due to the loading shown in *Figure 5.17f*, with $a = b = L$, as

Slopes and Deflections

$$v_{D2} = 0$$

Finally equation (5.52) gives the deflection at D due to the loading shown in *Figure 5.17e* as

$$v_{D3} = -\frac{WL}{12} \times \frac{L^2}{4EI} = -WL^3/(48EI)$$

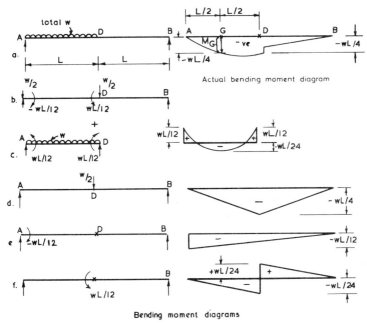

Figure 5.17

The deflection at point D under the actual loading (*Figure 5.17a*) is therefore given by

$$v_D = v_{D1} + v_{D2} + v_{D3}$$

$$v_D = -\frac{WL^3}{12EI} + 0 - \frac{WL^3}{48EI} = -\frac{5WL^3}{48EI}$$

5.18. Further examples

Example 1. The portions AB and BC of the cantilever shown in *Figure 5.18a* have flexural rigidity $2EI$ and EI respectively. Calculate the deflection of the free end C where the beam is subject to a point load W.

Answer: Because the cross sectional property of the beam changes at B, it is treated as if it consists of two members AB and BC connected together at joint B. The deflected form of the beam is shown in *Figure 5.18b* where it is indicated that: $M_{AB} = WL$, $M_{BA} = -WL/2$, $M_{BC} = WL/2$ and the total deflection v_C at C is equal to the sum of the deflections v_{BA} at B and the deflection v_{CB} of point C relative to point B.

Because end A is fixed $\theta_A = 0$ and at B we have $\theta_{BA} = \theta_{BC} = \theta_B$. Taking $k_1 = k_{AB}$ as $2EI/(0.5L) = 4EI/L$ and $k_2 = k_{BC}$ as $EI/(0.5L) = 2EI/L$ we obtain k_2/k_1 as $1/2$.

Applying the slope deflection equations to member 1

$$M_{AB} = -12k_1 v_{BA}/L + 2k_1 \theta_B = WL$$
$$M_{BA} = -12k_1 v_{BA}/L + 4k_1 \theta_B = -WL/2$$

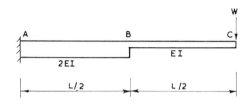

a. Beam and loading

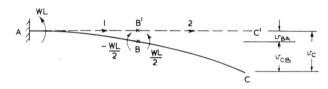

b. Deflected form of the beam

Figure 5.18

These give

$$\left. \begin{array}{l} \theta_B = -3WL/4k_1 \\ \\ v_{BA} = -5WL^3/(96EI) \end{array} \right\} \quad (5.53)$$

and

Applying the slope deflection equations to member 2

$$\left. \begin{array}{l} M_{BC} = -12k_2 v_{CB}/L + 4k_2 \theta_B + 2k_2 \theta_C = WL/2 \\ M_{CB} = -12k_2 v_{CB}/L + 2k_2 \theta_B + 4k_2 \theta_C = 0 \end{array} \right\} \quad (5.54)$$

with $k_2/k_1 = 1/2$ and using the value of θ_B given by equation (5.53), equations (5.54) give

$$v_{CB} = -13WL^3/(96EI)$$

Thus

$$v_C = v_{BA} + v_{CB} = -5WL^3/(96EI) - 13WL^3/(96EI)$$
$$= -3WL^3/(16EI)$$

Example 2. Calculate the deflection of the free end C of the bent bar shown in *Figure 5.19*. *EI* is constant throughout.

Answer: The small contraction of AB due to the axial force W can be neglected. The deflected form of the bar and the values of the bending moments are shown in *Figure 5.19b*. End A is fixed and thus $\theta_{AB} = 0$ and because the joint at B is rigid $\theta_{BA} = \theta_{BC} = \theta_B$. With $k = EI/L$, the slope deflection equations for member AB are

$$M_{AB} = 2k\theta_B - 6kv_B/L = WL$$

and
$$M_{BA} = 4k\theta_B - 6kv_B/L = -WL$$

These give
$$4k\theta_B = -4WL \qquad (5.55)$$

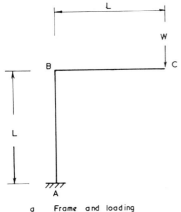
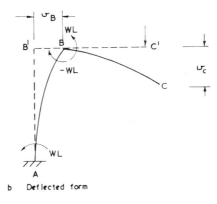

a Frame and loading

b Deflected form

Figure 5.19

The slope deflection equations for BC are
$$M_{BC} = 4k\theta_B + 2k\theta_C - 6kv_C/L = WL$$
$$M_{CB} = 2k\theta_B + 4k\theta_C - 6kv_C/L = 0$$

These and equation (5.55) give
$$v_C = -4WL^3/(3EI)$$

Example 3. The beam ACB is pinned at A and supported at C by the pin ended tie CD as shown in *Figure 5.20a*. The cross sectional area of the tie is A and the second moment of area of the beam is I. Calculate the deflection of the free end B where a load W is acting. E is the same for the beam and the bar.

Answer: As the beam deflects downwards a tensile force p develops in the tie CD which extends it by an amount $v_{CA} = C'C$. The force p is calculated by taking moments about A, thus
$$W \times 2L = p \times L$$
$$\therefore \qquad p = 2W$$

The extension of the tie is calculated from the stiffness equation of a pin ended member, thus
$$p = EA\, v_{CA}/L$$
$$\therefore \qquad v_{CA} = pL/EA = 2WL/(EA)$$

Since the tie is connected to the beam at C, the downward deflection of the beam there is also equal to v_{CA}. When considering the beam it is thus known that the deflection at C is $-2WL/EA$. This deflection is considered negative because the already known movement of point C relative to A is clockwise. With $k = EI/L$, the slope deflection equations for AC thus become
$$M_{AC} = 4k\theta_A + 2k\theta_C + (6k/L) \times (2WL/EA) = 0$$

$$M_{CA} = 4k\theta_C + 2k\theta_A + (6k/L) \times (2WL/EA) = -WL$$

These give

$$k\theta_C = -(WL/3) - (2kW/EA) \qquad (5.56)$$

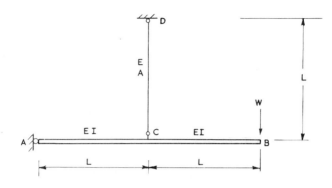

a. Beam and tie

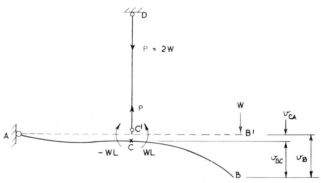

b. Deflected form of the system

Figure 5.20

The deflection v_{BC} of B relative to C is unknown and therefore assumed to be positive. The slope deflection equations for CB are

$$M_{CB} = 4k\theta_C + 2k\theta_B - 6kv_{BC}/L = WL$$
$$M_{BC} = 4k\theta_B + 2k\theta_C - 6kv_{BC}/L = 0$$

These and equation (5.56) give

$$v_{BC} = -(2WL^2/EI)[(L/3) + (EI/LEA)]$$

The total deflection v_B at B is thus $v_{CA} + v_{BC}$

i.e.

$$v_B = -(2WL^2/EI)[(L/3) + (I/LA)] - (2WL/EA)$$

Example 4. In the rigidly jointed frame shown in *Figure 5.21a*, the beam BC carries a uniformly distributed load of total value W. Draw the bending moment diagram for the frame. What are the bending moments at A, B and C when $k_1 = k_2$?

Answer: The frame and its load shown in *Figure 5.21a* is equivalent to the two systems shown in *Figures 5.21b* and *c*. The bending moment diagram for the beam in *Figure 5.21c* is shown in *Figure 5.21d*.

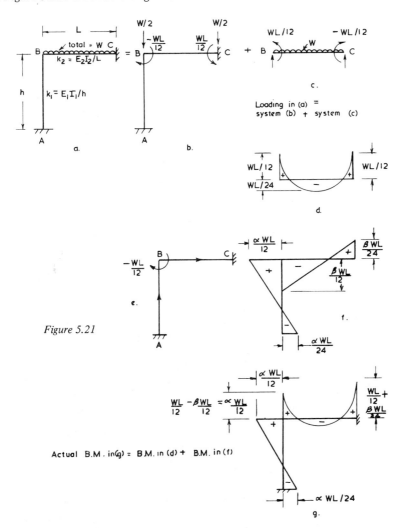

Figure 5.21

Referring to the system shown in *Figure 5.21b*, because joint C is fixed, the application of the vertical load W/2 and the moment WL/12 to this joint has no effect on the frame. The vertical load W/2 acting at B only causes a negligible vertical deflection at B. The loading system shown in *Figure 5.21b* is therefore the same as that shown in *Figure 5.21e* where the frame is subjected to an external moment −WL/12 at joint B. For this

$$\theta_A = \theta_C = \delta_{Bv} = \delta_{Bh} = 0$$

where δ_{Bv} and δ_{Bh} are the vertical and the horizontal movements at joint B. The slope deflection equations for the frame and loading shown in *Figure 5.21e* therefore reduce to

$$M_{AB} = 2k_1\theta_B$$
$$M_{BA} = 4k_1\theta_B$$
$$M_{BC} = 4k_2\theta_B$$
$$M_{CB} = 2k_2\theta_B$$
(5.57)

The second and the third of these equations give

$$M_{BC} = M_{BA} k_2/k_1 \qquad (5.58)$$

For the equilibrium of joint B, the sum of the bending moments M_{BC} and M_{BA} must add up to the external moment applied at B, thus

$$M_{BA} + M_{BC} = -WL/12$$

This and equation (5.58) give

$$M_{BA} = [k_1/(k_1 + k_2)] \times -WL/12 = -\alpha(WL/12) \qquad (5.59)$$
$$M_{BC} = [k_2/(k_1 + k_2)] \times -WL/12 = -\beta(WL/12) \qquad (5.60)$$

The first two of equations (5.57) then give

$$M_{AB} = -\alpha(WL/24) \qquad (5.61)$$

and the last two of these equations give

$$M_{CB} = -\beta(WL/24) \qquad (5.62)$$

The negative signs indicate that these moments are all clockwise. For BC, therefore, tension is developed at bottom fibres at B and at the top fibres at C while in member AB tension develops on the right at A and on the left at B. The bending moment diagram for the frame and loading shown in *Figure 5.21e* (and thus *Figure 5.21b*) is shown in *Figure 5.21f*. Adding these moments to those shown in *Figure 5.21d* gives the actual bending moment diagram for the frame. This is shown in *Figure 5.21g*.

For the special case when $k_1 = k_2$; $\alpha = \beta = 0.5$, *Figure 5.21c* and equations (5.59) to (5.62) give

$$M_{AB} = -WL/48 \qquad \text{(clockwise)}$$
$$M_{BA} = -WL/24 \qquad \text{(clockwise)}$$
$$M_{BC} = -\frac{WL}{24} + \frac{WL}{12} = WL/24 \qquad \text{(anticlockwise)}$$
$$M_{CB} = -\frac{WL}{48} - \frac{WL}{12} = -5WL/48 \qquad \text{(clockwise)}$$

The process of obtaining the final bending moment diagram for a frame by adding the results of the two separate systems can be incorporated directly to the slope deflection equations. Thus including the fixed end moments $WL/12$ at B and $-WL/12$ at C for the beam BC shown in *Figure 5.21c*, the slope deflection equations for the members of the frame become

$$M_{AB} = 2k_1\theta_B$$
$$M_{BA} = 4k_1\theta_B$$
$$M_{BC} = 4k_2\theta_B + WL/12$$
$$M_{CB} = 2k_2\theta_B - WL/12$$
(5.63)

6
Influence lines

6.1. Introduction

The previous chapters have been concerned with forces, moments, deflections etc. anywhere in a structure, when a load or set of loads has been acting at a specific position. In this chapter we study the influence of moving loads on the values of a force, a moment etc. at a specific point. Consider, for instance, the case of the bending moment M_D at the midspan D of a simply supported bridge beam AB. It is easy to calculate M_D when a load of 1 kN is at a specific point such as E. However, if the load is moving from A to B, the value of M_D changes continuously. A graph of M_D against the distance x from end A of the moving load is called the 'influence line' for M_D.

6.2. Influence line for support reactions

Consider the case of a simply supported beam AB of span L as shown in *Figure 6.1*. When a unit moving load acts on the beam at a distance x from A, the reactions

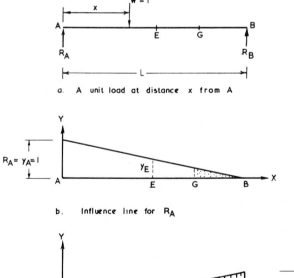

Figure 6.1

a. A unit load at distance x from A

b. Influence line for R_A

c. Influence line for R_B

141

R_A and R_B are

$$y = R_A = 1 \times (L - x)/L = (L - x)/L$$

and

$$y' = R_B = 1 - R_A = x/L$$

(6.1)

Both R_A and R_B are linearly related to the distance x from A of the moving load. A graph of x against R_A or R_B will therefore be a straight line. The graph of R_A against x is the influence line for R_A. Its ordinate y_1 at any point, where $x = x_1$, along the beam measures the value of the reaction R_A when the load is at a distance x_1 from A.

When $x = 0$, the first of equations (6.1) gives $R_A = y_A = 1$ and when $x = L$, it gives $y_B = R_A = 0$. The graph of R_A against x is shown in *Figure 6.1b*. This is the influence line for R_A. Similarly when $x = 0$, $y'_A = R_B = 0$ and when $x = L$, $y'_B = R_B = 1$. The influence line for R_B is shown in *Figure 6.1c*.

When the unit moving load is at E, the value of the reaction R_A is equal to the ordinate y_E of the influence line shown in *Figure 6.1b*. When the value of the moving load at E is not unity but W, the value of the reaction R_A is, by the principle of superposition, equal to $y_E W$.

By using the influence line for R_A it is possible to obtain the value of R_A for any system of loading. For instance, if $W_1, W_2, \ldots W_E, \ldots W_N$ are acting on the beam and $y_1, y_2, \ldots y_E, \ldots y_N$ are the corresponding ordinates of the influence line for R_A (*Figure 6.1b*) then by the principle of superposition

$$R_A = y_1 W_1 + y_2 W_2 + \ldots + y_E W_E + \ldots + y_N W_N = \sum_{E=1}^{N} y_E W_E \quad (6.2)$$

or

$$R_A = [y_1 \; y_2 \ldots y_E \ldots y_N] \{W_1 \; W_2 \ldots W_E \ldots W_N\} \quad (6.3)$$

or

$$R_A = \mathbf{Y}^T \mathbf{W}$$

where $\mathbf{Y}$ is the column vector of the ordinates at the N points where the loads are acting; the row vector $\mathbf{Y}^T$ is the transpose of $\mathbf{Y}$.

Similarly, denoting $y'_1 \; y'_2 \ldots y'_E \ldots y'_N$ as the ordinates of the influence line for R_B at points where $\{W_1 \; W_2 \ldots W_E \ldots W_N\}$ are acting, we obtain the reaction at B as

(6.4)

$$R_B = \Sigma y'_E W_E = \mathbf{Y}'^T \mathbf{W}$$

When a portion GB of the beam carries a uniform load of intensity w per unit length, each reaction is obtained by multiplying w by the corresponding area under the proper influence line; the areas by which w is to be multiplied are shown shaded in *Figures 6.1b and c*.

The influence lines for R_A and R_B for a beam DABE with overhangs, as shown in *Figure 6.2a* are constructed similarly. Considering R_A, for instance, when the unit load is at A, $R_A = 1$ and when the unit load is at B, $R_B = 1$ and $R_A = 0$. Thus point a, whose ordinate Aa is 1, is joined to point B (*Figure 6.2b*) and the straight line is extended in both directions to d and e. The line daBe is the influence line for R_A. This indicates that when the unit load is between D and A, $R_A > 1$ and when the unit load is between B and E, R_A is negative. The influence line for R_B is also constructed similarly and it is shown in *Figure 6.2c*.

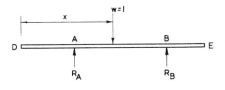

a.   Beam and the moving load

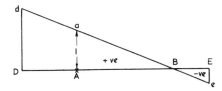

b.   Influence line for R_A

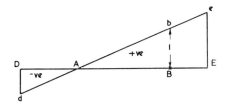

c.   Influence line for R_B

Figure 6.2

6.3. Influence line for shearing force

For a simply supported beam, such as DAB, shown in *Figure 6.3*, the influence line for the shearing force at a section such as E is obtained by considering the influence lines for the reactions R_A and R_B first. When the moving unit load is at A, $R_A = 1$ but $R_B = 0$. On the other hand when the unit load is at B, $R_B = 1$ and $R_A = 0$.

Considering section E, when the moving load is between E and B, the shearing force F_E at E is positive and numerically equal to R_A. On the other hand, when the moving load is between A and E, F_E is negative and numerically equal to R_B.

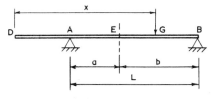

a.   Beam and the moving load.

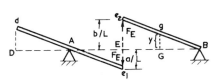

b.   Part of beam displaced at E

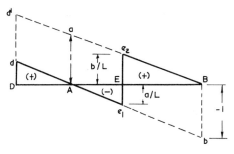

c.   Influence line for shearing force at E

Figure 6.3

Hence the influence line for F_E is given by the diagram bound by the heavy lines, shown in *Figure 6.3c*, which is obtained from the influence lines for R_A and $-R_B$. The diagram indicates that when the moving load is between D and A, the shearing force at E is positive. It is evident in *Figure 6.3c*, that $Ee_2 = b/L$, $Ee_1 = a/L$ and $e_1 e_2 = 1$.

The principle of 'virtual' work can be used to draw the influence line for the shearing force at any point. Consider section E again. Suppose the beam is cut at E and the two parts are displaced, as rigid bodies, by the shearing forces F_E as shown in *Figure 6.3b*. Because of this displacement the work done by F_E is $F_E \times (e_2 e_1)$. The work done by the unit load at G is $-1 \times y$; this work is negative because the movement of G to g is in the opposite direction to that of the unit load. Now since the beam here is assumed to act as a rigid body, it is considered that it is strained by a negligible amount and therefore it stores no strain energy. The work done by the shearing forces F_E is thus balanced by the work done by the unit load, i.e.

$$e_1 e_2 F_E - 1 \times y = 0$$

$$\therefore \quad F_E = y/e_1 e_2$$

Thus the diagram in *Figure 6.3b* can be taken as the diagram for the influence line for F_E provided that $e_1 e_2$ is drawn so that it is equal to one unit, Ee_2 is b/L and Ee_1 is a/L. Notice in *Figure 6.3c*, the similar triangles aAB and $e_2 EB$ give $aA/AB = Ee_2/b$, thus $1/L = Ee_2/b$ and hence $Ee_2 = b/L$. Similarly $Ee_1 = a/L$.

6.4. Example

The beam shown in *Figure 6.4* is simply supported at D, A, B and E. The parts are hinged at G and K. What uniform load system of intensity w kN/m makes the

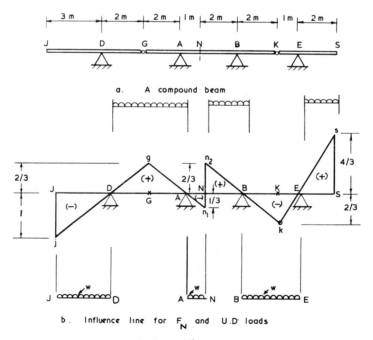

Figure 6.4

shearing force at section N, maximum and positive? What uniform load system makes the shearing force at N numerically maximum but negative?

Answer: Cutting the beam at N and displacing the two parts so that $Nn_2 = NB/AB = 2/3$ and $Nn_1 = NA/AB = 1/3$, the two parts of the beam take the shape given by jgAn_1 and n_2BkES. This is also the shape of the influence line diagram.

For maximum positive shearing force F_N, the parts DA, NB and ES are loaded as shown above the influence line diagram. The maximum positive F_N is obtained by multiplying w by the areas of the triangles DgA, Nn_2B and ESs, i.e.

$$F_{N,max} = w\,(0.5 \times 4 \times 2/3 + 0.5 \times 2 \times 2/3 + 0.5 \times 2 \times 4/3) = 10w/3 \text{ kN}.$$

The numerical maximum negative F_N is obtained when the parts JD, AN and BE are loaded as shown below the influence line diagram, and

$$-F_{N,max} = -w\,(0.5 \times 3 \times 1 + 0.5 \times 1 \times 1/3 + 0.5 \times 3 \times 2/3) = -8w/3 \text{ kN}.$$

6.5. Influence line for bending moments

Consider a simply supported beam AB which is carrying a unit load moving from A to B as shown in *Figure 6.5a*. The bending moment at section E, distance a from the support at A, changes as the unit load moves along the beam. However, wherever the load may be, it causes tension at the bottom fibres of the beam and the bending moment diagram will have negative ordinates. Therefore, the ordinate of the bending moment at E will also be always negative.

When the unit load is to the right of point E, the ordinate of the bending moment diagram at this section is given by $-R_A a$. On the other hand, when the unit load is to the left of point E, the ordinate of the bending moment diagram at E is given by $-R_B b$. Thus the influence line diagram for M_E is obtained from the influence line diagrams for $-R_A$ and $-R_B$. For the part of the beam to the right of E, the influence line for M_E is obtained by multiplying the ordinates of the influence line for $-R_A$ by the length a. Likewise, for the part of the beam to the left of E, the influence line for M_E is obtained by multiplying the ordinates of the influence line for $-R_B$ by the length b. The complete diagram for the influence line for M_E is the traingle AeB, shown bounded by the thick lines in *Figure 6.5b*. The numerically largest ordinate of the diagram is at E and is equal to $Ee = ab/L$. Thus when a single load W moves from A to B its worst effect on M_E occurs when the load is actually at E. The value of M_E will then be $-abW/L$. For the case of a moving uniformly distributed load, the worst effect on M_E occurs when the moving load covers the whole span AB. The value of M_E will then be $-wab/2$ where $ab/2$ is the area of the influence line diagram AeB and w is the intensity of the uniform load per unit length. For the case when E is at the midspan of the beam, $a = b = L/2$ and $M_E = -wL^2/8$.

The principle of virtual work can also be used to draw the influence line for the bending moment at any point such as E. This is done by inserting a fictitious hinge at E and assuming that the two parts AE and BE can rotate about this hinge as unstrained rigid bodies. As soon as the unit load moves on the beam, it forces the two parts AE and BE to rotate as a mechanism. AE rotates about point A by an amount $-\alpha_1$, say, and BE rotates about B by an amount $+\alpha_2$, anticlockwise rotations being positive. Point E moves to e and the rotation of the hinge there will be $-\alpha_1$ for eA and $+\alpha_2$ for eB. The total rotation of the hinge will therefore be $\theta = \alpha_1 + \alpha_2$ as shown in *Figure 6.5c*. When the unit load is at distance x from A, it will move down by an amount y and the work done is $1 \times y$. The moment M_E in EB rotates the hinge by α_2 and therefore the work done by it is $\alpha_2 M_E$. The moment M_E in EA rotates the hinge by α_1 and the work done by it is $M_E \times \alpha_1 = \alpha_1 M_E$. The total work done in rotating the hinge at E is therefore

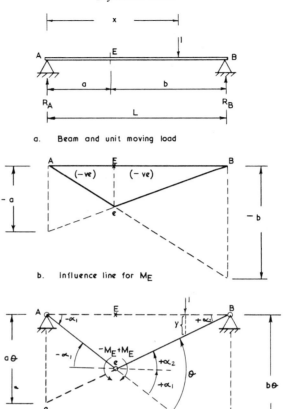

Figure 6.5

$(\alpha_1 + \alpha_2) M_E = M_E \theta$. The virtual work equation then becomes

$$1 \times y + M_E \theta = 0 \tag{6.5}$$

$$\therefore \quad M_E = -y/\theta \tag{6.6}$$

If θ, for convenience, is taken as unity, then equation (6.6) gives the bending moment M_E at E when the unit moving load is at a distance x from A. Thus the shape of the mechanism AeB is, for $\theta = 1$, exactly the same as the influence line diagram for the bending moment at E.

6.6. Example

Draw the influence line diagram for the bending moment M_N at section N of the beam shown in *Figure 6.4a*.

Answer: The beam is reproduced in *Figure 6.6a*. When a hinge is inserted at N, the beam acts as a mechanism as shown by jgnks. The ordinate $Nn = -ab/L = -1 \times 2/3 = -2/3$. The ordinates of the other points directly follow from the geometry of the mechanism. These are indicated in *Figure 6.6b*.

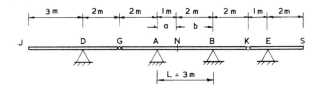

a   The compound beam of figure 6.4.a

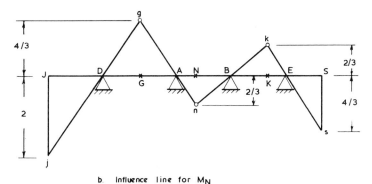

b. Influence line for M_N

Figure 6.6

6.7. The dangerous section

As a unit load moves along a beam, the values of the shearing force and the bending moment at a given section change. These values also change for another section somewhere else along the beam. There is therefore a single section in the beam whose shearing force is higher than any other section, no matter where the moving load is. This section is called the dangerous section for shearing force. The maximum shear force occurring there is called the absolute maximum shearing force which should be used as a basis for the design of the beam. Similarly the section at which the absolute maximum (or the absolute minimum but numerically the largest) bending moment occurs is called the dangerous section for bending moment. This absolute maximum bending moment is used as a basis for the design of the beam.

In the case of a single point load moving along a simply supported beam, *Figure 6.7a*, it is clear that, if various points, such as E, of varying distance b from B are considered, then as b increases from zero at B to L at A, the shearing force at E also increases until its absolute positive maximum is reached at A. Thus A is the dangerous section for the positive shearing force. The value of the absolute maximum shearing force is $1 \times W$ where W is the value of the load. Similarly point B is the dangerous section for the negative shearing force.

The dangerous section for bending moment in the same beam when a single load is moving on the beam, is the midspan point G. It was stated that due to a unit moving load the worst bending moment at a section is $-ab/L$ and it occurs when the unit load is on the section. Clearly (see *Figure 6.7c*), when the unit load is at midspan G the worst bending moment occurs in the beam. Its absolute value is $-0.5L \times 0.5L/L = -L/4$ which is the ordinate Gg.

For the case of a uniformly distributed load of intensity w, the dangerous section for bending moment is also point G, when the load covers the whole span and $M_G = -wL^2/8$. The dangerous section for the positive shear force is at A and

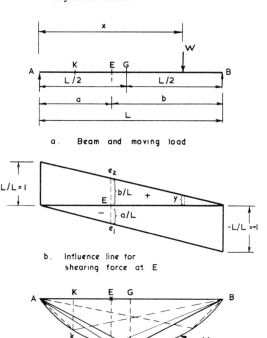

Figure 6.7

the absolute maximum shearing force is $wL/2$ when the load covers the span AB. The absolute minimum shearing force is $-wL/2$ and the dangerous section for this is B.

6.8. The dangerous section for several loads

Consider a simply supported beam AB carrying a number of moving loads as shown in *Figure 6.8*. The distance between one load and the next is fixed. Let R be the resultant of all the loads on the beam. The distance of R from A is x. The worst bending moment occurs under one of the loads. This will be at a point where the shearing force changes sign as explained in Chapter 3. Let W_i be the load which is at a distance s from the resultant R. The reaction R_A is obtained by taking moments about B

$$-R_A L + R(L - x) = 0$$

$$\therefore \quad R_A = R(L - x)/L$$

The bending moment M_i under the load W_i is

$$M_i = -R_A(x - s) + M$$

i.e.

$$M_i = M - R(L - x)(x - s)/L = M - R(Lx - Ls - x^2 + sx)/L \quad (6.7)$$

where M is the sum of the moments about point i of all the loads to the left of W_i.

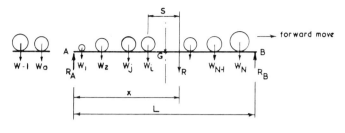

Beam carrying several loads

Figure 6.8

Consider that if the loads move forward by a short distance, i.e. if x increases by a small amount, the last load W_N does not leave the beam and a new load, such as W_0 does not move on the beam. Under these circumstances, the shearing force changes its sign when $dM_i/dx = 0$ i.e., using equation (6.7), when

$$dM_i/dx = -R(L - 2x + s)/L = 0$$

which happens when

$$L - 2x + s = 0$$

i.e. when

$$s/2 = x - L/2 \tag{6.8}$$

This proves that the midspan point G bisects the distance between the resultant R and the load W_i which is on the dangerous point. Thus as soon as R passes the centre line of the beam by a distance $x - L/2$, the dangerous point is behind R by a distance $s = 2x - L$.

6.9. The significance of influence lines

It is now clear how, as the loads move along a structure, such as a bridge, the values of functions such as reactions, shearing force, etc. change; not only from point to point but also they change at a particular point because the loads are moving. It is also clear that there are dangerous points in the structure which have to be discovered and taken into consideration as bases of design. These facts are very significant in structural engineering. However, with the advent of computers and the use of matrix methods, structural engineers use influence line diagrams less than before. This is because a single analysis of a structure by a matrix method gives the results for several load cases. These results indicate the dangerous points and how the load cases representing one set of moving loads are influencing the structure.

6.10. Influence lines for pin jointed frames

The method of constructing the influence diagrams as given above can be used to draw the influence lines for various other functions in different structures. For instance, it can be used to draw the influence line for the normal force at a certain section of a three pinned arch or for the axial force in a member of a pin jointed frame. In the latter case, consider the variation of the forces p_1, p_2 and p_3 in members GK, ED and GD of the pin jointed frame shown in *Figure 6.9*. A unit load is moving from A to B along the bottom chord AEDMB. Cutting the frame

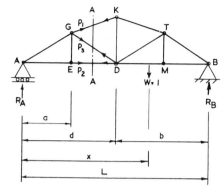

a. A unit load moving from A to B

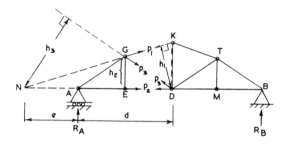

b. The two parts of the frame

Figure 6.9

by section A–A and considering the portion to the left of this section, *Figure 6.9b*, the force p_1 is found by taking moments about point D. The load $W = 1$ is to the right of D. Thus

$$-dR_A - p_1 h_1 = 0$$

∴
$$p_1 = -dR_A/h_1 \text{ (compressive)} \qquad (6.9)$$

This indicates that the compressive force p_1 is proportional to the bending moment $-dR_A$ at D of a corresponding simply supported beam AB of the same span L. Hence the influence line of p_1 is obtained by dividing the ordinates of the influence line for the bending moment at D by h_1. This is also true for the case when the unit load is to the left of D. The influence line for p_1 is shown as Ad_1B in *Figure 6.10b*.

The force p_2 in member ED is found by taking moments about point G, thus

$$p_2 = aR_A/h_2 \text{ (tensile)} \qquad (6.10)$$

Hence the influence line for p_2 is obtained by dividing by $-h_2$ the ordinates of the influence line for the bending moment at G. This is shown as AgB in *Figure 6.10c*.

To construct the influence line for the force p_3 in member GD, consider first that the unit load is to the right of joint D. Taking moments about point N in *Figure 6.9b* where the extensions of the lines KG and DE meet, we obtain

$$p_3 h_3 - R_A e = 0$$

∴
$$p_3 = eR_A/h_3 \text{ (tensile)} \qquad (6.11)$$

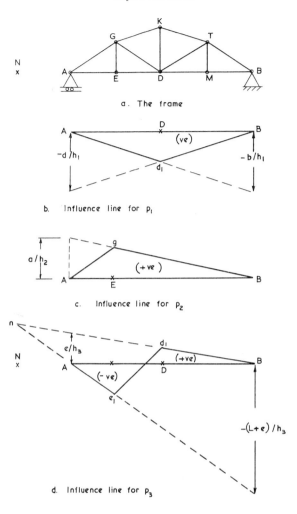

Figure 6.10

Hence while the load is between D and B, the influence line for p_3 is obtained by multiplying the ordinates of the influence line for R_A by e/h_3.

Next consider the unit load to be between A and E. Taking moments of the forces acting on the right portion of the frame about point N

$$R_B (L + e) + p_3 h_3 = 0$$

$$\therefore \quad p_3 = -R_B (L + e)/h_3 \text{ (compressive)} \tag{6.12}$$

Hence the influence line for p_3 is, in this case, obtained by multiplying the ordinates of the influence line for R_B by $-(L + e)/h_3$.

When the unit load is between E and D, we assume that the actual loading of the frame is via cross beams attached to the panel points E, D and M. This enables us to join the two ends of the parts of the influence line for p_3 by a straight line $e_1 d_1$ (why?) as shown in *Figure 6.10d*. The complete diagram is shown as $Ae_1 d_1 B$.

6.11. Examples

Example 1. A vehicle crosses a bridge which has a span of 10 m. The effective concentrated load on the back axle of the vehicle is 15 kN while that on the front axle is 10 kN. The distance between the axles is 4m. Find the dangerous section of the beam for bending moments. What is the numerical value of the absolute minimum bending moment?

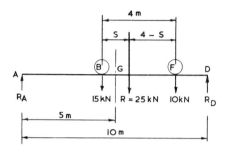

Figure 6.11

Answer: The bridge is shown in *Figure 6.11*.

The resultant of the two loads is $R = 15 + 10 = 25$ kN. Taking moments about the centre of the front axle F

$$15 \times 4 = 25(4 - s)$$
$$\therefore \quad s = 1.6 \text{ m}$$

The dangerous point is therefore $s/2 = 0.8$m to the left of the centre line at G and it is under the back axle B. The numerical value of the bending moment there is M_B. Taking moments about support D

$$-R_A \times 10 + R(5 - 0.8) = 0$$
$$\therefore \quad R_A = 25 \times 4.2/10 = 10.5 \text{ kN}$$
$$M_B = 10.5(5 - 0.8) = 44.1 \text{ kN m}$$

Example 2. Construct the influence line for the force in member KD of the pin jointed frame shown in *Figure 6.12*. The unit load is moving along the bottom chord.

Answer: Cutting the frame with section A–A and taking moments of the forces, acting on the part to the left of A–A, about point D

$$-R_A d - p_1 h_1 = 0$$
$$\therefore \quad p_1 = -dR_A/h_1 \quad (6.13)$$

Resolving the forces horizontally at joint K, it is found that the force in member KG is also p_1.

Resolving the forces vertically at joint K

$$-2p_1 \cos\theta - p_2 = 0$$
$$\therefore \quad p_2 = -2p_1 \cos\theta$$

and from equation (6.13)

$$p_2 = 2dR_A \cos\theta/h_1$$
$$p_2 = 2 \times 5 \times 0.3046\, R_A/2.286 = 1.328\, R_A$$

Influence Lines 153

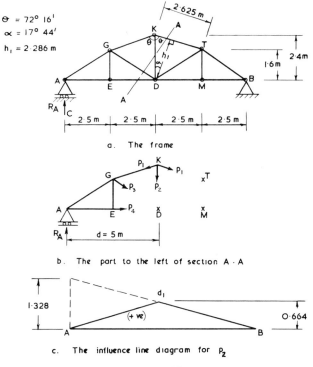

Figure 6.12

Thus when the unit force is to the right of point D the influence line for the force p_2 in member KD is obtained by multiplying the ordinates of R_A by 1.328. When the force is to the left of point D, the influence line for p_2 is obtained by multiplying the ordinates of R_B by 1.328. The influence line diagram for p_2 is shown in Figure 6.12 as Ad_1B.

Exercises on Chapter 6

1. The train of loads shown in Figure 6.13 crosses the simply supported beam from A to B. Calculate the maximum shearing force and bending moment at point C.

Ans. 175.25 kN, 5722.5 kN m.

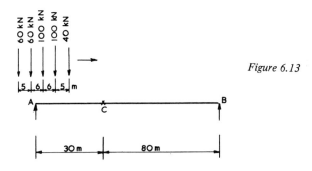

Figure 6.13

154   *Influence Lines*

2. Draw the influence line diagrams for the shearing force and the bending moment at point E for the compound beam shown in *Figure 6.14*.

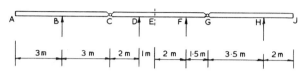

Figure 6.14

3. Draw the influence lines for the forces in members AC and AB for the pin jointed frame shown in *Figure 6.15* for a unit load moving along DACE. What are the maximum forces in these members when a uniform load of 144 kN and length 48m crosses from A to E?

Ans. 185.3 kN, −55.4 kN.

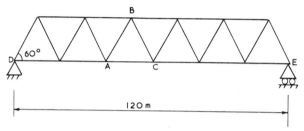

Figure 6.15

4. The weight of the bridge shown in *Figure 6.16* is 0.5 kN/m. The diagonals are unsuitable for sustaining compressive forces. For a uniform moving load of 1.25 kN/m, calculate the number of bays to be counterbraced.

Ans. The central five.

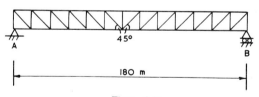

Figure 6.16

5. A symmetrical parabolic three pinned arch is 100 m long and has a rise of 10 m. Calculate the maximum bending moment at its quarter points for two moving loads of 10 kN and 15 kN, at 10 m.

Ans. 159.3 kN m.

7
Flexural properties of structures

7.1. Introduction
A flexural element, which is a member subject to bending moments and shearing forces, has a number of properties such as flexibility, stiffness, strain energy and plasticity. For a flexural element these are different from those of a member subject to an axial force. In this chapter these properties are presented.

7.2. The flexural stiffness
The slope deflection equations (5.14) relate the bending moments at the ends of a member to its deformations, i.e. to its sway and end rotations. These are therefore stiffness equations for the member. A further equation can be derived which gives the shearing force in the member in terms of its deformation. Considering *Figure 5.2* again and taking moments about B', we obtain

$$F_{AB}L + M_{AB} + M_{BA} = 0 \qquad (7.1)$$

Substitutions from the slope deflection equations (5.14) for M_{AB} and M_{BA} into equation (7.1) gives

$$F_{AB} = (12EI/L^3)v - (6EI/L^2)\theta_{AB} - (6EI/L^2)\theta_{BA} \qquad (7.2)$$

Equations (7.2) and (5.14) are written in matrix form as

$$\begin{bmatrix} F_{AB} \\ M_{AB} \\ M_{BA} \end{bmatrix} = \begin{bmatrix} 12EI/L^3 & -6EI/L^2 & -6EI/L^2 \\ -6EI/L^2 & 4EI/L & 2EI/L \\ -6EI/L^2 & 2EI/L & 4EI/L \end{bmatrix} \begin{bmatrix} v_{AB} \\ \theta_{AB} \\ \theta_{BA} \end{bmatrix} \qquad (7.3a)$$

or simply

$$\mathbf{P}_{AB} = \mathbf{k}_{AB}\boldsymbol{\delta}_{AB} \qquad (7.3b)$$

The 3 × 3 matrix $\mathbf{k}_{AB}$ is the stiffness matrix of the member. Thus it is noticed that the stiffness of a flexural member is not defined by a single number but by a matrix with nine stiffness coefficients. Each coefficient of course has a definite meaning. For instance, if a propped cantilever AB, *Figure 7.1*, is subjected to an external moment M_{AB} at A the rotation at B and the sway of the member will

Figure 7.1

both be zero while the rotation at A is θ_{AB} given by

$$M_{AB} = (4EI/L)\theta_{AB}$$

and

$$M_{AB}/\theta_{AB} = 4EI/L \tag{7.4}$$

Thus the stiffness or the stiffness coefficient $4EI/L$ is the moment required to be applied at the hinged end of a propped cantilever to produce a unit rotation at that end while all the other deformations of the member are prevented. Under such conditions, $\theta_{BA} = 0$, $v_{AB} = 0$ and the last of equations (7.3a) or the original slope deflection equations (5.14) gives

$$M_{BA}/\theta_{AB} = 2EI/L \tag{7.5}$$

Equation (7.2), on the other hand gives

$$F_{AB}/\theta_{AB} = -6EI/L^2 \tag{7.6}$$

The stiffness or the stiffness coefficient $2EI/L$ can therefore be defined as the moment produced at the second end B of the propped cantilever per unit rotation at the first end. Similarly, the stiffness coefficient $-6EI/L^2$ is the shear force produced in the propped cantilever per unit rotation of the first end. However, it is much easier to look upon the nine elements of the stiffness matrix k as a single entity and consider the matrix as one that defines all the forces in the member in terms of all its deformations. Notice that the term 'force' here is used with its generalised meaning which includes forces and moments.

Defining, for simplicity

$$\begin{aligned} b &= 12EI/L^3 \\ d &= -6EI/L^2 \\ e &= 4EI/L \\ f &= 2EI/L = e/2 \end{aligned} \tag{7.7}$$

Equations (7.3a) become

$$\begin{bmatrix} F_{AB} \\ M_{AB} \\ M_{BA} \end{bmatrix} = \begin{bmatrix} b & d & d \\ d & e & f \\ d & f & e \end{bmatrix} \begin{bmatrix} v_{AB} \\ \theta_{AB} \\ \theta_{BA} \end{bmatrix} \tag{7.3c}$$

and

$$k_{AB} = \begin{bmatrix} b & & \text{symmetrical} \\ d & e & \\ d & f & e \end{bmatrix} \tag{7.8}$$

For a rigidly jointed plane frame with N members, equations such as (7.3) can be written for each member. All of these together will take the form

$$\begin{bmatrix} P_1 \\ P_2 \\ P_3 \\ P_{AB} \\ P_N \end{bmatrix} = \begin{bmatrix} k_1 & & & & 0 \\ & k_2 & & & \\ & & k_3 & & \\ & & & k_{AB} & \\ 0 & & & & k_N \end{bmatrix} \begin{bmatrix} \delta_1 \\ \delta_2 \\ \delta_3 \\ \delta_{AB} \\ \delta_N \end{bmatrix} \quad (7.9a)$$

or simply $\quad P = k\delta \quad$ (7.9b)

Matrix **k** is constructed by arranging the stiffness matrices of the members so that their leading diagonals form the leading diagonal of **k**. This matrix is called the member stiffness matrix of the unassembled structure.

7.3. Examples

Example 1. Define the stiffness coefficient b. A fixed ended beam AB has a span of 10 m, its modulus of elasticity is 200 kN/mm² and its second moment of area is 10^6 mm⁴. What forces applied to the ends of the beam cause end B to subside by 50 mm?

Answer: For a member AB, with $\theta_{AB} = \theta_{BA} = 0$, equation (7.2) gives the shearing force as

$$F_{AB} = (12EI/L^3)v = bv$$

Thus

$$b = F_{AB}/v$$

and is the shear force per unit sway of a member without end rotations. The answer to the second part of the questions, therefore is

$$F_{AB} = \frac{12 \times 200 \times 10^6}{10^{12}} \times 50 = 0.12 \text{ kN.}$$

Example 2. A simply supported pin ended beam AB is subjected to equal end moments $M_{AB} = M_{BA}$. The beam is 10 m long with $E = 200$ kN/mm² and $I = 10^6$ mm⁴. What is its stiffness?

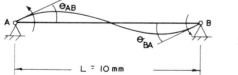

Figure 7.2

Answer: The beam is shown in *Figure 7.2*, and it is in double curvature. Since the sway of the member is zero, the slope deflection equations become

$$M_{AB} = 4k\theta_A + 2k\theta_B$$
$$M_{BA} = 2k\theta_A + 4k\theta_B$$

where $k = EI/L$. Since $M_{AB} = M_{BA}$, it follows that

$$4k\theta_A + 2k\theta_B = 2k\theta_A + 4k\theta_B$$

∴ $\qquad \theta_A = \theta_B$

Thus $M_{AB} = 6k\theta_A = 6EI\theta_A/L$

158 *Flexural Properties of Structures*

and the stiffness K of the beam is

$$K = M_{AB}/\theta_{AB} = 6EI/L$$

$$\therefore K = \frac{6 \times 200 \times 10^6}{10 \times 10^3} = 120\,000 \text{ kN mm}$$

or

$$K = 120 \text{ kN m}$$

7.4. The flexibility

Consider a simply supported pin ended member AB of length L subject to two moments, M_{AB} at A and $-M_{BA}$ at B as shown in *Figure 7.3*. The slope deflection equations with $v = 0$ and $k = EI/L$ for this member are

$$M_{AB} = 4k\,\theta_{AB} - 2k\,\theta_{BA}$$

$$-M_{BA} = 2k\,\theta_{AB} - 4k\,\theta_{BA}$$

Solving these equations for θ_{AB} and θ_{BA}

$$\left.\begin{array}{l}\theta_{AB} = L(2M_{AB} + M_{BA})/6EI \\ \theta_{BA} = L(M_{AB} + 2M_{BA})/6EI\end{array}\right\} \qquad (7.10a)$$

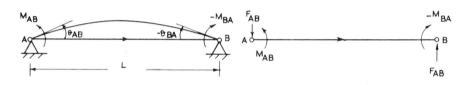

a.   A member in single curvature   b.   Shear force in the member

Figure 7.3

These are known as the flexibility equations of a member. They define the end rotations of a member in terms of the bending moments at these ends. These equations can be written in matrix form as

$$\begin{bmatrix}\theta_{AB} \\ \theta_{BA}\end{bmatrix} = \begin{bmatrix}L/3EI & L/6EI \\ L/6EI & L/3EI\end{bmatrix}\begin{bmatrix}M_{AB} \\ M_{BA}\end{bmatrix} = (L/6EI)\begin{bmatrix}2 & 1 \\ 1 & 2\end{bmatrix}\begin{bmatrix}M_{AB} \\ M_{BA}\end{bmatrix} \qquad (7.10b)$$

or simply as

$$\boldsymbol{\theta}_i = \mathbf{f}_i\,\mathbf{M}_i \qquad (7.10c)$$

where for member i, $\boldsymbol{\theta}_i = \{\theta_{AB}\ \ \theta_{BA}\}_i$ and $\mathbf{M}_i = \{M_{AB}\ \ M_{BA}\}_i$. The 2×2 matrix $\mathbf{f}$ is the flexibility matrix of member AB. To define the flexibility

Flexural Properties of Structures

coefficient $L/6EI$ and $L/3EI$, consider that $M_{BA} = 0$ in *Figure 7.3*, then equations (7.10a) or (7.10b) give

$$\theta_{AB} = (L/3EI) \times M_{AB}$$

i.e.

$$\theta_{AB}/M_{AB} = L/3EI$$

and

$$\theta_{BA} = (L/6EI) \times M_{AB} = \tfrac{1}{2}\theta_{AB}$$

$$\theta_{BA}/M_{AB} = L/6EI$$

(7.11)

Thus the flexibility or the flexibility coefficient $L/3EI$ is the rotation produced at end A by a unit moment applied there. The flexibility $L/6EI$, on the other hand, is the rotation produced at the second end of a member by a unit moment applied at the first end.

Equations similar to (7.10) can be written for each member of a rigidly jointed plane frame. These take the form

$$\begin{bmatrix} \theta_1 \\ \theta_2 \\ \vdots \\ \theta_i \\ \vdots \\ \theta_N \end{bmatrix} = \begin{bmatrix} f_1 & & & & 0 \\ & f_2 & & & \\ & & \ddots & & \\ & & & f_i & \\ & & & & \ddots \\ 0 & & & & f_N \end{bmatrix} \begin{bmatrix} M_1 \\ M_2 \\ \vdots \\ M_i \\ \vdots \\ M_N \end{bmatrix}$$

(7.12a)

or simply

$$\boldsymbol{\theta} = \mathbf{f}\,\mathbf{M}$$ (7.12b)

Matrix $\mathbf{f}$ is called the member flexibility matrix of the unassembled structure. It is constructed by arranging the flexibility matrices of the individual members so that their leading diagonal forms the leading diagonal of $\mathbf{f}$.

7.5. Examples

Example 1. Construct matrix $\mathbf{f}$ for the cantilever ABC shown in *Figure 5.18*.
Answer: $L_{AB} = L_{BC} = 0.5L$; $(EI)_{AB} = 2EI$, $(EI)_{BC} = EI$.

$$\mathbf{f} = \begin{bmatrix} f_{AB} & 0 \\ 0 & f_{BC} \end{bmatrix} = (0.5L/EI) \begin{bmatrix} 1/6 & 1/12 & 0 & 0 \\ 1/12 & 1/6 & 0 & 0 \\ 0 & 0 & 1/3 & 1/6 \\ 0 & 0 & 1/6 & 1/3 \end{bmatrix}$$

Example 2. Construct the member stiffness and flexibility equations for the structure shown in *Figure 7.4*. The beam ABC has constant I and the area of the pin ended member BD is A. The value of E is constant throughout.

Flexural Properties of Structures

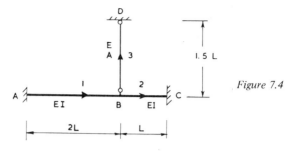

Figure 7.4

Answer: For AB with length $2L$; $b_1 = 12EI/(2L)^3$, $d_1 = -6EI/(2L)^2$, $e_1 = 4EI/(2L)$.
For member BC; $b_2 = 12EI/L^3 = 8b_1$, $d_2 = -6EI/L^2 = 4d_1$, $e_2 = 4EI/L = 2e_1$.
For member BD, $a_3 = EA/(1.5L)$.
Because ends A and C are fixed with $\theta_{AB} = \theta_{CB} = 0$, the rows and columns of matrix k corresponding to these rotations can be disregarded. Thus $\mathbf{P} = \mathbf{k}$ become

$$\begin{bmatrix} F_{AB} \\ M_{BA} \\ F_{BC} \\ M_{BC} \\ P_{BD} \end{bmatrix} \begin{bmatrix} b_1 & d_1 & 0 & 0 & 0 \\ d_1 & e_1 & 0 & 0 & 0 \\ 0 & 0 & 8b_1 & 4d_1 & 0 \\ 0 & 0 & 4d_1 & 2e_1 & 0 \\ 0 & 0 & 0 & 0 & a_3 \end{bmatrix} \begin{bmatrix} v_{AB} \\ \theta_{BA} \\ v_{BC} \\ \theta_{BC} \\ \delta L_{BD} \end{bmatrix}$$

with $g = L/(6EI)$, the flexibility equations are

$$\begin{bmatrix} \theta_{AB} \\ \theta_{BA} \\ \theta_{BC} \\ \theta_{CB} \\ \delta L_{BD} \end{bmatrix} \begin{bmatrix} 4g & 2g & 0 & 0 & 0 \\ 2g & 4g & 0 & 0 & 0 \\ 0 & 0 & 2g & g & 0 \\ 0 & 0 & g & 2g & 0 \\ 0 & 0 & 0 & 0 & 1.5L/(EA) \end{bmatrix} \begin{bmatrix} M_{AB} \\ M_{BA} \\ M_{BC} \\ M_{CB} \\ P_{BD} \end{bmatrix}$$

Example 3. How is it possible to reduce the stiffness, i.e. increase the flexibility, of any of the horizontal members shown in *Figure 7.4*?

Answer: The stiffness coefficients contain EI/L, EI/L^2 or EI/L^3 while the flexibility coefficients contain L/EI.
∴ The stiffness can be reduced by

  (1) reducing the modulus of elasticity E.
  (2) reducing the second moment of area I.
  (3) increasing the length L.
  (4) While keeping E, I and L constants, the stiffness of each member is reduced by applying equal and opposite external horizontal forces at A and C. which causes compression in the member. Such a force helps

the moments to rotate the beam further. Thus more rotation is produced by a unit moment. Compare, for instance, the two beams in *Figure 7.5*. Also see Chapter 10.

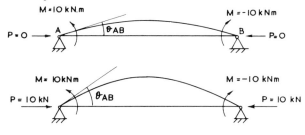

Figure 7.5

7.6. Strain energy in pure bending

1. Consider an element of a beam in pure bending as shown in *Figure 7.6a*. The material of the element has a non-linear $M - \theta$ relationship as in *Figure 7.6b*.

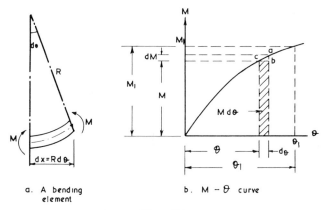

a. A bending element

b. $M - \theta$ curve

Figure 7.6

As the external moment M increases from M to $M + dM$, the rotation in the element increases from θ to $\theta + d\theta$. The work done by the external moment is

$$dw = (M + dM/2) \, d\theta = M d\theta + dM d\theta/2$$

Since both dM and $d\theta$ are small, the area of the triangle abc = $dM d\theta /2$ in *Figure 7.6b* is small and can be neglected, thus the work done is

$$dw = M d\theta \qquad (7.13)$$

This work is stored in the element as strain energy which for elastic, linear or non-linear material, can be lost during the removal of the external moment. Thus the strain energy dU stored in the element is

$$dU = M d\theta$$

and the total strain energy stored in the element, as M increases from zero to M_1 and θ increases from zero to θ_1, is

$$U = \int_0^{\theta_1} M d\theta \qquad (7.14)$$

2. In the linear elastic region, where M is related linearly to θ, equation (7.14) gives

$$U = 0.5 M_1 \theta_1 \qquad (7.15)$$

and since the beam is in pure bending, the bending moment M_1 remains constant along its length L. The deformed shape of the beam is in the form of a circular arc of curvature M/EI and the angle subtended by the arc is

$$\theta_1 = M_1 L/(EI) \qquad (7.16)$$

Thus using equation (7.15), the total strain energy in the beam is given by

$$\left.\begin{array}{c} U = M_1^2 \, L/(2EI) \\ \\ U = EI \, \theta_1^2/(2L) \end{array}\right\} \qquad (7.17)$$

or

3. In the linear elastic region consider that the bending moment in the beam changes from one point to another because the beam is not subject to pure bending but some other type of loading. Assuming that over a short length dx of the beam the bending moment remains constant and equal to M_x. The change in the slope of the beam is $d\theta = dx/R = (d^2 y/dx^2) dx$, because $1/R = d^2 y/dx^2$.
Equations (7.17) when applied to the small length dx become

$$\left.\begin{array}{c} dU = M_x^2 \, dx/(2EI) \\ \\ dU = (EI/2)(d^2 y/dx^2)^2 \, dx \end{array}\right\} \qquad (7.18)$$

and

This is provided that the strain energy due to the shearing of the beam is small and can be neglected, which is the case when the beam is not deep.
Integrating equations (7.18) over the length L of the beam

$$\left.\begin{array}{c} U = \int_0^L M_x^2 \, dx/2EI \\ \\ U = (EI/2) \int_0^L (d^2 y/dx^2)^2 \, dx \end{array}\right\} \qquad (7.19)$$

and also

7.7. Examples

Example 1. Compare the strain energy stored in three rectangular bars of exactly the same length L, cross section and linear elastic properties. The first bar is under direct tensile force p, the second is in pure bending and the third is fixed at one end as a cantilever and carries a point load p at the other. The tensile stress in the bars should not exceed the limiting value σ.

Answer: The bars are shown in *Figure 7.7*. Let B and d be the width and the depth of the bars with $d > B$. From equation (2.39), the strain energy U_1 in the first bar is

$$U_1 = 0.5 \, BdL \, \sigma^2/E \qquad (7.20)$$

From the first of equations (7.17), the strain energy U_2 in the second bar is

$$U_2 = M^2 L/2EI$$

But
$$M = \sigma z = \sigma Bd^2/6 \text{ and } I = Bd^3/12$$
$$\therefore \quad U_2 = 0.5\,BdL\,\sigma^2/3E \tag{7.21}$$

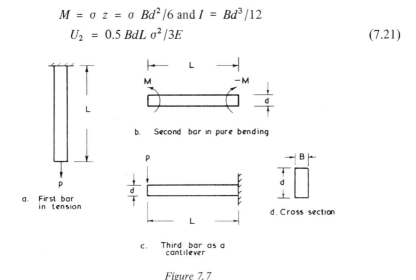

Figure 7.7

From the first of equations (7.19), the strain energy U_3 in the cantilever is

$$U_3 = \int_0^L M_x^2\,dx/(2EI)$$

The bending moment M_x at a distance x from the free end is px

$$\therefore \quad U_3 = \int_0^L p^2 x^2\,dx/(2EI) = p^2 L^3/6EI \tag{7.22}$$

The maximum bending stress at the top fibres of the beam at the fixed end should not exceed σ given by

$$pL = \sigma z = \sigma Bd^2/6$$

Substituting for pL in equation (7.22), we obtain

$$U_3 = 0.5BdL\,\sigma^2/9E \tag{7.23}$$

Thus
$$U_1 = 3U_2 = 9U_3$$

i.e. the strain energy can be best stored in the bar in direct tension and the least work is required to make the stress in the cantilever reach the limiting stress.

Example 2. In a pure bending experiment on a mild steel bar, it was found that at the limit of proportionality M was 1250 kN mm and θ was 0.09 rad. After this limit it was decided to approximate the $M - \theta$ relationship to a hyperbolic function of the form

$$M = 1670\,\theta/(0.003 + \theta) \tag{7.24}$$

which is shown graphically in *Figure 7.8*. Failure was noticed to take place at an ultimate bending moment of 1670 kN mm. The angle θ just before failure took

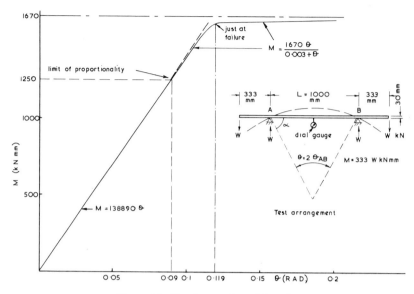

Figure 7.8. M-θ diagram for mild steel

place was noticed to be 0.119 rad. The bar was 1000 mm long with a square cross section 30 mm × 30 mm. Calculate:

(1) The extreme fibre stress at the limit of proportionality and the dial gauge reading of *Figure 7.8* at this limit.
(2) The total strain energy stored in the bar when it failed.
(3) The value of E.
(4) Make suggestions for altering the experimental arrangement shown in *Figure 7.8*.
(5) Plot a graph for the stiffness of the bar against M. What conclusions can be made from this graph?

Note: The values given in this example and equation (7.24) agree very well with the real values of a medium strength mild steel bar. Equation (7.24) is generally in the form $M = b\theta/(a + \theta)$, where b is the ultimate bending moment and a is obtained from the values of M and θ at the limit of proportionality.

Answer:
(1) From $M/I = \sigma/y$, the value of σ at the limit of proportionality is

$$\sigma = \frac{My}{I} = \frac{1250 \times 15 \times 12}{30 \times 27000} = 0.277 \text{ kN/mm}^2 (= 18 \text{ tons/in}^2)$$

The gauge reading δ is

$$\delta = R - R\cos(0.5\theta) = R[1 - \cos(0.5\theta)]$$

Now $R = L/\theta$ and with $\theta = 0.09$ rad

$$R = 1000/0.09$$

∴ $$\delta = 1000/0.09 \,[1 - \cos(180 \times 0.045/\pi)]$$

i.e. $$\delta = 11.11 \text{ mm}.$$

(2) The total strain energy U is, using equations (7.14) and (7.24)

$$U = \frac{1250 \times 0.09}{2} + \int_{0.09}^{0.119} \frac{1670\,\theta}{0.003 + \theta}\,d\theta$$

$$U = 56.25 + 47.13 = 103.38 \text{ kN mm}$$

(3) From $\theta = ML/EI$, we obtain

$$E = ML/I\theta$$

$$\therefore \quad E = \frac{1250 \times 1000 \times 12}{30 \times 27000 \times 0.09} = 206 \text{ kN/mm}^2 \ (29.88 \times 10^6 \text{ lb/in}^2)$$

(4) Nearer failure the weights W may fall off the beam. The arrangement shown in *Figure 7.9* does not suffer from this weakness.

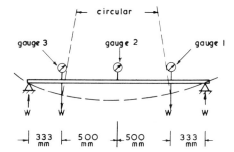

Figure 7.9. *Improved test arrangement*

(5) From *Figure 7.8*

$$\alpha + \theta_{AB} = 90°$$

and

$$\alpha + \theta/2 = 90$$

$$\therefore \quad \theta = 2\theta_{AB}$$

The slope deflection equation for the bar is

$$M_{AB} = M = (4EI/L)\theta_{AB} - (2EI/L)\theta_{AB}$$

$$M = (2EI/L)\theta_{AB}$$

The stiffness $K = M/\theta_{AB} = 2EI/L$

and

$$K = M/0.5\theta = 2M/\theta$$

Up to the limit of proportionality

$$K = 1250/0.045 = 27777.8 \text{ kN mm.}$$

and the flexibility $F = 1/27\,777.8$ (kN mm)$^{-1}$ = 3.6×10^{-5} (kN mm)$^{-1}$

After the limit of proportionality the $M - \theta$ relationship is nonlinear and the stiffness is given by $2dM/d\theta$. From equation (7.24)

$$M = 1670\,\theta/(0.003 + \theta)$$

$$\therefore \quad dM/d\theta = (1670 \times 0.003)/(0.003 + \theta)^2$$

and

$$K = 2\,dM/d\theta = 10.02/(0.003 + \theta)^2$$

when $\theta = 0.1$, $K = 944.48$ when $\theta = 0.119$, $K = 673.2$ and when $M = 1670$, $\theta = \infty$, thus $2\,dM/d\theta = K = 0$

The graph of the stiffness K against the applied moment M is shown in *Figure 7.10*. Up to the limit of proportionality the material is linear elastic and its stiffness is constant at a high value of 27 777.8 kN mm per radian of the angular rotation θ_{AB}. After this limit K decreases very rapidly until at failure the bar's stiffness is depleted to zero. Of course at the linear elastic range the flexibility of the bar is very low then increases rapidly. At failure the flexibility of the bar reaches infinity. The bar is said to be infinitely flexible at this stage. Indeed zero stiffness (infinite flexibility) is, as was also stated in Chapter 2, the only criterion for failure.

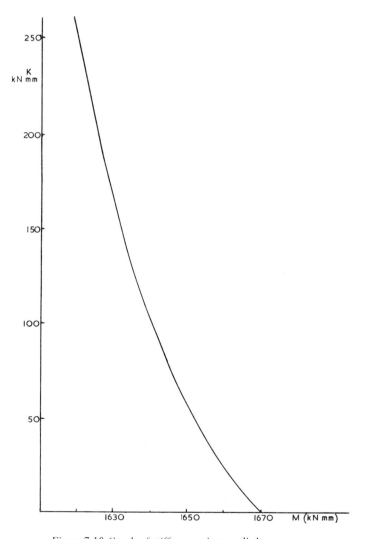

Figure 7.10. Graph of stiffness against applied moment

Flexural Properties of Structures

Example 3. A simply supported beam is 4 m long with $E = 200$ kN/mm^2 and $I = 10^6$ mm^4. Calculate the deflection of the beam when:

(a) A weight of 12 kN is placed gently on its midspan.
(b) When the same weight is suddenly applied to the beam at its midspan.
(c) When the weight is dropped from the height of 2 m on the midspan of the beam.

Answer:

(a) From equation (5.25) the central deflection of the beam is

$$v = WL^3/(48EI) \tag{7.25}$$

$$v = 12 \times 64 \times 10^9/(48 \times 200 \times 10^6) = 80 \text{ mm.}$$

The negative sign is dropped for simplicity. To answer parts (b) and (c) it is more convenient to derive a general expression for case (c) and then use it for case (b) when the height of the falling weight from the beam is zero.

Let W be the weight, y its height from the beam and v its total deflection. When the weight falls through a total distance $(y + v)$ the work it does is $W(y + v)$. When the beam is in its extreme deflected position, the weight exerts a force of W' downwards on the beam and the strain energy in the beam is $W'v/2$. Since the work done by the weight is stored as strain energy in the beam, it follows that

$$W(y + v) = W'v/2 \tag{7.26a}$$

Hence

$$W' = 2W(y + v)/v \tag{7.26b}$$

To calculate the deflection produced by a load W' acting at the midspan of the beam, we use equation (7.25) and find that the magnitude of this deflection is

$$v = W'L^3/48EI$$

and it is a downward movement.

Using equation (7.26) for W'

$$v = \frac{WL^3}{48EI} \times \frac{2(y + v)}{v}$$

Now $WL^3/48EI$ is the magnitude of the deflection v_g produced by the weight W when acting gradually and it is also a downward movement, then

$$v = 2v_g(y + v)/v$$

which gives

$$v = v_g + \sqrt{(v_g^2 + 2v_g y)} \tag{7.27}$$

and when $y = 0$, for case (b)

$$v = v_g + \sqrt{(v_g^2)} = 2v_g \tag{7.28}$$

Thus for case (b), the value of $v = 2 \times 80 = 160$ mm down. Finally for case (c) with $y = 2000$ mm

$$v = 80 + \sqrt{(6400 + 2 \times 80 \times 2000)}$$
$$v = 80 + \sqrt{(326\,400)}$$
$$v = 80 + 571.3 = 651.3$$

Thus the value of $v = -651.3$ mm, i.e. downwards. This deflection is so great that it is likely that the falling weight causes plasticity or even failure of the beam.

It is noticed in equation (7.26) that the work done by the falling weight W is $Wy + Wv$. This is because W is constant throughout. On the other hand the work done by the contact force W' is $W'v/2$. This is because when the weight touches the beam its contact force is zero but as it pushes the midspan point of the beam down it begins to exert a force on it which increases up to W' when the beam is at its extreme position. The total work done by W' is therefore $0.5\, W'v$.

Example 4. Calculate the vertical deflection at the free end A of the circular arc shown in *Figure 4.6*.

Answer: If we assume that the diameter of the cross section is small compared to the radius of the arc, it is possible to neglect any deflection due to shear. The deflection at A is then due to the bending and torsion of the member. An expression similar to equation (7.19) can be derived for the strain energy U_t due to torsion. 1. This is of the form

$$U_t = \int_0^L T^2\, dx/(2GJ)$$

where T is the torque at any point on a beam and GJ is known as the torsional rigidity of the section. The total strain energy due to bending and torsion of a beam is thus given by

$$U = \int_0^L M^2\, dx/(2EI) + \int_0^L T^2\, dx/(2GJ)$$

Referring to *Figure 4.6*, an element of length $Rd\theta$ at E is subject to a bending moment $WR \sin \theta$ and a torque $WR(1 - \cos \theta)$ as given by equations (4.14) and (4.15). The total strain energy in the circular member is therefore given by

$$U = \int_0^{\pi/2} W^2 R^2 \sin^2 \theta \times Rd\theta/(2EI)$$

$$+ \int_0^{\pi/2} W^2 R^2 (1 - \cos \theta)^2 \times Rd\theta/(2GJ)$$

Thus

$$U = \frac{\pi W^2 R^3}{8EI} + \frac{(3\pi - 8)W^2 R^3}{8GJ}$$

The vertical deflection under the load at A is v and the work done by the load is $Wv/2$. Equating this work to the total strain energy

$$\frac{\pi W^2 R^3}{8EI} + \frac{(3\pi - 8)W^2 R^3}{8GJ} = \frac{Wv}{2}$$

$$\therefore \quad v = \frac{\pi WR^3}{4EI} + \frac{(3\pi - 8)WR^3}{4GJ}$$

7.8. Strain energy in pure shear

When lateral forces are applied to structural members both shearing forces and bending moments are developed in the members. The strain energy stored in a member due to pure bending was presented in the last two sections. To evaluate

Flexural Properties of Structures

the strain energy due to the shearing forces, consider a portion of a beam subject to a pure shearing force F as shown in *Figure 7.11*. The vertical shearing force acting on the free surface abcd causes this surface to move upwards by an amount v

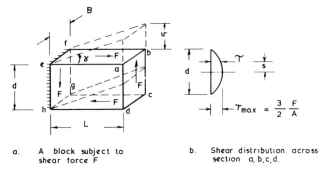

a. A block subject to shear force F

b. Shear distribution across section a, b, c, d.

Figure 7.11

relative to its fixed surface efgh. Work is therefore done by this force which is stored in the beam as strain energy. The relation between F and v can be represented by a graph similar to that shown for pure bending (in *Figure 7.6*). By the same reasoning used in section 7.6, the increment in strain energy dU_F is given by

$$dU_F = F dv \tag{7.29}$$

and if F increases from zero to F_1 while v increases from zero to v_1, the total strain energy will be

$$U_F = \int_0^{v_1} F dv \tag{7.30}$$

In the case of a linear elastic material, the $F - v$ diagram is a straight line and the total strain energy is

$$U_F = F_1 v_1 / 2 \tag{7.31}$$

In *Figure 7.11*, it is noticed that the shear strain γ is given by

$$\gamma = v/L \tag{7.32}$$

and if A is the area of the surface abcd, the average shear stress τ on this surface is

$$\tau = F/A \tag{7.33}$$

and

$$G = \tau/\gamma \tag{7.34}$$

where G is the shear modulus. Thus from equations (7.32), (7.33) and (7.34)

$$v = (L/GA)F \tag{7.35}$$

Substituting for v from equation (7.35) into equation (7.31), when $v = v_1$ and $F = F_1$, we obtain

$$U_F = F_1^2 L / 2GA \tag{7.36}$$

Similarly the strain energy in shear can also be expressed as

$$U_F = AGv_1^2 / 2L \tag{7.37}$$

In terms of the average shearing stress τ, U_F becomes

$$U_F = 0.5\tau^2 \, AL/G \tag{7.38}$$

and in terms of strain γ

$$U_F = G\gamma^2 \, AL/2 \tag{7.39}$$

7.9. The effect of shear stress distribution

In the last section it was assumed that the shearing stress on the section can be averaged to $\tau = F/A$ as given by equation (7.33). In reality the distribution of the shearing stress across the section is not uniform but parabolic as shown in *Figure 7.11b*. For a rectangular section the maximum stress τ_{max} at the neutral axis is $1.5F/A$ and varies according to the function

$$\tau = (F/2I)(0.25d^2 - s^2) \tag{7.40}$$

where I is the second moment of area of the section, d is its depth and s is the distance from the neutral axis. The derivation of this expression can be found in any book on strength of materials.

For a steel section with $E/G = 2.5$, the exact expression for the vertical deflection v becomes

$$v = (FL^3/3EI) \times (6/8) \times (d^2/L^2) \tag{7.41}$$

as compared to equation (7.35) which gives

$$v = (FL^3/3EI) \times (5/8) \times (d^2/L^2)$$

Equation (7.38), for the strain energy changes to

$$U = \int_{-d/2}^{d/2} (F^2 L/8GI^2)[(d^2/4) - s^2]^2 \, B ds \tag{7.42}$$

Because of this parabolic shear stress distribution, plane sections such as abcd do not remain plane after bending but warp.

If the maximum shear stress $\tau_{max} = 1.5F/A$ is assumed to be acting on the whole section, the vertical deflection due to shear becomes

$$v = (FL^3/3EI) \times (7.5/8) \times (d^2/L^2)$$

Of course the correct value of v is given by equation (7.41).

7.10. Example

A rectangular cantilever is 10 m long and carries a point load of $W = 2$kN at its free end. The value of $E = 200$ kN/mm^2 and G is 80 kN/mm^2. If the width of the cross section is 100 mm and its depth is 400 mm calculate the strain energy due to shear and compare it with that due to bending. What is the total deflection of the free end?

Answer: From equation (7.42)

$$U_F = \int_{-d/2}^{d/2} (W^2 LB/8GI^2)[d^2/4) - s^2]^2 ds = 3W^2 L/5GBd$$

$$U_F = \frac{3 \times 4 \times 10^4}{5 \times 80 \times 4 \times 10^4} = 3/400 \text{ kN mm}$$

From equation (7.22), the strain energy in bending is
$$U_B = 4 \times 10^{12} \times 12/(6 \times 200 \times 4^3 \times 10^8) = 2500/400 \text{ kN mm}.$$
Thus U_B is $833\frac{1}{3} U_F$

The deflection of the free end due to shear is from equation (7.41)
$$v_F = (-WL^3/3EI) \times (6d^2/8L^2)$$
From equation (5.17a), the deflection of the free end due to bending is
$$v_B = -WL^3/(3EI)$$
The total deflection is therefore
$$v = v_F + v_B = (-WL^3/3EI)[1 + (6d^2/8L^2)]$$
$$\therefore \quad v = -100[1 + (12/10\,000)]/16 = -6.258 \text{ mm}.$$

It is evident from this example that the deflection of the beam due to shear is very small compared to that due to bending. However the deflection due to shear is proportional to d^2/L^2 and for deep beams such as shear walls with $d \triangleq L$ it could become significant.

7.11. The work equation

Consider a beam, a bent bar or a rigidly jointed structure as shown in *Figure 7.12*. Each one has a total of M joints and N members. A typical joint is j and a typical member is i. Each joint such as j is subject to external horizontal and vertical loads H_j and V_j and an external moment M_j. Due to the external forces acting at all the joints, let the displacement of a typical joint be x_j, y_j and θ_j. The external forces cause the development of member forces and member deformations. Let the forces in a typical member i be F_i, M_{1i} and M_{2i} and the corresponding deformations

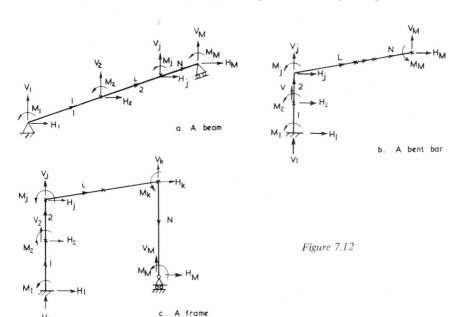

Figure 7.12

of the member be v_i, θ_{1i} and θ_{2i}. Suffixes 1 and 2 refer to the first and second ends of the members while the suffix i refers to the member number.

By the principle of conservation of energy, the total work done by the external forces and moments is equal to the total work done by the member forces. Thus

$$(H_1 x_1 + V_1 y_1 + M_1 \theta_1 + \ldots H_j x_j + V_j y_j + M_j \theta_j + \ldots$$
$$+ H_M x_M + V_M y_M + M_M \theta_M) = (F_1 v_1 + M_{11} \theta_{11} + M_{21} \theta_{21} + \ldots$$
$$+ F_i v_i + M_{1i} \theta_{1i} + M_{2i} \theta_{2i} + \ldots + F_N v_N +$$
$$+ M_{1N} \theta_{1N} + M_{2N} \theta_{2N}) \tag{7.43a}$$

i.e.

$$\sum_{j=1}^{M} H_j x_j + V_j y_j + M_j \theta_j = \sum_{i=1}^{N} F_i v_i + M_{1i} \theta_{1i} + M_{2i} \theta_{2i} \tag{7.43b}$$

This equation can be written in matrix form as

$$[H_1 V_1 M_1 \ldots H_j V_j M_j \ldots H_M V_M M_M] \{x_1 y_1 \theta_1 \ldots x_j y_j \theta_j \ldots x_M y_M \theta_M\}$$
$$= [F_1 M_{11} M_{21} \ldots F_i M_{1i} M_{2i} + \ldots F_N M_{1N} M_{2N}] \times \{v_1 \theta_{11} \theta_{21}$$
$$\ldots v_i \theta_{1i} \theta_{2i} \ldots v_N \theta_{1N} \theta_{2N}\} \tag{7.44a}$$

or simply

$$\mathbf{W}^T \mathbf{X} = \mathbf{P}^T \boldsymbol{\delta} \tag{7.44b}$$

where the row vector $\mathbf{W}^T$ is the transpose of the external load vector $\{\mathbf{W}\}$ and $\mathbf{P}^T$ is the transpose of the column vector of member forces $\mathbf{P}$. The column vector $\mathbf{X}$ consists of the joint deflections and rotations while the column vector $\boldsymbol{\delta}$ consists of the member deformations.

The work equation (7.44) is often used in calculating the deflection V at a point in a structure in a specified direction. To do this a unit load is applied to the structure in the direction of V. The work done by this load is thus $1 \times V$. The member forces caused by the applied unit load are $\mathbf{P}_u^T$ and equation (7.44b) becomes

$$1 \times V = \mathbf{P}_u^T \boldsymbol{\delta} \tag{7.44c}$$

At a given point, a member of a plane frame may be subject to an axial force p, a shear force F and a bending moment M. The resulting member deformations consist of an elongation $\delta = p\,ds/EA$, a rotation $\theta = M\,ds/EI$ and a shear deformation $\psi = fF\,ds/GA$. (Here f is a factor which depends on the distribution of the shear stresses over the cross section. For a rectangular section for instance this factor is equal to 1.2.) In these expressions ds is the length of an element of the member.

The unit load, on the other hand, produces at the same point in the member an axial force p_u, a shear force F_u and a bending moment M_u. When equation (7.44c) is applied to every point in the frame, it takes the form

$$V = \Sigma \int \frac{p_u p}{EA} ds + \Sigma \int \frac{M_u M}{EI} ds + \Sigma \int \frac{F_u F_f}{GA} ds \tag{7.44d}$$

where the integration is over the length of each member and the summation covers all the members.

Equation (7.44b) is valid even if the displacements are 'virtual', i.e. imaginary. It is then known as the 'virtual work equation' and is used to define the state of equilibrium in a structure. The virtual work equation was applied in sections 6.3 and 6.5 and will be used in section 7.12 and throughout Chapter 8.

7.12. Examples

Example 1. Calculate the central deflection of a simply supported beam loaded (a) centrally with W and (b) uniformly with w/unit length.

Answer: (a) the beam is shown in *Figure 5.4a* in section 5.6. At a distance x from A, the bending moment due to the central load W is $M = -Wx/2$. Applying a positive (upward) unit load at midspan, the bending moment due to this load at distance x from A is $0.5x$. Considering the deflection of the beam due to the sole effect of bending, equation (7.44d) reduces to

$$V = \int_0^L M_u \, M dx/EI = 2 \int_0^{L/2} M_u M dx/EI$$

$$\therefore \quad V = (2/EI) \int_0^{L/2} 0.5x \times -0.5 Wx dx$$

$$= (-0.5 W/EI) \int_0^{L/2} x^2 dx = -WL^3/(48EI)$$

(b) This beam is shown in *Figure 5.7* in section 5.10. At D, distance x from A, the bending moment due to the actual U.D. load is $0.5w(x^2 - Lx)$. The bending moment at D due to a unit upward positive force is $0.5x$.

Equation (7.44d) gives

$$V = 2 \int_0^{L/2} M_u M dx/EI = 2 \int_0^{L/2} 0.5x \times 0.5w(x^2 - Lx) dx/EI$$

$$= (0.5w/EI) \int_0^{L/2} (x^3 - Lx^2) dx = -5wL^4/(384EI)$$

Example 2. With $V_B = M_B = 0$, calculate the rotation and the deflections of end B in the semicircular beam shown in *Figure 4.4*, section 4.4.

Answer: The arc is now under a single horizontal force H_B. The radius R is assumed to be large compared to the cross sectional dimensions. Therefore, deflections other than those due to bending can be neglected.

Measuring ϕ anticlockwise from the horizontal, the bending moment at a point E (*Figure 4.4c*) due to the horizontal force H_B is $M = H_B R \sin \phi$. The length of an element of the arc is $R d\phi$. To calculate the rotation θ_B at B a unit positive moment $M_B = 1$ is applied there and in equation (7.44d) V is replaced by θ_B. The bending moment at E due to the unit moment at B is $M_u = 1$, thus

$$\theta_B = 2 \int_0^{\pi/2} H_B R \sin \phi \times 1 \times R d\phi/EI$$

$$= (2H_B R^2/EI) \int_0^{\pi/2} \sin \phi d\phi = 2H_B R^2/EI$$

which is positive, i.e. anticlockwise.

To calculate the horizontal deflection h_B at B a unit positive horizontal force is applied there. This causes a bending moment $M_u = 1 \times R \sin \phi$ at E and equation (7.44d) now gives

$$h_B = 2 \int_0^{\pi/2} H_B R \sin \phi \times R \sin \phi \times R d\phi / EI$$

$$= (2H_B R^3 / EI) \int_0^{\pi/2} \sin^2 \phi \, d\phi = 0.5\pi H_B R^3 / EI$$

which is to the right.

Finally to calculate the vertical deflection V_B at B, a unit upward force is applied there. This causes a bending moment $M_u = R(1 - \cos \phi)$ at E.

Equation (7.44d) now becomes

$$V_B = 2 \int_0^{\pi/2} H_B R \sin \phi \times R(1 - \cos \phi) \times R d\phi / EI$$

$$= (2H_B R^3 / EI) \int_0^{\pi/2} (\sin \phi - \sin \phi \cos \phi) d\phi$$

$$= (2H_B R^3 / EI) \times (0.5) = H_B R^3 / EI \text{ up.}$$

Example 3. Calculate the bending moment M_A at support A for the propped cantilever shown in *Figure 7.13*. The beam is subject to a vertical load W at D.

Answer: The equations of static equilibrium of forces are not by themselves sufficient to calculate the reactions R_A, R_B and M_{AD}. This beam is therefore hyperstatic. The slope deflection equations (5.14) or (7.3) are equally valid for hyperstatic structures as they are for isostatic ones. These equations are now applied to the parts AD and DB of the propped cantilever. It is noticed that $\theta_{AD} = 0$, because end A is fixed and because end B is resting on a simple support, $M_{BD} = 0$. Denoting $k = EI/(0.5L) = 2EI/L$, the slope deflection equations become

$$M_{AD} = 2k\theta_D - 12kv/L \quad (7.45)$$

$$M_{DA} = 4k\theta_D - 12kv/L \quad (7.46)$$

$$M_{BD} = 4k\theta_B + 2k\theta_D + 12kv/L = 0 \quad (7.47)$$

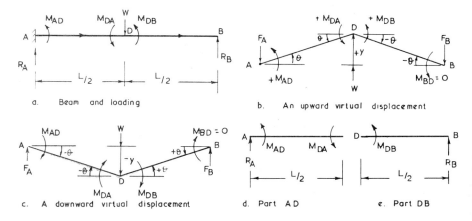

Figure 7.13

Hence
$$2k\theta_B = -k\theta_D - 6kv/L$$
and therefore
$$M_{DB} = 3k\theta_D + 6kv/L \tag{7.48}$$

To solve for the unknowns θ_D and v, two further equations are needed. One is obtained by considering the equilibrium of joint D which gives $M_{DA} + M_{DB} = 0$, and using equations (7.46) and (7.48), we obtain
$$k\theta_D = 6kv/7L \tag{7.49}$$

Another equation, often known as the sway equation is obtained by equating the work done by the external load W to the work done by the internal moments. This is now derived by three slightly different methods.

Method 1: Suppose that AD and DA are kept rigid while an upward virtual displacement y is given to joint D as shown in *Figure 7.13b*. If y is small AD rotates through an anticlockwise angle $\theta = y/0.5L$. On the other hand DB rotates through a clockwise angle $-\theta = -y/0.5L$. Such a displacement can only be achieved by an upward force W as shown in the figure. The work done by this force is Wy, that done by the shear force F_A and F_B is zero because they do not move and that done by M_{BD} is also zero because $M_{BD} = 0$.

The work done by M_{AD} is $M_{AD}\theta = M_{AD}y/0.5L$. that done by M_{DA} is $M_{DA}\theta = M_{DA}y/0.5L$ and that done by M_{DB} is $M_{DB} \times -\theta = -M_{DB}y/0.5L$. It is noticed that in this system the external load and the bending moments are all assumed to be positive. The virtual work equation (7.43) or (7.44) reduces to
$$Wy = (M_{AD}y + M_{DA}y - M_{DB}y)/0.5L$$

$$0.5WL = M_{AD} + M_{DA} - M_{DB} \tag{7.50a}$$

Method 2: Under similar conditions as for method 1, a downward virtual displacement $-y$, by a downward force $-W$, causes the member rotations shown in *Figure 7.13c*. It is noticed that because W has reversed its direction, the member forces and bending moments have also reversed their direction. The virtual work equation now becomes
$$-W \times -y = (-M_{AD} \times -\theta) + (-M_{DA} \times -\theta) + (-M_{DB} \times +\theta)$$
with $\theta = y/0.5L$, equation (7.50) is obtained.

This indicates that the direction of the virtual displacement is in fact immaterial. Furthermore, since the virtual displacement y appears on both sides of the virtual work equation, it is always cancelled out, thus even its magnitude is immaterial.

Method 3: Consider now the equilibrium of parts AD and DB with all the moments assumed positive. For AD, taking moments about D (*Figure 7.13d*)
$$-0.5R_AL + M_{AD} + M_{DA} = 0$$
$$\therefore \quad 0.5R_AL = M_{AD} + M_{DA} \tag{7.51}$$

For DB, taking moments about D (*Figure 7.13e*)
$$0.5R_BL + M_{DB} = 0$$
$$\therefore \quad 0.5R_BL = -M_{DB} \tag{7.52}$$

Adding equation (7.51) to (7.52) and realising that $R_A + R_B = W$, we obtain
$$0.5WL = M_{AD} + M_{DA} - M_{DB} \tag{7.50b}$$
which is exactly the same as equation (7.50a) obtained earlier.

In method 3, the sway equation (7.50) was obtained by considering the equilibrium of the parts of the beam while the first two methods used the concept of work. This indicates, once again, that the virtual work equation is merely an expression of the state of equilibrium. In other words, the virtual work equation is a method for deriving the equilibrium equation for a structure and vice versa.

Substituting from equations (7.45), (7.46), (7.48), and (7.49), equation (7.50) gives

$$kv = -7WL^2/384$$

i.e.

$$v = -7WL^3/(768\ EI) \qquad (7.53)$$

Thus using equations (7.49) and (7.53), equation (7.45) gives

$$M_{AD} = 3WL/16 \qquad (7.54)$$

This example also demonstrates the use of the slope deflection equations to analyse hyperstatic problems. In these, it is necessary to derive as many joint equilibrium equations as there are unknown joint rotations θ. It is also necessary to derive as many sway equations as there are unknown joint deflections v. Further details on this are given in Chapter 12.

7.13. Plasticity

So far attention has been focused on the elastic analysis of structures. The stress in the material of the structure was assumed to remain elastic, below the limit of proportionality. If the stress in a member is increased slightly above this limit, plasticity begins to develop and the strains increase dramatically as shown by the portion ab in *Figure 7.14*. For a typical annealed mild steel the ratio of the strain ϵ_s at b and the yield strain ϵ_y is of the order of 10 to 20. After point b, it is said that the material is strain hardening. Plasticity is in fact the state of yielding of the material.

Consider a cross-section of a structural member which is subject to a gradually increasing moment. While the section is elastic the stress distribution across the section is linear as shown in *Figure 7.15a*. At a certain value of the bending moment the extreme outer fibres begin to yield and plasticity is initiated in these fibres. This takes place at a stress σ_y as shown in *Figure 7.15b*.

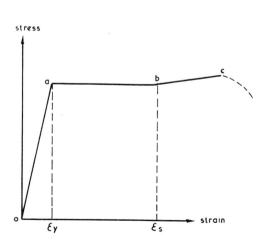

Figure 7.14

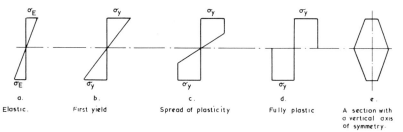

Figure 7.15

Increasing the moment further causes the spread of plasticity to the inner fibres and the stress distribution across the section takes the form shown in *Figure 7.15c*. At some stage the bending moment reaches a certain value M_p, known as the fully plastic moment, and the entire section becomes plastic. The diagram for the stress distribution takes the form shown in *Figure 7.15d*. From then onwards, the section becomes unable to sustain an increasing moment. At this constant moment and fibre stress σ_y, large strains develop across the section and large deformations of the structural member takes place. It is said that the section has become fully plastic and has developed a plastic hinge.

7.14. Evaluation of the fully plastic moment

For a symmetrical section the plastic moment of resistance which is also equal to M_p, can be calculated from a knowledge of the dimensions of the section and the value of the yield stress. In the case of the section shown in *Figure 7.15e*, the stresses above and below the horizontal axis of symmetry add up to a pair of forces each being equal to $A\sigma_y/2$. Here A is the area of the cross section. If d is the depth of this section, the moment $0.5A\sigma_y d/2$ of these equal and opposite forces balance the fully plastic moment acting on the section, thus

$$M_p = A\sigma_y d/4 \tag{7.55}$$

For instance, for a rectangular section of width B and depth d, *Figure 7.16*, the resultant force acting on each half is $\sigma_y Bd/2$. These forces act at the centroid of each portion and thus are $d/2$ apart. Hence for this section

$$M_p = 0.5\sigma_y dBd/2 = Bd^2 \sigma_y/4 \tag{7.56}$$

Unlike an elastic section, where the centroidal axis is the neutral axis, a plastic section has its equal area axis as the neutral axis. This is because the force acting on an element dA of an area is $\sigma_y dA$ and the moments of these forces about the neutral axis are zero, thus:

$$\int \sigma_y dA = \sigma_y \int dA = 0$$
$$\therefore \int dA = 0 \tag{7.57}$$

This means that the area above the neutral axis is equal to that below it.

7.15. Example

Calculate the value of M_p for the I section shown in *Figure 7.16*.

Answer: Resultant of forces acting on each flange = $Bt_2 \sigma_y$
where Bt_2 is the area of the flange.

Distance between the forces acting on the flanges = $(d - t_2)$.

Force acting on each half of the web = $0.5(d - 2t_2) t_1 \sigma_y$.
Distance between forces acting on the web = $(d - 2t_2)/2$
Thus for the I section

$$M_p = Bt_2\sigma_y(d - t_2) + 0.5(d - 2t_2) t_1 \sigma_y \times (d - 2t_2)/2$$
$$M_p = [Bt_2(d - t_2) + 0.25t_1(d - 2t_2)^2] \sigma_y \qquad (7.58)$$

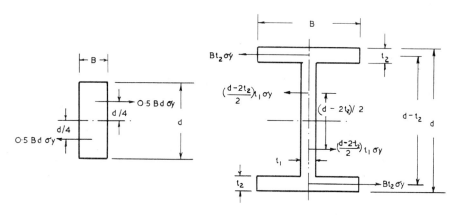

a. Forces on a fully plastic rectangular section

b. Forces on a fully plastic I section

Figure 7.16

When $t_1 = 0$, $M_p = Bt_2(d - t_2) \sigma_y$ which is the fully plastic moment of two parallel plates of width B, thickness t_2 with the distance between the outer fibres being d.

7.16. The shape factor

The fully plastic moment of a section is the product of the yield stress and a quantity dependent on the dimensions of the section, i.e.

$$M_p = z_p \sigma_y \qquad (7.59)$$

where z_p is known as the plastic section modulus of the section. For a rectangular section, for instance, equation (7.56) gives z_p as $BD^2/4$.

The plastic section modulus is analogous to the elastic section modulus z and the ratio $\alpha = z_p/z$ is known as the shape factor of the section. For a rectangular section

$$\alpha = z_p/z = (Bd^2/4)/(Bd^2/6) = 1.5$$

Similarly for an I section, the shape factor α is 1.15 and for a pair of thin parallel plates the shape factor is unity.

When the outer fibres of a section first yield, the elastic bending moment is M_y which is equal to $z\sigma_y$, thus

$$\alpha = z_p/z = \sigma_y z_p/\sigma_y z = M_p/M_y \qquad (7.60)$$

It follows that the shape factor is the ratio between the fully plastic moment of the section and the moment at first yield.

7.17. Examples

Example 1. The I section in *Figure 7.16* has $B = 100$ mm, $t_1 = 5$ mm, $t_2 = 10$ mm and $d = 100$ mm. Calculate the values of the plastic and elastic section moduli and the second moment of area. If the yield stress is $\sigma_y = 0.30$ kN/mm, calculate the fully plastic hinge moment.

Answer: Comparing equations (7.58) and (7.59) it is found that

$$z_p = Bt_2 (d - t_2) + 0.25 t_1 (d - 2t_2)^2$$

$$\therefore \quad z_p = 100 \times 10(100 - 10) + 0.25 \times 5(100 - 20)^2 = 98 \times 10^3 \text{ mm}^3$$

For this section $\alpha = 1.15 = z_p/z$, the elastic section modulus is thus given by

$$z = z_p/\alpha = 98\,000/1.15 = 85.217 \times 10^3 \text{ mm}^3$$

The second moment of area I of the section is $zd/2$

$$I = 100 \times 85.217 \times 10^3/2 = 4.261 \times 10^6 \text{ mm}^4$$

With $\sigma_y = 0.30$

$$M_p = \sigma_y z_p = 0.3 \times 98 \times 10^3 = 29.4 \text{ kN m.}$$

Example 2. It is intended to increase the fully plastic moment of the section in the last example by 33 kN m by welding two parallel plates to the top and bottom flanges. Calculate the thickness t_3 of these plates.

Answer: From equation (7.58) with $t_1 = 0$ the fully plastic hinge moment of two parallel plates is

$$M_p = Bt_2(d - t_2)\sigma_y$$

where d is the distance between the outer fibres and t_2 is the thickness of the plates. Here $d = 100 + 2t_3$ and the thickness of the plates is t_3 and not t_2

$$\therefore M_p = Bt_3(100 + 2t_3 - t_3)\sigma_y$$

Thus

$$33 \times 10^3 = 100\, t_3(100 + t_3) \times 0.3$$

which gives

$$t_3 = 10 \text{ mm.}$$

Exercises on Chapter 7

1. Construct the member stiffness matrix for the structure shown in *Figure 7.17*. $E_1 = E_2 = 200$ kN/mm^2, $E_3 = 100$ kN/mm^2. $I_1 = 10^6$ mm^4, $I_2 = 10^5$ mm^4 and $I_3 = 10^4$ mm^4.

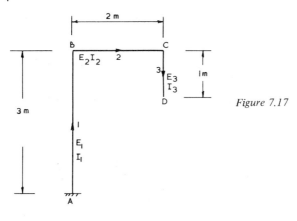

Figure 7.17

180 Flexural Properties of Structures

2. Construct the member flexibility matrix for the structure of the last example.

3. Calculate the total bending strain energy for the beam shown in *Figure 7.18*. Take $E = 200$ kN/mm² and $I = 10^9$ mm⁴.

Ans. 7.7 kN mm.

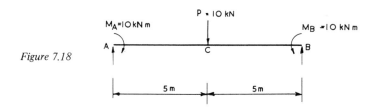

Figure 7.18

4. A reinforced concrete shear wall is 5 m deep, 0.5 m thick and 15 m high. Calculate the total deflection at its free end where it is subject to a horizontal load of 100 kN. Take $E = 15$ kN/mm² and $G = 6$ kN/mm².

Ans. 1.56 mm.

5. Calculate the vertical and the horizontal deflection of the free end of the curved member shown in *Figure 7.19*. Take $E = 200$ kN/mm² and $I = 10^6$ mm⁴.

Ans. 0.271 mm, 0.025 mm.

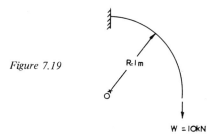

Figure 7.19

6. Calculate the plastic section modulus of a circular section of radius 10 mm.

Ans. 1333 mm³.

7. The cantilever AC is resting on the simply supported beam DE at its midpoint as shown in *Figure 7.20*. At B the cantilever carries a point load W. Calculate the force between the beam and the cantilever and the stiffness of the simply supported beam against vertical movement. Both beams have the same EI.

Ans. $R_C = 5W/17$, $k_C = R_C/v_C = 48EI/L^3$.

8. In *Figure 7.21* two cantilevers AB and CD are connected together by a spring of axial stiffness k. A load W is acting at B. If the deflection at D is half that at B, calculate the stiffness of the spring and the bending moment at A and C.

Ans. $k = 3EI/L^3$, $M_A = 2WL/3 = 2M_C$.

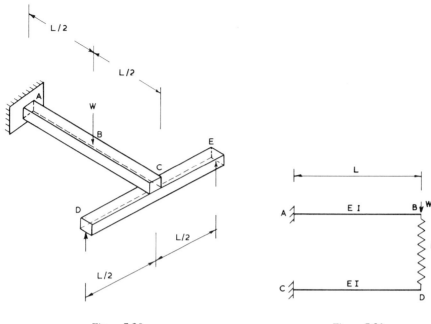

Figure 7.20

Figure 7.21

9. The beam ABC shown in *Figure 7.22* is fixed at C and simply supported at B. For AB, $EI/L = k_1$ while for BC it is k_2. Calculate the stiffnesses of BA, BC and joint B (a) when A is fixed as shown in *Figure 7.22a* and (b) when A is pinned as shown in *Figure 7.22b*.

Ans. (a) $k_{BA} = M_{BA}/\theta = 4k_1$, $k_{BC} = 4k_2$, $k_B = 4(k_1 + k_2)$;
(b) $k_{AB} = 3k_1$, $k_{BC} = 4k_2$, $k_B = 3k_1 + 4k_2$.

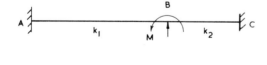

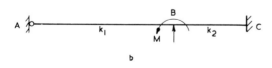

Figure 7.22

10. End B of the beam AB shown in *Figure 7.23* is forced to deflect relative to A by an amount v. Calculate the stiffness $k = M_{AB}/v$ (a) when end B is fixed and no rotation takes place at A or B (*Figure 7.23a*); (b) when end B is pinned and the rotation at A is zero (*Figure 7.23b*).

Ans. (a) $-6EI/L^2$; (b) $-3EI/L^2$.

182   *Flexural Properties of Structures*

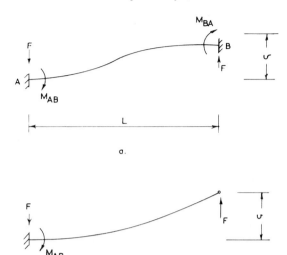

Figure 7.23

8

The plastic theory

8.1. Introduction

One assumption made in the analysis of structures by the plastic theory is that the strains and deformations in the structure at the limit of proportionality are small compared to those at the onset of strain hardening. The theory therefore neglects the elastic strains and deformations. It assumes that the structure remains undeformed until suddenly a number of fully plastic hinges develop at discrete sections in the structure which convert it into a collapsing mechanism. At incipient collapse, which is the instant when the structure is just converted into a mechanism with the formation of the last hinge, the structure is therefore isostatic. At this stage, the bending moments throughout the structure can be calculated by using the static equilibrium of forces alone. The plastic theory should therefore be included with topics on isostatic structures. This is not because of the simplicity of the theory but principally because it is merely dependent upon the state of equilibrium of forces.

8.2. The load factor

Consider a simply supported beam which is subject to a point load at its midspan as shown in *Figure 8.1a*. Let the initial value of this load be P_w. If this load is gradually increased then its value at any other stage can be expressed as λP_w, where λ is known as the load factor. For values of the load above the initial, λ is more

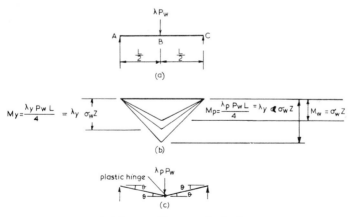

Figure 8.1. Beam subject to point load
(a) beam and loading
(b) bending moment diagram
(c) collapse mechanism

than one. As the load factor increases the maximum bending moment in the beam also increases. Eventually at some load factor λ_y, yielding will begin at the outside fibres of the beam. The corresponding maximum elastic bending moments at the initial load and at first yield are shown in *Figure 8.1b* and are given respectively as

$$M_w = P_w L/4 = \sigma_w Z \qquad (8.1)$$
$$M_y = \lambda_y P_w L/4 = \sigma_y Z = \lambda_y \sigma_w Z$$

Further increases of the load factor cause yield to spread sideways along the beam and inwards, towards the neutral axis until eventually at a load factor λ_p the whole section at midspan B yields and a plastic hinge is developed there. The beam is thus converted into a mechanism with three hinges, a plastic hinge at the midspan and two real hinges at the supports. The load factor λ_p cannot be increased any further and the beam collapses with the bending moment at B remaining constant and equal to M_p. This moment is given by

$$M_p = \lambda_p P_w L/4 = \sigma_y Z_p \qquad (8.2)$$

i.e.

$$\lambda_p P_w = 4M_p/L = 4\sigma_y Z_p/L$$

Furthermore, from equations (7.60) (8.1) and (8.2)

$$\sigma_y Z_p = \lambda_y \sigma_w Z_p = \lambda_y \alpha \sigma_w Z$$

It follows that

$$M_p = \lambda_y \alpha \sigma_w Z \qquad (8.3)$$

Thus

$$M_p/M_w = \lambda_p = \lambda_y \alpha \sigma_w Z/\sigma_w Z = \lambda_y \alpha \qquad (8.4)$$

This means that the load factor at collapse is given by the product of the load factor at first yield and the shape factor of the section.

8.3. Proportional loading

Consider the same simply supported beam but subject to two point loads with initial values P_w at midspan and $3P_w$ at a quarter span as shown in *Figure 8.2a*. When gradually increasing these loads, a proportional loading procedure of

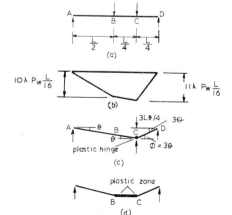

Figure 8.2. Beam subject to proportional loading
(a) beam and loading
(b) bending moment diagram
(c) collapse mechanism
(d) mechanism with constant loading along BC

the beam is achieved if both loads are increased by the same load factor λ and the ratio of the two loads remain constant. The bending moment diagram for this beam during the elastic range is shown in *Figure 8.2b* where it is apparent that yield first takes place at C under the heavier load. Eventually a plastic hinge develops at C and the beam collapses as a mechanism. This is shown in *Figure 8.2c*.

The load factor λ_p at collapse is once again obtained by equating the maximum moment in the beam to M_p, thus

$$M_p = 11\lambda_p P_w L/16$$

and

$$\lambda_p = 16M_p/11LP_w$$

The state of collapse with a plastic hinge at C was obtained by increasing the loads proportionally. If this procedure had not been adopted the collapse stage would have been obtained in a totally different manner. For instance if the loads at C and B are increased by applying load factors λ and λ' respectively, with $\lambda' = 1.5\lambda$ say, then the bending moments at B and C would both be equal to $3\lambda P_w L/4$. The bending moment over the entire portion BC would also be constant and equal to $3\lambda P_w L/4$. At collapse therefore, this entire zone yields and acts as a plastic hinge as shown in *Figure 8.2d*. For values of λ' higher than 1.5λ, the bending moment at B becomes greater than that at C and at collapse the plastic hinge develops at B instead of C.

This exposé shows clearly that the state of collapse and the position of the plastic hinge is defined entirely by the manner of loading. The plastic theory assumes that the loads acting on a structure are increased proportionally up to collapse which takes place with the formation of a unique mechanism. In reality structures are not subject to proportional loading. For instance a bridge designed to carry a moving train in gale winds may carry the train with no wind or a breeze. In the case of building frames it is more realistic to expect that some of the vertical live loads are already acting, before the frame is subjected to the wind loads.

8.4. The use of the virtual work equation

It has been stated earlier that when the collapse mechanism develops, a structure is statically determinate and the equations of static equilibrium can be utilised to calculate the collapse load factor. It is convenient to derive the equilibrium equation corresponding to a mechanism by the virtual work method.

For instance the mechanism corresponding to the collapse of a simply supported beam loaded at midspan is shown in *Figure 8.1c*. The virtual work equation for this mechanism can be derived by allowing an incremental rotation θ to the collapsing beam. The point of application of the load will move down by an amount $0.5L\theta$ and the work done by this load is $0.5L\theta\, \lambda_p P_w$. This work is absorbed by the plastic hinge during the rotation of each half of the beam by θ. Thus the work absorbed by the hinge is $2\theta M_p$. Equating the work done by the applied load to that absorbed by the hinge rotation, we obtain

$$0.5\lambda_p P_w L\theta = 2M_p\theta$$

Hence

$$\lambda_p P_w = 4M_p/L$$

or

$$M_p = \lambda_p P_w L/4$$

It is seen that the quantity θ appears on both sides of the virtual work equation and thus it does not appear either in the expression for the collapse load or for M_p. This is why the value of the incremental rotation is immaterial.

The virtual work equation can be used to derive the load factor that gives rise to any mechanism. In the case of a fixed-ended beam, *Figure 8.3*, three plastic hinges are required to develop a mechanism. These hinges form at ends A and C of the beam and under the applied load at B. These positions can be easily found from the elastic bending moment diagram, which shows that the bending moments at these points are higher than those anywhere else along the beam.

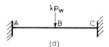

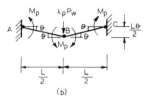

Figure 8.3. A fixed ended beam
(a) beam and loading
(b) collapse mechanism

Once again an incremental rotation θ given to the system will make the applied load do work $\lambda_p P_w L\theta/2$. From *Figure 8.3b*, it is noticed that the work absorbed by each support hinge is $M_p\theta$, while that absorbed by the midspan hinge is $2M_p\theta$. Hence the virtual work equation is

$$M_p\theta + M_p\theta + 2M_p\theta = \lambda_p P_w L\theta/2$$

i.e.

or
$$\left.\begin{array}{l} \lambda_p P_w = 8M_p/L \\ \\ M_p = \lambda_p P_w L/8 \end{array}\right\} \quad (8.5)$$

Similarly for a propped cantilever supporting a load λP_w at its midspan, the reader can verify that $\lambda_p P_w = 6M_p/L$ and $M_p = L\lambda_p P_w/6$. Collapse takes place with two hinges one under the load and the other at the fixed support.

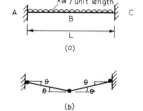

Figure 8.4. Fixed ended beam carrying
uniformly distributed load
(a) beam and loading
(b) collapse mechanism

In the case of a fixed ended beam carrying a uniformly distributed load of λw per unit length, w being the initial value of the load, collapse also takes place with three plastic hinges A, B and C shown in *Figure 8.4*. For an incremental rotation θ, the total work absorbed by these hinges is also $4M_p\theta$. The midspan point B moves down by $L\theta/2$ and thus the average movement of the total applied load downwards is $L\theta/4$. The work done by the load is thus $\lambda_p wL \times L\theta/4$ and the virtual work equation becomes

$$4M_p\theta = \lambda_p wL^2\theta/4$$

giving

$$\lambda_p w = 16M_p/L^2$$

or

$$M_p = \lambda_p wL^2/16$$

The collapse mechanism for the simply supported beam in *Figure 8.2* is shown in *Figure 8.2c*. An incremental rotation of portion AC by θ will make point B move down by $L\theta/2$ and point C by $3L\theta/4$. Thus the rotation ϕ of portion CD is $(3L\theta/4)/(L/4)$; i.e. $\phi = 3\theta$. The work done by the point load at B is thus $\lambda_p P_w L\theta/2$ and that by the load at C is $3\lambda_p P_w \times 3L\theta/4$. The work absorbed by the hinge at C is $M_p(\theta + \phi)$ and the virtual work equation is

$$4M_p\theta = \lambda_p P_w L\theta/2 + 3\lambda_p P_w \times 3L\theta/4$$

∴

$$\lambda_p P_w = 16M_p/11L$$

or

$$M_p = 11\lambda_p P_w L/16$$

In the above instance it was easy to find the position of the plastic hinges and hence the collapse mechanism. Sometimes the exact positions of all the hinges are not apparent. For instance, consider the propped cantilever shown in *Figure 8.5* which is carrying a uniformly distributed load λw per unit length. This beam becomes a mechanism with two hinges. One hinge develops at end A where the positive ordinate of the bending moment is maximum. The second hinge develops in the span at some point C where the negative ordinate of the moment is numerically maximum. To find the position of this hinge, let us assume that it is at a distance x from the simply supported end as shown in *Figure 8.5c*. Giving the portion BC an incremental rotation θ, point C moves down by $x\theta$. Hence the average movement of the uniform load down is $x\theta/2$ and the hinge rotation at A is $x\theta/(L-x)$. At collapse the work done by the load is thus $\lambda_p wLx\theta/2$ and the

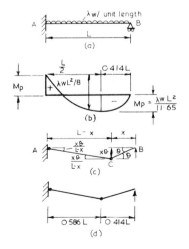

Figure 8.5. Beam with a simply supported end
(a) beam and loading
(b) elastic bending moment diagram
(c) a possible collapse mechanism
(d) the actual collapse mechanism

work absorbed by the hinges at A and C are $x\theta M_p/(L-x)$ and $M_p[\theta + x\theta/(L-x)]$ respectively. The virtual work equation is thus given by

$$\lambda_p wLx\theta/2 = M_p\theta + 2M_p x\theta/(L-x)$$

i.e.

$$\lambda_p = (2M_p/Lw)\,[(L+x)/(Lx-x^2)]$$

The beam collapses when λ_p is a minimum. That is to say as the load factor increases collapse takes place with a mechanism that can develop with the smallest load. The second hinge can be anywhere in the span so long as the load that causes it is the least. This is not only common sense, but in fact it is also in accordance with the 'kinematic theorem' which will be stated later. Thus at collapse

$$\frac{d\lambda}{dx_p} = \frac{2M_p}{wL}\left[\frac{xL - x^2 - (L+x)(L-2x)}{(Lx-x^2)^2}\right] = 0$$

which gives $x = 0.414L$. Hence the collapse load factor is given by

$$\left.\begin{array}{c}\lambda_p w = 11.65 M_p/L^2 \\ \\ M_p = \lambda_p w L^2/11.65\end{array}\right\} \qquad (8.6)$$

i.e.

The actual collapse mechanism is shown in *Figure 8.5d*. The procedure just described for finding the position of a plastic hinge in a beam which is uniformly loaded can be applied to other structures such as continuous beams and frames.

8.5. Collapse of continuous beams

In order to calculate the collapse load of a continuous beam with several spans, the virtual work equation is used to calculate the collapse load of each span separately. The collapse load of the beam is then selected as the lowest load that causes the collapse of the weakest span in the beam.

Consider, for instance, the uniform continuous beam ABCDEF of *Figure 8.6*. The spans AB and BD are of length L, while span DF is $2L$. AB carries a total uniform working load W, BD carries a working load W at its midpoint at C and DF carries a load of $2W$ acting at a distance of $L/2$ from support D. As the loads are increased proportionally by multiplying each span load by the same load factor λ, the first collapse mechanism may develop at any of the three spans. These

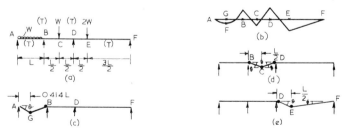

Figure 8.6. Collapse of a continuous beam
(a) beam and loading
(b) elastic bending moment diagram
(c) mechanism in AB alone
(d) mechanism in BD
(e) mechanism in DF

mechanisms are shown in *Figures 8.6c, d* and *e*. For the mechanism to develop in span AB (*Figure 8.6c*), a hinge has to form at B, where the positive ordinate of the bending moment is largest in this span, and another hinge has to form somewhere such as G in the span where the negative ordinate of the moment is the highest. This type of collapse is similar to that of the propped cantilever of *Figure 8.5* and hence point G is $0.414L$ from the simple support A. Once again the load factor is given by equation (8.6), thus

$$\left. \begin{array}{l} \lambda_p = 11.65 M_p/LW \\ M_p = \lambda_p WL/11.65 \end{array} \right\} \quad (8.7)$$

For the mechanism to develop in span BD, the bending moments at B, C and D should be numerically equal and they must all have numerical value M_p. This is similar to the case of the fixed-ended beam of *Figure 8.3* and hence the collapse load factor is given by equation (8.5), i.e.

$$\left. \begin{array}{l} \lambda_p = 8M_p/WL \\ M_p = \lambda_p WL/8 \end{array} \right\} \quad (8.8)$$

Finally for the mechanism to develop in span DF, the numerical values of M_D and M_E should be equal to M_p. This mechanism is shown in *Figure 8.6e*. The collapse load factor λ_p with this mechanism can be derived by using the virtual work equation. Giving an incremental rotation θ to the portion DE, then point E moves down by $\theta L/2$ and the work done by the load $2W\lambda_p$ is $\lambda_p WL\theta$. The rotation ϕ of the simply supported end F is $0.5L\theta/1.5L$, i.e. $\phi = \theta/3$. The hinge rotation at E is therefore $\theta + \theta/3$ and the work absorbed by this hinge is $(\theta + \theta/3)M_p$. The work absorbed by the hinge at D is $M_p\theta$ and the virtual work equation becomes

$$M_p\theta + M_p(\theta + \theta/3) = \lambda_p WL\theta$$

Hence

$$\left. \begin{array}{l} \lambda_p = 7M_p/3WL \\ M_p = 3\lambda_p WL/7 \end{array} \right\} \quad (8.9)$$

The continuous beam collapses when only one of these mechanisms develops. The lowest of the load factors given by equations (8.7), (8.8), and (8.9) is the load factor that causes collapse. Since

$$11.65 M_p/WL > 8M_p/WL > 7M_p/3WL$$

it follows that collapse takes place by the mechanism of *Figure 8.6e* with $\lambda_p = 7M_p/3WL$ given by equation (8.9).

8.6. Selection of mechanisms

In the case of continuous beams, although there is the possibility of the development of a separate mechanism in each span, the final collapse takes place with only one of these mechanisms. In the case of frames, however, there may be the possibility of several separate member mechanisms. Collapse may take place by any one of these individual mechanisms. On the other hand hinges may be selected, throughout the frame, that render the frame or part of it into a mechanism. In a frame collapsing in this manner, no member will have its own individual mechanism. In order to find the collapse load factor λ_p all these mechanisms have to be investigated to find out the one that can develop with the lowest load factor. To achieve this aim, it is necessary to list all the possible positions of hinges in a frame and select from this list two or more hinges that can give rise to the development of a mechanism. From among these mechanisms the collapse mechanism is then selected

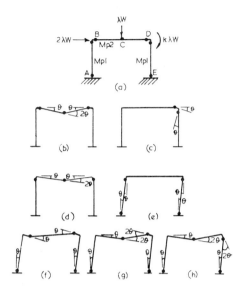

Figure 8.7. Separate mechanics in a portal frame
(a) possible position of hinges
(b) beam mechanism
(c) joint mechanism
(d) beam mechanism
(e) sway mechanism
(f) sway mechanism
(g) combined mechanism
(h) combined mechanism

according to the 'kinematic theorem' which states that: of all the mechanisms formed by assuming plastic hinge positions, the correct one for collapse is that which requires the minimum load.

For a member that is not loaded anywhere along its span, there is the possibility of a hinge forming at either end of such a member. If, however, the member is subject to some form of lateral loading, then an extra hinge can form along the span of the member. A mechanism in the frame is uniquely defined when all the incremental rotations in the frame can be algebraically expressed in terms of one of these rotations.

As an example consider the portal frame of *Figure 8.7*. The columns AB and ED are not loaded along their lengths, hinges can therefore develop only at their ends. On the other hand the beam BCD is laterally loaded at C and can therefore have three hinges. Altogether seven hinges can develop in this frame. In *Figure 8.7b* a beam-type collapse mechanism is shown with hinges B, C and D in the beam and an incremental rotation θ. The mechanism shown in *Figure 8.7d* is also beam type except that the hinges at B and D are developed in the columns instead of the ends of the beam. Whether the hinges form at the end of the beam or the column heads depends on the values of the fully plastic hinge moments M_p^2 of the beam and M_p^1 of the columns. If $M_p^2 > M_p^1$ then the hinges develop in the columns, otherwise they form at the ends of the beam. In *Figure 8.7c* a joint mechanism is developed. Such a mechanism is possible at a joint where an external moment is acting and it takes place with a plastic hinge at the end of all the members meeting at that joint. Mechanisms e and f in the figure are sway type mechanisms. These can generally develop when the side load acting on the frame is large. The last two mechanisms are a combined type mechanism with a hinge developing under the point load at C and the frame moving sideways. In fact, a combined mechanism can be obtained by superimposing a beam type mechanism on a sway mechanism.

The Plastic Theory

For instance the hinge rotations at end B of the beam in figures b and f are in opposite directions. Once these two mechanisms are added together, the hinge rotation at B reduces to zero. The hinge rotations at D are in the same direction in both these mechanisms and therefore they add up, resulting in a rotation of 2θ at D. In this manner mechanism g is obtained.

For each one of these mechanisms a virtual work equation can be written and the load factor to cause collapse by each mechanism can be calculated. The lowest of these factors is the actual collapse load factor. It is not necessary to try every mechanism, since if M_{p^2} is different from M_{p^1} then the hinges at B and D may form either in the beam or the column. Of course the mechanisms satisfying the actual conditions are the ones that are possible to develop. If $M_{p^1} = M_{p^2}$, then the load factors given by mechanisms b and d are equal. Similarly the load factors for mechanisms e and f will be equal as will those for mechanisms g and h.

Let us assume, by way of an example, that $M_{p^2} = 1.5 M_{p^1}$, $AB = L$, $BC = 3L/4 = CD$ and $k = 0$. Since $M_{p^1} < M_{p^2}$ mechanisms d, e and h are the only ones that are possible to develop. In the case of mechanism d, the reader can derive the virtual work equation as

$$2M_{p^1}\theta + 3M_{p^1}\theta = 3\lambda_p WL\theta/4, \quad (M_{p^2} = 1.5 M_{p^1});$$

$$\therefore \quad \lambda_p = 20 M_{p^1}/3WL$$

and

$$M_{p^1} = 3\lambda_p WL/20 \tag{8.10}$$

The virtual work equation for mechanism e is

$$4M_{p^1}\theta = 2\lambda_p WL\theta$$

$$\therefore \quad \lambda_p = 2 M_{p^1}/WL$$

and

$$M_{p^1} = \lambda_p WL/2 \tag{8.11}$$

Finally, for mechanism h the virtual work equation is

$$2\theta M_{p^1} + 2\theta M_{p^1} + 2\theta M_{p^2} = 2\lambda_p WL\theta + 3\lambda_p WL\theta/4$$

This gives

$$\lambda_p = 28 M_{p^1}/11 WL$$

i.e.

$$M_{p^1} = 11\lambda_p WL/28 \tag{8.12}$$

From equations (8.10) through (8.12) the smallest value of λ_p is $2M_{p^1}/WL$ given by mechanism e which is therefore the collapse mechanism.

8.7. Example

The pitched roof frame of *Figure (8.8)* is fixed at A and E and is subject to a vertical load of $2\lambda W$ at the apex C and a horizontal load of $3\lambda W$ at D. The members of the frame are made out of the same cross section with a fully plastic hinge moment M_p. Calculate the collapse load.

Answer: This frame may collapse with one of the three mechanisms shown in the figure. In the case of the sway mechanism (*Figure 8.8b*) hinges form at A, B, D and E. The hinges at B and D may form either in the columns or in the rafter. For an incremental rotation θ, as shown in the figure, the work absorbed by the hinges

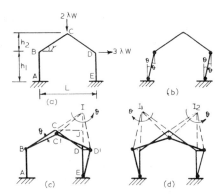

Figure 8.8. Pitched roof frame
(a) frame and loading
(b) sway mechanism
(c) frame mechanism
(d) symmetrical frame mechanism

is $4M_p\theta$. The vertical load at C does not move down and hence does no work. The horizontal load at D moves by $h_1\theta$ and hence the work done by it is $3\lambda W h_1 \theta$.

The virtual work equation is thus

$$4M_p\theta = 3\lambda_p W h_1 \theta$$

i.e.

or
$$\left.\begin{array}{l} \lambda_p = 4M_p/(3Wh_1) \\ \\ M_p = 3\lambda_p W h_1/4 \end{array}\right\} \quad (8.13)$$

In the case of the frame mechanism (*Figure 8.8c*) hinges develop at B, C, D and E. In order to find the hinge rotations, member CD is rotated by θ about the instantaneous centre I. From the geometry of the frame, it is evident that $CJ = L/2$, $JD = h_2 = IJ$ and $CI = CB$, thus $CC' = IC\,\theta$ and the vertical movement of the load $2\lambda W$ is $CC' \cos \gamma = L\theta/2$. The hinge rotation at B is $CC'/BC = \theta$. The horizontal movement at D is $2h_2\theta$ and the hinge rotation at E is therefore $2h_2\theta/h_1$. Since members CB and CD both rotate by θ, the hinge rotation at C is, therefore, 2θ. On the other hand the hinge at D rotates by $\theta + 2h_2\theta/h_1$. The virtual work equation is thus given by

$$M_p\left[\theta + 2\frac{h_2}{h_1}\theta + 2\theta + \left(\theta + \frac{2h_2\theta}{h_1}\right)\right] = \lambda_p WL\theta + 3\lambda_p W 2h_2 \theta$$

Hence for $k = h_2/h_1$ we obtain

$$\lambda_p = 2M_p(1 + k)/[W(0.5L + 3h_2)]$$

and
$$M_p = W\lambda_p(0.5L + 3h_2)/[2(1 + k)]$$
(8.14)

The reader can derive the virtual work equation for the symmetrical frame mechanism of *Figure 8.8d*. This gives

$$\lambda_p = 4M_p(1 + k)/[W(L + 3h_2)]$$

and
$$M_p = W\lambda_p(L + 3h_2)/[4(1 + k)]$$
(8.15)

The lowest value of λ_p calculated by equations (8.13), (8.14) and (8.15) is the load factor at collapse. This depends upon the geometry of the frame. For instance, if $L = 3h_2$ and $k = h_2/h_1 = 1$, equation (8.13) gives $\lambda_p = 1.33\, M_p/(Wh_2)$, equation (8.14) gives $\lambda_p = 0.89\, M_p/(Wh_2)$ while equation (8.15) gives $\lambda_p = 1.33\, M_p/(Wh_2)$. In this case the actual collapse load is $0.89\, M_p/(Wh_2)$ and the collapse mechanism is that shown in *Figure 8.8c*.

8.8. Design by rigid-plastic theory

It has been pointed out that when a rigid-plastic collapse of a structure is about to take place, the structure is isostatic. This simplifies the design procedure and reduces it to that of a simple analysis. In general, the problem of design is the reverse of the problem of an analysis. In the case of an analysis, the loads, dimensions and sectional properties are known and it is required to find the member forces and bending moments. In the problem of design, the sectional properties are required so that the structure can carry the applied loads. For isostatic structures an analysis can be carried out without a knowledge of sectional properties. Once this is done, suitable sections can be selected to withstand the member forces and moments. For this reason in the preceding sections, the result of an analysis was always expressed either to predict the load factor or to calculate the fully plastic moments of the section.

Various codes of practice decide upon the load factor λ_p at collapse in a definite manner. For instance, at present, the load factor at collapse under vertical loadings alone is 1.75, while that under combined vertical and wind loads is 1.4. Once the type of loading and the corresponding load factor is decided upon, a rigid-plastic analysis of a structure determines the sectional properties of the members. in the case of the simply supported beam of *Figure 8.1* for instance, the virtual work equation gave the value of M_p as $\lambda_p P_w L/4$. For a load factor $\lambda_p = 1.75$, once the applied working loads P_w and the span L of the beam are decided upon, the required M_p for the section can be calculated. This value of M_p together with the yield stress σ_y of the material are used to calculate Z_p of the section from

$$Z_p = M_p/\sigma_y$$

A section is then selected with an actual Z_p greater than that calculated. This same procedure can be used to design propped cantilevers and fixed-ended beams.

In the case of continuous beams it is possible to obtain more than one acceptable design depending on the type of mechanisms that are allowed to develop at collapse and the type of sections to be used.

8.9. Example

Design the continuous beam of section 8.5 shown in *Figure 8.6*.

Answer: There are three possible types of mechanisms that can develop when the beam collapses. The value of the fully plastic hinge moment for these mechanisms are given by equations (8.7), (8.8) and (8.9).

If it is required to make the beam from the same uniform section then it is necessary to make this section sufficiently strong to withstand the largest M_p given by these three equations. Since $3\lambda_p WL/7 > \lambda_p WL/8 > \lambda_p WL/11.65$, it follows that the selected section should have $M_p > 3\lambda_p WL/7$. The bending moment diagram for the beam with $M_p = 3\lambda_p WL/7$ is shown in *Figure 8.9b*, and the collapse mechanism is shown in *Figure 8.6e*. Nowhere between A and D does the bending moment reach the value of M_p. Thus so long as the section selected is sufficiently large to withstand a moment $3\lambda_p WL/7$ at D and E, it will be safe to carry the loads on all the spans.

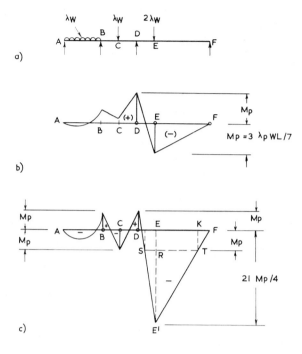

Figure 8.9. Bending moment diagrams for continuous beam
(a) beam and loading
(b) bending moment diagram when mechanism develops in DF
(c) bending moment diagram when mechanism develops in BD

A lighter design of this beam can be obtained by forcing the beam to collapse with a mechanism in span BD and another one in span DF. From equation (8.8) $M_p = \lambda_p WL/8$ obtained by allowing a mechanism to develop in span BD (*Figure 8.6d*). Three hinges would develop at B, C and D. The moments at the supports B and D will be M_p, and the moment at C will be M_p. In *Figure 8.9c*, the bending moment diagram for the beam is shown with these moments indicated. Now if

$$M_p = \lambda_p WL/8$$

then

$$\lambda_p W = 8M_p/L$$

and with the load acting on span AB, the bending moment midway between A and B is

$$M = \lambda_p WL/8 - 0.5M_p = 0.5M_p$$

Thus if the same section was used for the beam to withstand $M_p = \lambda_p WL/8$, then such a section would not collapse by the mechanism shown in *Figure 8.6c*, developing in span AB. Everywhere in this span the bending moment is less than M_p as shown in *Figure 8.9c*.

For span DF, however, the bending moment at E is

$$M_E = 2\lambda_p W \times 0.5L \times 1.5L/2L - 1.5M_p/2 = 21M_p/4$$

The support moment $M_D = M_p$. This indicates that the section selected to withstand the moments in the middle span is not sufficiently strong to carry the negative moments at E. In *Figure 8.9c* the straight line ST is drawn parallel to DF so that TK = M_p. The uniform section used for AD if continued to F will provide a plastic moment of resistance of M_p = TK all along DF. The portion of the beam vertically above ST will thus require reinforcement. This can be provided by a pair of parallel plates covering the uniform section between S and T and having a moment of resistance equal to RE', i.e. $17M_p/4$. A beam designed in this manner will collapse with mechanisms developing in spans BD and DF simultaneously.

8.10. Effect of axial load on M_p

In general the members of a structure are subject to axial forces as well as bending moments. The effect of the axial load in a member is to reduce its fully plastic moment. Consider a member subject to a moment and a force P with a yield stress σ_y. When the section of this member becomes fully plastic, this takes place by the combined effect of the moment and the axial force. Assuming that the axial force is acting centrally, the area of the section which yields because of the presence of P is

$$A_p = P/\sigma_y \tag{8.16}$$

The rest of the section yields by the acting moment which causes tension and compression on equal areas on either side of the plastic neutral axis. These stresses are represented diagrammatically in *Figure 8.10*. The central section of area A_p reaches the stress σ_y by the axial load as shown in *Figure 8.10b*. The section loses the fully plastic hinge moment M_{pp} of this central section and the bending stress distribution on the rest of the section is shown in *Figure 8.10c*. The final fully plastic stress distribution is the sum of these two.

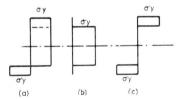

Figure 8.10. Effect of axial loads
(a) final stresses (=)
(b) stresses caused by P (+)
(c) stresses caused by the moment

This is shown in *Figure 8.10a*. Naturally the reduced plastic hinge moment M'_p of the section is given by

$$M'_p = M_p - M_{pp} \tag{8.17}$$

For example, let a rectangular section of width B and depth D be subject to an axial force P. The area A_p is given by $P/\sigma_y = Bd$, where d is the depth of the central portion yielded by the axial force, i.e.

$$d = P/B\sigma_y$$

The plastic section modulus of the central portion Z_{pp} is therefore given by

$$Z_{pp} = Bd^2/4 = P^2/(4B\sigma_y^2)$$

The reduced plastic section modulus is thus given by

$$Z'_p = Z_p - Z_{pp} = (BD^2/4) - (P^2/4B\sigma_y^2)$$

and the reduced plastic hinge moment $M'_p = \sigma_y Z'_p$ is given by

$$M'_p = \sigma_y [(BD^2/4) - (P^2/4B\sigma_y^2)]$$

$$\therefore \quad M'_p = M_p(1 - d^2/D^2)$$

In a similar manner it can be found that the reduced plastic hinge moment of the I section of Section 7.15, *Figure 7.16*, is given by

$$M'_p = M_p \left\{ 1 - 0.25 t_1 h^2 / [Bt_2(d - t_2) + 0.25(d - 2t_2)^2 t_1] \right\}$$

where h is the depth of the central portion of the web which is yielded by the effect of an axial load P. The value of h is again calculated using equation (8.16). In this expression for M'_p it is assumed that the central area, which is yielded by the axial load, does not spread to the flanges of the I section. Generally the reduction in fully plastic moment due to an axial force is more severe for an I section than for a rectangular section.

The application of the plastic theory is limited to beams and small structures such as building frames with one or two storeys. As a rule the axial forces in the members of these structures are small and therefore, their effects on reducing the fully plastic hinge moments of the members are also small. In tall building frames, with stanchions carrying heavy axial thrusts, the effect of thrusts on reducing the values of M_p in these stanchions can be considerable. This is particularly the case with the stanchions of the lower storeys. Furthermore such structures require elastic-plastic failure load analyses, where instability effects due to high axial loads are also important. In such an analysis, therefore, the effect of axial forces on reducing the fully plastic moment of the section becomes very significant. This is particularly so after the formation of a plastic hinge in a member. As the load factor increases, the axial forces in the members also increase and the combined effect of instability and reduction in M_p become major factors in producing failure in the structure.

8.11. Examples

Example 1. Design the frame of *Figure 8.11* which is subject to a working vertical load $2W$ at the midspan E of the beam AC and a horizontal load W at the beam level.

Answer: This frame may be converted into a mechanism in three different ways. A beam mechanism develops with three hinges, one at E under the vertical load and one at each end of the beam as shown in *Figure 8.11b*. Because the columns are made stronger than the beam, the plastic hinges tend to develop at the ends of the beam before those at the ends of the columns.

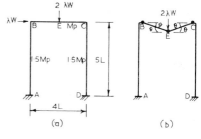
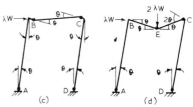

Figure 8.11. Various mechanisms of a portal
(a) frame and loading
(b) beam mechanism
(c) sway mechanism
(d) combined mechanism

As the structure deflects at factored loads $2\lambda w$ acting vertically and λw horizontally, the rotation of the plastic hinges at B and C, at a given instant, is θ while that of the hinge at E is 2θ. The load $2\lambda w$ moves down by an amount $2L\theta$ and the work done by this load is therefore $4\lambda wL\theta$. The horizontal load does no work and is not shown in *Figure 8.11b*.

The virtual work equation for the collapse mechanism, is thus

$$M_p\theta + M_p\theta + 2M_p\theta = 4\lambda wL\theta$$

Hence

$$M_p = \lambda wL \quad (8.18)$$

On the other hand a sway mechanism develops in this frame with four hinges at A, D, B and C as shown in *Figure 8.11c*. The virtual work equation for this collapse mechanism can be derived similarly as

$$3M_p\theta + 2M_p\theta = 5\lambda wL\theta$$

giving

$$M_p = \lambda wL \quad (8.19)$$

Finally a combined mechanism can develop with two hinges at A and D and two other hinges at E and C in the beam. This is shown in *Figure 8.11d*, for which the virtual work equation is

$$3M_p\theta + 4M_p\theta = 5\lambda wL\theta + 4\lambda wL\theta$$

Thus

$$M_p = 9\lambda wL/7 \quad (8.20)$$

Comparing equations (8.18), (8.19) and (8.20), it is noticed that the largest M_p is required by equation (8.20), for the combined mechanism. To prevent collapse it is therefore necessary to select a section for the beam with a plastic hinge moment which is at least as large as that given by equation (8.20). Accordingly the value of the plastic hinge moment for the columns must be 1.5 times larger.

Example 2. Calculate the plastic modulus of a circular section of radius R. A cantilever beam of span L is made from a solid circular section of diameter D_1 for half its span and another of diameter D_2 for the rest. What is the ratio D_1/D_2 which ensures the simultaneous development of plastic hinges at the support and at midspan? The beam is subject to a point load W at its free end.

Answer: In *Figure 8.12*: Area of circle = πR^2
Force on half area = $0.5\pi R^2 \sigma_y$
Distance of centroid of semicircle from base = $4R/3\pi$

$$Z_p = 0.5\pi R^2 \times (4R/3\pi) \times 2 = 4R^3/3.$$

In *Figure 8.12b*

$$M_{pA} = \lambda_p WL = 4R_1^3\sigma_y/3$$
$$M_{pD} = 0.5\lambda_p WL = 4R_2^3\sigma_y/3$$
∴ $$R_1^3/R_2^3 = \lambda_p WL/0.5\lambda_p WL = 2$$
∴ $$(0.5D_1)^3/(0.5D_2)^3 = 2$$
∴ $$D_1/D_2 = 1.26$$

Example 3. Design the two storey rigidly jointed frame shown in *Figure 8.13* by selecting the value of M_p. The load factor against collapse under vertical loads acting alone is $\lambda_1 = 1.75$ and that under combined vertical and horizontal load is $\lambda_2 = 1.4$.

198   The Plastic Theory

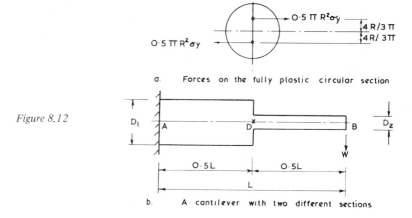

Figure 8.12

a. Forces on the fully plastic circular section

b. A cantilever with two different sections

Figure 8.13

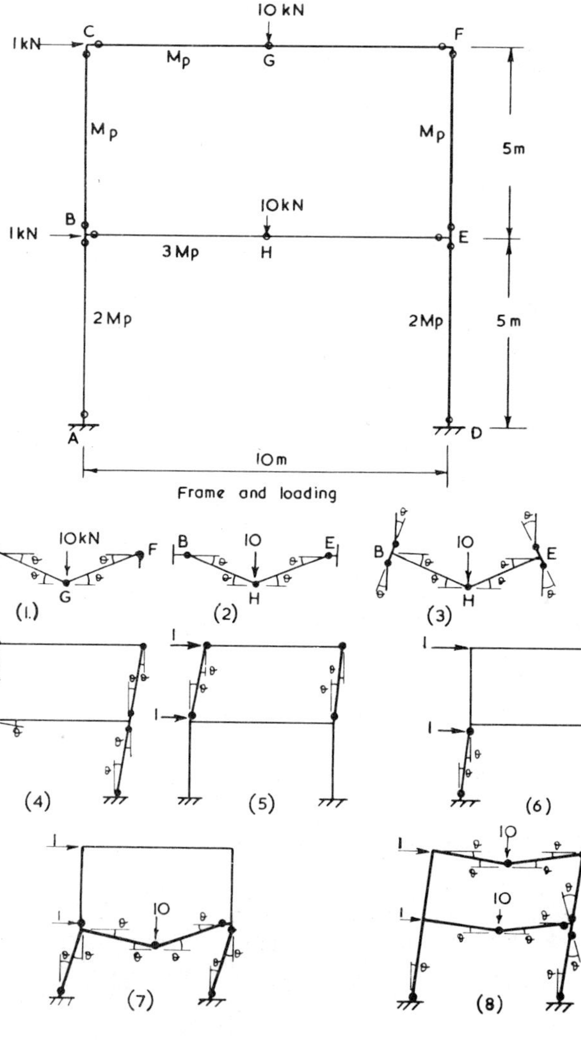

Frame and loading

The Plastic Theory 199

Answer: All the possible positions of the plastic hinges are shown as little circles in the figure. There are altogether 14 possible hinges that can take place. Out of these three beam type mechanisms as shown in 1, 2 and 3 in *Figure 8.13*, three sway mechanisms 4, 5, 6 and three combined mechanisms 7, 8 and 9 may develop. The development of mechanism 9 is left as an exercise for the reader. The loads causing these mechanisms are also shown. The virtual work equations for these are

(1) Mechanism in beam CGF

$$4M_p\theta = 10 \times 5\lambda_1\theta$$

∴ $M_p = 12.5\lambda_1$

(2) Mechanism in beam BHE

$$4 \times 3M_p\theta = 10 \times 5\lambda_1\theta$$

∴ $M_p = 4.167\lambda_1$

(3) In beam BHE

$$2M_p\theta + 2 \times 2M_p\theta + 3M_p \times 2\theta = 10 \times 5\lambda_1\theta$$

∴ $M_p = 4.167\lambda_1$

(4) Sway mechanism 4

$$2M_p \times 3\theta + 3M_p\theta + 3M_p\theta = 1 \times 5\lambda_2\theta + 1 \times 10\lambda_2\theta$$

∴ $M_p = 1.25\lambda_2$

(5) Sway of top storey

$$4M_p\theta = 1 \times 5\lambda_2\theta$$

∴ $M_p = 1.25\lambda_2$

(6) Sway of the bottom storey

$$2M_p \times 4\theta = 1 \times 5\lambda_2\theta + 1 \times 5\lambda_2\theta$$

∴ $M_p = 1.25\lambda_2$

(7) Combined mechanism for the bottom storey

$$2M_p \times 3\theta + M_p\theta + 6M_p\theta + 3M_p\theta = 1 \times 5\lambda_2\theta + 1 \times 5\lambda_2\theta + 10 \times 5\lambda_2\theta$$

∴ $M_p = 3.75\lambda_2$

(8) Combined mechanism for the whole frame

$$2M_p \times 3\theta + 3M_p \times 2\theta + 3M_p\theta + M_p\theta + 2M_p\theta + 2M_p\theta$$
$$= 1 \times 5\lambda_2\theta + 1 \times 10\lambda_2\theta + 10 \times 5\lambda_2\theta + 10 \times 5\lambda_2\theta$$

∴ $M_p = 5.75\lambda_2$

(9) Combined mechanism for the top storey

$$M_p = 9.166\lambda_2$$

Mechanism 1 gives the highest value of M_p. This is the deciding value of M_p for all the other section. Its value is

$$M_p = 12.5 \times 1.75 = 21.875 \text{ kN m}.$$

8.12. Limitation of the plastic theory

The plastic theory assumes that the material of the structure is ductile with large strains taking place while the material is flowing plastically. Thus with brittle materials such as concrete the theory is inapplicable.

The theory also assumes that once a hinge develops it undergoes a rotation of any magnitude so long as the bending moment remains constant at the fully plastic value. Once a hinge is introduced at a section it assumes that this hinge continues to rotate in the same direction. This is not always the case. For instance, in the case of a continuous beam under several point loads, which is similar to that of uniform loading, a hinge may develop under one of the loads and then stop to allow for another hinge to develop under a different point load.

The plastic theory assumes that a structure is always subject to a proportionally increasing load. This, as was stated earlier, is not always realistic.

The plastic theory neglects the deflections of the structure prior to collapse and particularly at the working loads. These deflections have to be calculated by the elastic theory.

As soon as the axial loads in the members of a structure are considered, the entire basic fundamentals of the plastic theory are totally undermined. This is because the virtual work equation, i.e. the equilibrium condition, has to be altered to allow for the axial loads. Axial loads also cause the bending moments at certain sections, not necessarily at joints or under point loads, to become higher than the fully plastic moments of these sections. The worst effect of the axial loads, however, is to make collapse take place before the formation of a mechanism at a load factor lower than that given by the plastic theory. Finally the axial loads may alter the mode of deformation in such a way that the sequence of hinge formation would be entirely different from those predicted by the plastic theory.

These are the reasons for applying the plastic theory only to steel beams and one or two storey steel frames. The theory itself is simple and attractive and its application to the collapse of plates has proved to be of value. This latter application is now presented.

8.13. The yield line theory in plates

Consider a uniformly loaded slab simply supported by two walls as shown in *Figure 8.14*. As the load increases, collapse takes place when a plastic hinge develops along a line in the slab known as the yield line. For other types of loading or support conditions, the yield line pattern will be more complicated and several such patterns may be possible. From the kinematic theorem, however, the pattern with the lowest collapse load is the collapse mechanism.

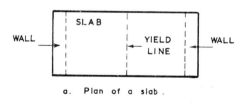

a. Plan of a slab.

Figure 8.14

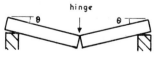

b. Collapse mechanism

The Plastic Theory

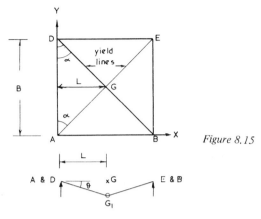

Figure 8.15

In deriving the virtual work equation, the work absorbed by the hinge along the yield lines should be calculated. Consider the slab shown in *Figure 8.15* with yield lines developed along the diagonals DB and AE. Giving each triangle a virtual rotation about its support so that the virtual displacement at G in the vertical direction is GG_1, then the rotation of the triangle DGA is θ, where

$$L\theta = GG_1$$

i.e. $\quad\quad \theta = GG_1/L$

If mp is the plastic hinge moment per unit length of a yield line then the work absorbed by the rotation of the triangle is w given as

$$w = B(GG_1/L)mp \tag{8.21}$$

where B is the base AD of the triangle. This is because the rotation of the triangle is about the base AD and the work absorbed at the yield lines AG and DG are obtained by multiplying mp by the projection of AG and DG on the Y axis. This projection is equal to

$$B = AG\cos a + DG\cos a \tag{8.22}$$

8.14. Rectangular reinforced concrete slabs

It was stated that the plastic theory does not apply to concrete frames. In the case of reinforced concrete slabs, if the thickness of the slab is small compared to its length and width then a yield line pattern can develop at collapse. For the case of a slab reinforced in the bottom of the slab only, tensile zones developed at the top of the slab cannot resist bending deformation and the plastic moment of resistance *mp* in these zones is zero. For a rectangular slab simply supported on all four sides and with bottom reinforcement, the yield line pattern at collapse will be similar to that shown in *Figure 8.16a*. On the other hand when the slab is reinforced top and bottom with equal areas of steel per unit area of the cross section in two perpendicular directions, then the pattern of the yield line will be similar to that shown in *Figure 8.16b*. The collapse load obtained from *Figure 8.16a* is smaller than that obtained from *Figure 8.16b*.

8.15. Example

A rectangular slab is reinforced at top and bottom with equal areas of steel and simply supported by four walls. The corners of the slab are held down. The slab carries a uniform load of intensity q per unit area. With a load factor of $\lambda = 2$ calculate the plastic hinge moment of the slab.

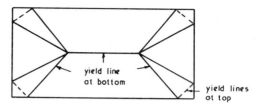
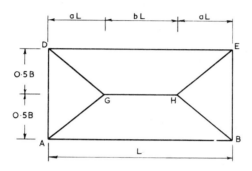

a. Yield lines for slabs with bottom reinforcements.

b. Yield lines for slabs with top and bottom reinforcements.

Figure 8.16

Answer: Giving a virtual displacement z to the yield line GH of *Figure 8.16b*, the work done by the load on the triangle DGA is

$$w_{DGA} = \lambda q \times 0.5BaL \times z/3$$

where $0.5BaL$ is the area of the triangle. It is assumed that the load acting on the triangle will be concentrated at its centroid which is at a distance $aL/3$ from the edge DA and hence moves down by $z/3$. Similarly the work done by the load on the triangle EHB is $\lambda q \times 0.5BaL \times z/3$.

The work done by the load acting on the trapezium GHED is

$$w_{GHED} = \lambda q \times 0.5BaL \times z/3 + \lambda q \times bL \times 0.5B \times z/2$$

wher $0.5BaL$ is the area of the two triangles at either side of the trapezium and $bL \times 0.5B$ is the area of the remaining rectangular part. The load acting on this rectangular part moves down by $z/2$. Similarly the work done by the load on AGHB is equal to w_{GHED}.

The work absorbed by the hinge rotation at yield lines GD and GA, using equation (8.21), is $Bmpz/aL$. That absorbed by the lines HE and HB is also $Bmpz/aL$. The work absorbed by the lines DGHE in rotating the trapezium DGHE about DE is $Lmp \times z/(0.5B)$. Similarly for the trapezium AGHB rotating about AB.

With $\lambda = 2$, $b = 1 - 2a$, the virtual work equations gives

$$mp = \frac{qL^2(3a - 2a^2)}{6(1 + 2aL^2/B^2)}$$

The value of a is selected so that the largest mp is selected and the slab is designed so that it withstands this value of mp, thus for $dmp/da = 0$ we obtain

$$a = 0.5\left[\sqrt{\left(\frac{B^4}{L^4} + \frac{3B^2}{L^2}\right)} - \frac{B}{L}\right]$$

and with this value of a, mp becomes

$$mp = \frac{qB^2}{12}\left[\sqrt{\left(\frac{B^2}{L^2} + 3\right)} - \frac{B}{L}\right]^2$$

Exercises on Chapter 8

1. A beam of constant flexural rigidity EI and fully plastic hinge moment M_p is fixed at its ends and carries a uniformly distributed load w per unit length. If this load is increased gradually, calculate the central deflection of the beam (a) at first yield, (b) when the first set of plastic hinges fully develop and (c) at incipient collapse. The span of the beam is L and its shape factor is α.

Ans. $M_pL^2/(32EI\alpha)$, $M_pL^2/(32EI)$, $M_pL^2/(12EI)$.

2. ABC is a uniform beam fixed at A and C and carries a proportionally increasing point load at B. $AB = 2BC = 2L$. The flexural rigidity of the beam is EI and its fully plastic hinge moment is M_p. Calculate the value of the applied load at collapse and the deflection under the load at incipient collapse.

Ans. $3M_p/L$, $2M_pL^2/(3EI)$.

3. A rectangular fixed base portal frame has a constant cross section with fully plastic hinge moment M_p and flexural rigidity EI. The frame is subject to a horizontal force W at the top of one of its columns and a vertical load W acting at the midspan of its beam. The column height is L and the span is $2L$. Calculate the value of W at collapse. What is the sway in the columns at incipient collapse?

Ans: $W = 3M_p/L$, $\delta_h = M_pL^2/3EI$.

4. A two bay rectangular portal frame has a height of 2m and a span of 2m per bay. The frame carries a horizontal working load of 11 kN at the top of one of the outer columns and a vertical load of 22 kN at the centre of each bay. Design the frame so that the fully plastic moment of the beam is twice that of the columns. Take the load factor at collapse as 1.4.

Ans. $M_p = 11.2$ kN m.

5. A pinned-base portal frame has a span of 4m, column height of 2m and an angle of pitch of $45°$. It carries a load of 20 kN at its apex. The frame has a constant cross section. Calculate the value of M_p if the frame collapses when the load is increased by a factor of 1.4.

Ans. $M_p = 22.4$ kN m.

6. ABCDE is a uniform beam of constant M_p and EI. The beam is simply supported at A, B and E and carries a point load of $1.3W$ at C and another of $0.7W$ at D. The span $AB = 3L$ while $BC = CD = DE = L$. Draw the bending moment diagram and show that the first plastic hinge forms at C. Calculate the value of W when a hinge actually

develops at this point. Derive the virtual work equations for the two possible mechanisms and calculate the values of W from each of these equations. Hence show that the plastic theory makes the wrong assumption that a hinge never takes place at C.

Ans. $1.32 M_p/L$, Hinges at B and C: $W = 1.82 M_p/L$, Hinges at B and D, $W = 1.48 M_p/L$.

7. A pinned-base pitched roof portal frame has a constant M_p of 1000 kN m. It is loaded with a vertical load of $22\lambda W$ at the apex and a horizontal force of $3\lambda W$ at the top of one of the columns. Calculate the value of λ at collapse. Take $W = 10$ kN, $L = 30$ m, $h_1 = 10$ m and $h_2 = 6$ m (see *Figure 8.8* for these dimensions).

Ans. 5.42.

9

Analysis of structures by the force method

9.1. Deflections of structures

So far we have been concerned mainly with the member forces in a structure. Once these are evaluated, the cross sections of the members can be selected so that the stresses developed in them are safe. It is often the case that although a structure is strong enough to carry the applied loads safely, the deflections at its joints turn out to be excessive to the extent that its serviceability is impaired. For this reason it is also necessary to evaluate these deflections and if excessive, alter the member cross sections in an effort to reduce the deflections to acceptable values.

Deflection calculations become particularly useful in the analysis of structures by the force method.

9.2. Relationship between member and joint deflections

Consider a pin jointed frame ABC with two members subject to a vertical and a horizontal force V and H respectively as shown in *Figure 9.1*. The inclination of

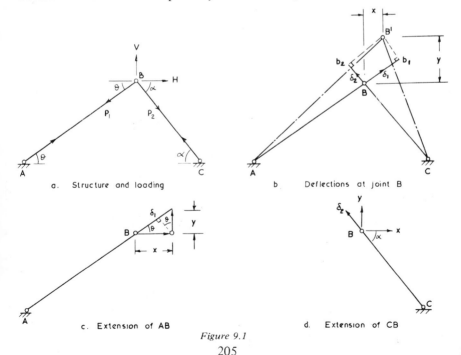

Figure 9.1

the members are θ and α as shown in the figure. Resolving the forces at B horizontally and vertically

$$H - p_1 \cos\theta + p_2 \cos\alpha = 0$$
$$V - p_1 \sin\theta - p_2 \sin\alpha = 0$$

Solving these two equations for p_1 and p_2, we obtain

$$p_1 = \frac{H \sin\alpha}{\sin(\alpha + \theta)} + \frac{V \cos\alpha}{\sin(\alpha + \theta)}$$

$$p_2 = \frac{-H \sin\theta}{\sin(\alpha + \theta)} + \frac{V \cos\theta}{\sin(\alpha + \theta)}$$

These two equations can be written in matrix form $\mathbf{P} = \mathbf{B}\,\mathbf{W}$ as

$$\begin{bmatrix} p_1 \\ p_2 \end{bmatrix} = \frac{1}{\sin(\alpha + \theta)} \begin{bmatrix} \sin\alpha & \cos\alpha \\ -\sin\theta & \cos\theta \end{bmatrix} \begin{bmatrix} H \\ V \end{bmatrix}$$

Here

$$\mathbf{B} = \frac{1}{\sin(\alpha + \theta)} \begin{bmatrix} \sin\alpha & \cos\alpha \\ -\sin\theta & \cos\theta \end{bmatrix}$$

and its transpose is

$$\mathbf{B}^T = \frac{1}{\sin(\alpha + \theta)} \begin{bmatrix} \sin\alpha & -\sin\theta \\ \cos\alpha & \cos\theta \end{bmatrix} \qquad (9.1)$$

Once the member forces p_1 and p_2 are calculated, the extension (or contraction) δ_1 and δ_2 of these members can be obtained from Hooke's law, thus

$$\left.\begin{aligned} \delta_1 &= \frac{L_1 p_1}{E_1 A_1} \\[4pt] \delta_2 &= \frac{L_2 p_2}{E_2 A_2} \end{aligned}\right\} \qquad (9.2)$$

Analysis of Structures by the Force Method

where L is the length of a member, E is its Young's modulus of elasticity and A is its area.

But for the constraint offered by CB, the extension of AB would displace B in the line AB produced by an amount δ_1 to b_1, *Figure 9.1b*. Similarly, but for AB, the elongation of CB would displace B along CB produced by an amount δ_2 to b_2. The bars, however, restrain one another and in the event both bars rotate about their supports A and C until their ends coincide at B'. These rotations are small and it can be regarded that the ends of the members move at right angles to AB and CB respectively as shown in *Figure 9.1b*.

Joint B moves horizontally by an amount x and vertically by an amount y. Since both members remain attached to each other at B, the extension δ_1 (*Figure 9.1c*) is given in terms of the joint displacements x and y as

$$\delta_1 = x \cos \theta + y \sin \theta$$

Similarly, from *Figure 9.1d*

$$\delta_2 = -x \cos \alpha + y \sin \alpha$$

Solving these two equations for x and y

$$x = \frac{\delta_1 \sin \alpha}{\sin(\alpha + \theta)} - \frac{\delta_2 \sin \theta}{\sin(\alpha + \theta)}$$

$$y = \frac{\delta_1 \cos \alpha}{\sin(\alpha + \theta)} + \frac{\delta_2 \cos \theta}{\sin(\alpha + \theta)}$$

and in matrix form

$$\mathbf{X} = \begin{bmatrix} x \\ y \end{bmatrix} = \frac{1}{\sin(\alpha + \theta)} \begin{bmatrix} \sin \alpha & -\sin \theta \\ \cos \alpha & \cos \theta \end{bmatrix} \begin{bmatrix} \delta_1 \\ \delta_2 \end{bmatrix} \quad (9.3)$$

$$\mathbf{X} = \mathbf{B}^T \boldsymbol{\delta} \quad (9.4)$$

Generally, for any structure $\mathbf{P} = \mathbf{BW}$ which gives

$$\mathbf{P}^T = \mathbf{W}^T \mathbf{B}^T$$

Post multiplying both sides of this equation by $\boldsymbol{\delta}$ we obtain $\mathbf{P}^T \boldsymbol{\delta} = \mathbf{W}^T \mathbf{B}^T \boldsymbol{\delta}$. But equation (7.44) gives $\mathbf{P}^T \boldsymbol{\delta} = \mathbf{W}^T \mathbf{X}$. Hence

$$\mathbf{P}^T \boldsymbol{\delta} = \mathbf{W}^T \mathbf{X} = \mathbf{W}^T \mathbf{B}^T \boldsymbol{\delta}$$

$$\therefore \mathbf{X} = \mathbf{B}^T \boldsymbol{\delta}$$

This means that once the member forces **P** are calculated, by equilibrium, and the member extensions **δ** are obtained, from Hooke's law, the joint deflections **X** can be readily evaluated from equations (9.4). Thus only a small effort is required to calculate these deflections.

9.3. Examples

Example 1. The angles θ and α in *Figure 9.1* are 30° and 45° respectively. Calculate the horizontal and the vertical deflections of joint B, of the frame shown in the figure when it is subjected, at B, to a horizontal force $H = 10$ kN and a downward vertical force $V = 5$ kN. The areas A_1 and A_2 are 100 mm² and 200 mm² respectively. Member AB is 1 m and E is 200 kN/mm².

Answer: Resolving the forces at B horizontally and vertically

$$H - 0.5\sqrt{3} \times p_1 + p_2/\sqrt{2} = 0$$
$$V - 0.5 p_1 - p_2/\sqrt{2} = 0$$

Solving for p_1 and p_2

$$\left. \begin{array}{l} p_1 = 0.732 H + 0.732 V \\ p_2 = -0.518 H + 0.896 V \end{array} \right\} \qquad (9.5a)$$

i.e.

$$\begin{bmatrix} p_1 \\ p_2 \end{bmatrix} = \begin{bmatrix} 0.732 & 0.732 \\ -0.518 & 0.896 \end{bmatrix} \begin{bmatrix} H \\ V \end{bmatrix} \qquad (9.5b)$$

Thus, with $H = 10$ and $V = -5$

$$p_1 = 0.732 \times 10 - 0.732 \times 5 = 3.66 \text{ kN (tension)}$$
$$p_2 = -0.518 \times 10 - 0.896 \times 5 = -9.66 \text{ kN (compression)}$$

Member 1 is 1000 mm and its extension δ_1 is

$$\delta_1 = \frac{1000 \times 3.66}{200 \times 100} = 0.183 \text{ mm}$$

Member BC is $0.5\sqrt{2} = 0.707$ m $= 707$ mm.

Its contraction δ_2 is

$$\delta_2 = \frac{707 \times -9.66}{200 \times 200} = -0.1707 \text{ mm}$$

Using equations (9.5), $X = B^T\delta$ becomes

$$\begin{bmatrix} x \\ y \end{bmatrix} = \begin{bmatrix} 0.732 & -0.518 \\ 0.732 & 0.896 \end{bmatrix} \begin{bmatrix} \delta_1 \\ \delta_2 \end{bmatrix}$$

∴ $x = 0.732 \times 0.183 + 0.518 \times 0.1707 = 0.222$ mm

and $y = 0.732 \times 0.183 - 0.896 \times 0.1707 = -0.019$ mm

The horizontal deflection x of point B is to the right and its vertical deflection is downwards.

Example 2. Calculate the horizontal deflection at joint C of the pin jointed frame shown in *Figure 9.2a*. All the members have area A of 100 mm² and $E = 200$ kN/mm² throughout.

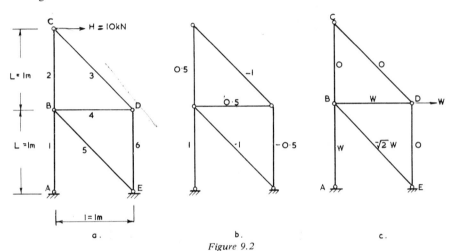

Figure 9.2

Answer: Resolving the forces at each joint horizontally and vertically, we obtain

$$\begin{bmatrix} p_1 \\ p_2 \\ p_3 \\ p_4 \\ p_5 \\ p_6 \end{bmatrix} = \begin{bmatrix} 2 \\ 1 \\ -\sqrt{2} \\ 1 \\ -\sqrt{2} \\ -1 \end{bmatrix} [H] = \begin{bmatrix} 20 \\ 10 \\ -14.14 \\ 10 \\ -14.14 \\ -10 \end{bmatrix} \text{ kN} \qquad (9.6)$$

and $B^T = [2 \quad 1 \quad -\sqrt{2} \quad 1 \quad -\sqrt{2} \quad -1]$

The member extensions are calculated by Hooke's law. These extensions are indicated in *Figure 9.2b*. From $X = B^T\delta$, the horizontal deflection of joint C is given as

$$x_c = [2 \quad 1 \quad -\sqrt{2} \quad 1 \quad -\sqrt{2} \quad -1] \{1 \quad 0.5 \quad -1 \quad 0.5 \quad -1 \quad -0.5\}$$

Thus $x_c = 2 + 0.5 + \sqrt{2} + 0.5 + \sqrt{2} + 0.5 = 6.328$ mm.

9.4. Deflection of unloaded joints

To obtain the deflection of an unloaded joint in a given direction, a load W is applied to that joint in the direction of the required deflection. The same procedure of sections 9.2 and 9.3 is then followed and when the required deflection is obtained in terms of the actual external loads and W, it is remembered that W is zero.

For instance, for the frame shown in *Figure 9.2*, if it is required to calculate the horizontal deflection at joint D, a force W is applied to joint D as shown in *Figure 9.2c*. The member forces due to W are indicated in this figure and equations $\mathbf{P} = \mathbf{B}\mathbf{W}$ due to the external force H and the extra force W now become

$$\begin{bmatrix} p_1 \\ p_2 \\ p_3 \\ p_4 \\ p_5 \\ p_6 \end{bmatrix} = \begin{bmatrix} 2 & 1 \\ 1 & 0 \\ -\sqrt{2} & 0 \\ 1 & 1 \\ -\sqrt{2} & -\sqrt{2} \\ -1 & 0 \end{bmatrix} \begin{bmatrix} H \\ W \end{bmatrix}$$

Thus

$$\mathbf{B}^T = \begin{bmatrix} 2 & 1 & -\sqrt{2} & 1 & -\sqrt{2} & -1 \\ 1 & 0 & 0 & 1 & -\sqrt{2} & 0 \end{bmatrix}$$

and equations $\mathbf{X} = \mathbf{B}^T \boldsymbol{\delta}$ are

$$\begin{bmatrix} x_C \\ x_D \end{bmatrix} = \begin{bmatrix} 2 & 1 & -\sqrt{2} & 1 & -\sqrt{2} & -1 \\ 1 & 0 & 0 & 1 & -\sqrt{2} & 0 \end{bmatrix} \begin{bmatrix} \boldsymbol{\delta} \end{bmatrix}$$

Where $\boldsymbol{\delta}$ is the vector of member extensions due to the actual external load H as indicated in *Figure 9.2b*.
Thus

$$x_D = [1 \ \ 0 \ \ 0 \ \ 1 \ \ -\sqrt{2} \ \ 0] \ \{1 \ \ 0.5 \ \ -1 \ \ 0.5 \ \ -1 \ \ -0.5\}$$

$$x_D = 1 + 0 \times 0.5 + 0 \times -1 + 1 \times 0.5 + \sqrt{2} \times 1 - 0.5 \times 0$$

$$x_D = 2.914 \text{ mm}.$$

9.5. Deflection of a structure with non-linear material characteristics

In the derivation of equations 9.3 or 9.4, no restrictions were imposed on the load-extension characteristics of the members of the pin jointed frames. In fact, the method used for evaluating the joint deflections can be applied to structures with non-linear material properties, provided that the displacements are sufficiently small to produce a negligible distortion of the shape of the structure. Matrix $\mathbf{B}$ and hence $\mathbf{B}^T$ does not alter and the non-linear relation $\boldsymbol{\delta} = \mathbf{fp}$ only influences the value of the extension δ of each member.

As an example, consider that the displacement δ in a member of the frame shown in *Figure 9.2* is given by the non-linear relationship

$$\delta = ap + bp^3 \tag{9.7}$$

where a and b are constants and p is the force in the member. Using equations (9.6) for the member forces, the member extensions are calculated, from equation (9.7) as

$$\delta_1 = 2aH + b(2H)^3 = 2aH + 8bH^3$$
$$\delta_2 = aH + b(H)^3 = aH + bH^3 \text{ and so on, thus}$$

$$\begin{bmatrix} \delta_1 \\ \delta_2 \\ \delta_3 \\ \delta_4 \\ \delta_5 \\ \delta_6 \end{bmatrix} = \begin{bmatrix} 2aH + 8bH^3 \\ aH + bH^3 \\ -\sqrt{2} \times aH - 2\sqrt{2} \times H^3 b \\ aH + bH^3 \\ -\sqrt{2} \times aH - 2\sqrt{2} \times bH^3 \\ -aH - bH^3 \end{bmatrix}$$

From $\mathbf{X} = \mathbf{B}^T \boldsymbol{\delta}$, the horizontal deflection x_c of joint C is calculated as

$$x_c = \begin{bmatrix} 2 & 1 & -\sqrt{2} & 1 & -\sqrt{2} & -1 \end{bmatrix} \begin{bmatrix} 2aH + 8bH^3 \\ aH + bH^3 \\ -\sqrt{2} \times aH - 2\sqrt{2} \times bH^3 \\ aH + bH^3 \\ -\sqrt{2} \times aH - 2\sqrt{2} \times bH^3 \\ -aH - bH^3 \end{bmatrix}$$

$$\therefore x_c = 11aH + 27bH^3$$

9.6. Examples

Example 1. Calculate the horizontal deflection at joints D and E for the frame shown in *Figure 9.3a*. The member deformation δ is given by the non-linear hyperbolic relationship

$$\delta = 0.003p/(1670 - p) \tag{9.8}$$

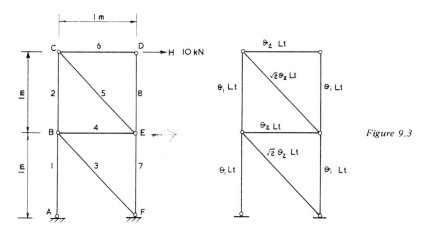

a. Frame and loading

b. Member extensions due to temperature rise

Figure 9.3

Answer: Applying a load W at E and resolving the forces at the joints horizontally and vertically, we obtain

$$\begin{bmatrix} p_1 \\ p_2 \\ p_3 \\ p_4 \\ p_5 \\ p_6 \\ p_7 \\ p_8 \end{bmatrix} = \begin{bmatrix} 2 & 1 \\ 1 & 0 \\ -\sqrt{2} & -\sqrt{2} \\ 1 & 1 \\ -\sqrt{2} & 0 \\ 1 & 0 \\ -1 & 0 \\ 0 & 0 \end{bmatrix} \begin{bmatrix} H \\ W \end{bmatrix}$$

For member 1 the deformation δ_1 due to the external force H is

$$\delta_1 = \frac{0.003 \times 2 \times 10}{1670 - 2 \times 10} = 0.000\,036 \text{ mm}$$

For member 2,

$$\delta_2 = \frac{0.003 \times 10}{1670 - 10} = 0.000\,018 \text{ mm}$$

For member 3,

$$\delta_3 = \frac{0.003 \times -14.14}{1670 + 14.14} = -0.000\,025 \text{ mm}$$

Thus

$$\delta = 10^{-6} \{36 \quad 18 \quad -25 \quad 18 \quad -25 \quad 18 \quad -18 \quad 0\} \text{ mm}$$

Equations $X = B^T\delta$ becomes

$$\begin{bmatrix} x_D \\ x_E \end{bmatrix} = \begin{bmatrix} 2 & 1 & -\sqrt{2} & 1 & -\sqrt{2} & 1 & -1 & 0 \\ 1 & 0 & -\sqrt{2} & 1 & 0 & 0 & 0 & 0 \end{bmatrix} 10^{-6} \begin{bmatrix} 36 \\ 18 \\ -25 \\ 18 \\ -25 \\ 18 \\ -18 \\ 0 \end{bmatrix} \quad (9.9)$$

$\therefore x_D = 10^{-6}(2 \times 36 + 1 \times 18 - \sqrt{2} \times -25 + 1 \times 18 - \sqrt{2} \times -25 + 1 \times 18 +$
$+ 1 \times 18)$

$$x_D = 214.7 \times 10^{-6} \text{ mm}$$

and

$$x_E = 10^{-6}(1 \times 36 - \sqrt{2} \times -25 + 1 \times 18)$$
$$= 89.36 \times 10^{-6} \text{ mm}$$

Example 2. How much does the horizontal deflection at D increase if the temperature of the frame rises by 60°? The coefficient of linear expansion of the vertical members is 4×10^{-6} per degree centigrade and that for the other members is 3×10^{-6}.
Answer: The increase in length of a member due to a temperature rise of $t°C$ is $\theta L t$ where L is the length and θ is the coefficient of linear expansion. With $\theta_1 = 4 \times 10^{-6}$ for the vertical members and $\theta_2 = 3 \times 10^{-6}$ for the other members, the member extensions δ are, as shown in *Figure 9.3b*, given as

$$\delta = \{0.24 \quad 0.24 \quad 0.255 \quad 0.18 \quad 0.255 \quad 0.18 \quad 0.24 \quad 0.24\}$$

Using these values, the joint deflection x_D is calculated from $X = B^T\delta$, where B^T is given by the first row of matrix B^T in equation (9.9). Thus

$$x_D = 2 \times 0.24 + 1 \times 0.24 - \sqrt{2} \times 0.255 + 1 \times 0.18 - \sqrt{2} \times$$
$$\times 0.255 + 1 \times 0.18 - 1 \times 0.24 = 0.12 \text{ mm}$$

9.7. Analysis of hyperstatic frames by the force method

The method for calculating the deflections of an isostatic frame is often used to analyse hyperstatic ones. Consider the frame shown in *Figure 9.4a* which is hyperstatic because it has more than three unknown reaction components. This frame can be rendered isostatic if one of the reaction components, such as V_D, is removed, as shown in *Figure 9.4b*. The vertical deflection y_{DW} at D due to the external loads, for this isostatic frame, is first calculated by the method given earlier in this chapter. The external loads are then removed and a load V_D is applied to the isostatic frame so that it acts in the same direction as y_{DW} as shown in *Figure 9.4c*. The frame is analysed again and the vertical deflection y_{DV} is calculated. This will be in terms of V_D.

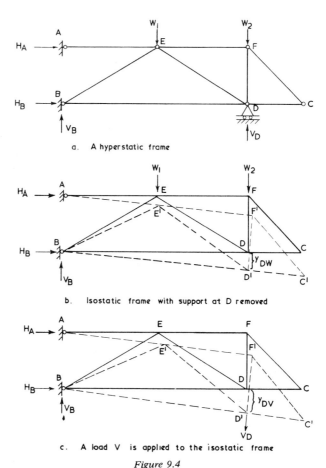

a. A hyperstatic frame

b. Isostatic frame with support at D removed

c. A load V is applied to the isostatic frame

Figure 9.4

Now in the original hyperstatic frame, *Figure 9.9a*, the deflection at D is zero. Thus when the reaction at D is acting, the sum of the above two deflections must add up to zero, hence

$$y_{DW} + y_{DV} = 0 \qquad (9.10)$$

This equation is solved to calculate the value of the reaction V_D. Once this is done, the member forces are calculated in the usual manner.

The same procedure is used when there is more than one redundant force. The frame shown in *Figure 9.5a*, for instance, has two redundant reaction components. Consider that the components H_C and V_D are chosen as redundants. To find their values, they are removed to obtain the isostatic frame shown in *Figure 9.5b*. This

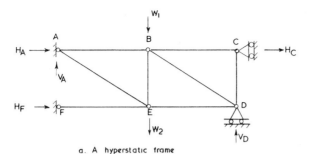

a. A hyperstatic frame

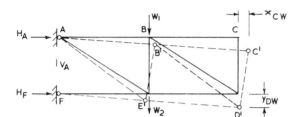

b. Isostatic frame with supports at C and D removed.

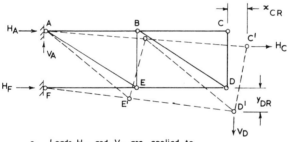

c. Loads H_C and V_D are applied to the isostatic frame

Figure 9.5

is analysed and the value of the deflections x_{CW} and y_{DW} due to the external loads are calculated. Next the external loads are removed and the isostatic frame is subjected to two forces H_C and V_D as shown in *Figure 9.5c*. The deflections x_{CR} and y_{DR} due to H_C and V_D are calculated. These will be in terms of H_C and V_D. The values of the forces H_C and V_D are then calculated from the fact that in the original hyperstatic frame, the horizontal deflection at C and the vertical deflection at D are zero. Thus

and
$$\left.\begin{aligned} x_{CW} + x_{CR} &= 0 \\ y_{DW} + y_{DR} &= 0 \end{aligned}\right\} \qquad (9.11)$$

This method is known as 'the force method' because the unknowns are the redundant forces.

9.8. Examples

Example 1. Calculate the member forces in the hyperstatic pin jointed frame shown in *Figure 9.6a*. EA is constant throughout.

Answer: Select V_D as the redundant reaction component. Remove the support at D to obtain the isostatic frame shown in *Figure 9.6b*. For this frame, to calculate

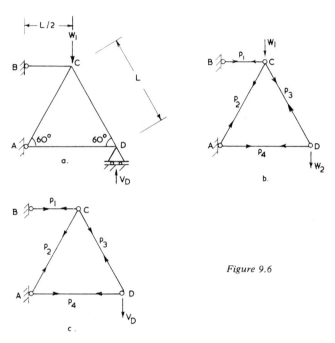

Figure 9.6

the vertical deflection y_{DW} at D, a force W_2 is applied there and it is later remembered that the value of this force is zero.

At D resolving vertically

$$p_3 \times 0.5\sqrt{3} - W_2 = 0$$

∴
$$p_3 = 2W_2/\sqrt{3}$$

At D resolving horizontally

$$p_4 + 0.5\,p_3 = 0$$

∴
$$p_4 = -W_2/\sqrt{3}$$

At C resolving vertically

$$W_1 + 0.5\sqrt{3} \times p_2 + 0.5\sqrt{3} \times p_3 = 0$$

∴
$$p_2 = -2W_1/\sqrt{3} - 2W_2/\sqrt{3}$$

At C resolving horizontally

$$p_1 + 0.5 p_2 - 0.5 p_3 = 0$$

$$\therefore \quad p_1 = W_1/\sqrt{3} + 2W_2/\sqrt{3}$$

Thus **P** = **B W** becomes

$$\begin{bmatrix} p_1 \\ p_2 \\ p_3 \\ p_4 \end{bmatrix} = (1/\sqrt{3}) \begin{bmatrix} 1 & 2 \\ -2 & -2 \\ 0 & 2 \\ 0 & -1 \end{bmatrix} \begin{bmatrix} W_1 \\ W_2 \end{bmatrix} \qquad (9.12)$$

and $\mathbf{B}^T = (1/\sqrt{3}) \begin{bmatrix} 1 & -2 & 0 & 0 \\ 2 & -2 & 2 & -1 \end{bmatrix}$

The member deformations **δ** are obtained from Hooke's law with $W_2 = 0$, thus

$$\delta_1 = (0.5L/EA) \times W_1/\sqrt{3} = W_1 L/(2\sqrt{3} \times EA)$$

$$\delta_2 = (L/EA) \times -2W_1/\sqrt{3} = -2W_1 L/(\sqrt{3} \times EA)$$

$$\delta_3 = \delta_4 = 0$$

The vertical deflections y_{CW} and y_{DW} are calculated from $\mathbf{X} = \mathbf{B}^T \boldsymbol{\delta}$, thus

$$\begin{bmatrix} y_C \\ y_D \end{bmatrix} W = (1/\sqrt{3}) \begin{bmatrix} 1 & -2 & 0 & 0 \\ 2 & -2 & 2 & -1 \end{bmatrix} \times LW_1/(\sqrt{3} \times EA) \times \begin{bmatrix} 0.5 \\ -2 \\ 0 \\ 0 \end{bmatrix}$$

$$\therefore \quad y_{DW} = (1/\sqrt{3}) \times [LW_1/(\sqrt{3} \times EA)][2 \times 0.5 + (-2) \times (-2)]$$

$$y_{DW} = 5W_1 L/(3EA)$$

The external load W_1 is now removed and a force V_D is applied at D to the isostatic frame as shown in *Figure 9.6c*. For this, resolving at D vertically

$$0.5\sqrt{3} \times p_3 - V_D = 0$$

$$\therefore \quad p_3 = 2V_D/\sqrt{3}$$

At D resolving horizontally

$$p_4 + 0.5p_3 = 0$$

$$\therefore \quad p_4 = -V_D/\sqrt{3}$$

At C resolving vertically

$$0.5\sqrt{3} \times p_2 + 0.5\sqrt{3} \times p_3 = 0$$

$$\therefore \quad p_2 = -2V_D/\sqrt{3}$$

At C resolving horizontally

$$p_1 + 0.5p_2 - 0.5p_3 = 0$$

$$\therefore \quad p_1 = 2V_D/\sqrt{3}$$

Thus **P = B W** becomes

$$\begin{bmatrix} p_1 \\ p_2 \\ p_3 \\ p_4 \end{bmatrix} = (1/\sqrt{3}) \begin{bmatrix} 2 \\ -2 \\ 2 \\ -1 \end{bmatrix} \begin{bmatrix} V_D \end{bmatrix} \text{ and } \mathbf{B}^T = (1/\sqrt{3}) \, [2 \ -2 \ 2 \ -1]$$

The member deformations δ due to V_D are

$$\delta_1 = 0.5L/EA \times 2V_D/\sqrt{3} = V_D L/(\sqrt{3} \times EA)$$

$$\delta_2 = L/EA \times -2V_D/\sqrt{3} = -2V_D L/(\sqrt{3} \times EA)$$

$$\delta_3 = L/EA \times 2V_D/\sqrt{3} = 2V_D L/(\sqrt{3} \times EA)$$

$$\delta_4 = L/EA \times -V_D/\sqrt{3} = -V_D L/(\sqrt{3} \times EA)$$

Thus from $\mathbf{X} = \mathbf{B}^T \boldsymbol{\delta}$

$$y_{DV} = (1/\sqrt{3}) \, [2 \ -2 \ 2 \ -1] \times V_D L/(\sqrt{3}EA) \begin{bmatrix} 1 \\ -2 \\ 2 \\ -1 \end{bmatrix}$$

Analysis of Structures by the Force Method

which gives

$$y_{DV} = [V_D L/(3EA)] (2 \times 1 - 2 \times -2 + 2 \times 2 - 1 \times -1)$$

$$y_{DV} = 11 V_D L/(3EA)$$

Now in the hyperstatic frame shown in *Figure 9.6a*, the vertical deflection at D is zero, thus

$$y_{DW} + y_{DV} = 0$$

i.e. $5W_1 L/(3EA) + 11 V_D L/(3EA) = 0$

Hence $V_D = -5W_1/11$ (i.e. upwards).

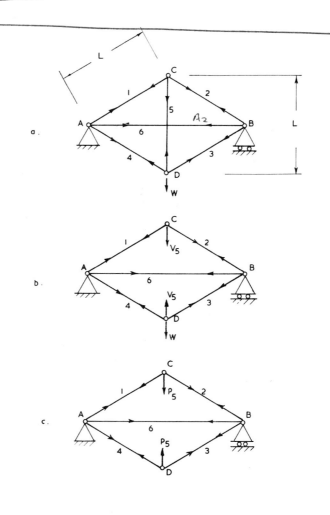

Figure 9.7

To calculate the member forces in the hyperstatic frame, *Figure 9.6a*, the value $V_D = -5W_1/11$ is substituted for W_2 in equations (9.12). These give

$$p_1 = (1/\sqrt{3})(1 \times W_1 + 2 \times -5W_1/11) = +W_1/(11\sqrt{3})$$

$$p_2 = (1/\sqrt{3})(-2 \times W_1 - 2 \times -5W_1/11) = -12W_1/(11\sqrt{3})$$

$$p_3 = (1/\sqrt{3})(0 \times W_1 + 2 \times -5W_1/11) = -10W_1/(11\sqrt{3})$$

and

$$p_4 = (1/\sqrt{3})(0 \times W_1 - 1 \times -5W_1/11) = +5W_1/(11\sqrt{3})$$

Example 2. Calculate the force in member 5 of the frame shown in *Figure 9.7a*. Members 1 to 5 have area A_1 while member 6 has area A_2. E is constant.
Answer: This frame is externally isostatic having only three reaction components. However, internally it is hyperstatic because it has 6 members and only 4 joints. Equation (1.9) is not satisfied and $m - (2j - 3) = 1$. There is therefore one redundant member. Since the force in member 5 is required, the force p_5 in this member is taken as redundant.

The procedure to find p_5 is to remove member 5, to obtain the isostatic frame shown in *Figure 9.7b*. Two equal and opposite forces V_5 are applied to the frame and the relative movement y_{DW} between joints C and D is calculated remembering that $V_5 = 0$. Next the external load is removed and the frame is analysed under the action of forces p_5 acting at C and D as shown in *Figure 9.7c*. The relative movement y_D between joints C and D are again calculated. the value of the force p_5 in member 5 is obtained by the fact that $y_{DW} + y_{Dp}$ in the hyperstatic frame is zero. This is because member 5 prevents any relative movement from taking place between joints C and D except that due to the deformation in member 5 which is taken into consideration during the calculation.

The analysis of the isostatic frame in *Figure 9.7b* gives

$$\begin{bmatrix} p_1 \\ p_2 \\ p_3 \\ p_4 \\ p_5 \\ p_6 \end{bmatrix} = \begin{bmatrix} 0 & -1 \\ 0 & -1 \\ 1 & -1 \\ 1 & -1 \\ 0 & 1 \\ -\sqrt{3}/2 & \sqrt{3} \end{bmatrix} \begin{bmatrix} W \\ V_5 \end{bmatrix}$$

Thus $\mathbf{B}^T = \begin{bmatrix} 0 & 0 & 1 & 1 & 0 & -\sqrt{3}/2 \\ -1 & -1 & -1 & -1 & 1 & \sqrt{3} \end{bmatrix}$

With $V_5 = 0$, the member deformations become

$$\delta_1 = 0, \; \delta_2 = 0, \; \delta_3 = WL/EA_1, \; \delta_4 = WL/EA_1, \; \delta_5 = 0,$$
$$\delta_6 = (L\sqrt{3}/EA_2) \times (-0.5\sqrt{3} \times W) = -3WL/(2EA_2)$$

Analysis of Structures by the Force Method

The equation $X = B^T \delta$ then becomes

$$\begin{bmatrix} y_D \\ y_{DW} \end{bmatrix} = \begin{bmatrix} 0 & 0 & 1 & 1 & 0 & -0.5\sqrt{3} \\ -1 & -1 & -1 & -1 & 1 & \sqrt{3} \end{bmatrix} (LW/E) \begin{bmatrix} 0 \\ 0 \\ 1/A_1 \\ 1/A_1 \\ 0 \\ -3/2A_2 \end{bmatrix}$$

which gives

$$y_{DW} = \frac{WL}{E}\left(-\frac{1}{A_1} - \frac{1}{A_1} - \frac{3\sqrt{3}}{2A_2}\right) = -\frac{WL(4A_2 + 3\sqrt{3} \times A_1)}{2A_1 A_2 E}$$

The analysis of the isostatic frame shown in Figure 9.7c gives

$$\begin{bmatrix} p_1 \\ p_2 \\ p_3 \\ p_4 \\ p_5 \\ p_6 \end{bmatrix} = \begin{bmatrix} -1 \\ -1 \\ -1 \\ -1 \\ 1 \\ \sqrt{3} \end{bmatrix} [p_5]$$

Notice that the force p_5 in member 5 is included in the equations $P = BW$. In this manner its deformation is included in the calculations.
The member deformations are

$$\delta_1 = -p_5 L/EA_1 = \delta_2 = \delta_3 = \delta_4, \quad \delta_5 = p_5 L/EA_1$$

$$\delta_6 = \frac{\sqrt{3} \times L}{EA_2} \times \sqrt{3} \times p_5 = 3p_5 L/EA_2$$

Thus

$$y_{Dp} = \begin{bmatrix} -1 & -1 & -1 & -1 & 1 & \sqrt{3} \end{bmatrix}(p_5 L/E)\begin{bmatrix} -1/A_1 \\ -1/A_1 \\ -1/A_1 \\ -1/A_1 \\ 1/A_1 \\ 3/A_2 \end{bmatrix}$$

Hence

$$y_{Dp} = \frac{p_5 L(5A_2 + 3\sqrt{3} \times A_1)}{A_1 A_2 E}$$

Thus

$$y_{DW} + y_{Dp} = \frac{-WL(4A_2 + 3\sqrt{3} \times A_1)}{2A_1 A_2 E} + \frac{p_5 L(5A_2 + 3\sqrt{3} \times A_1)}{A_1 A_2 E} = 0$$

$$p_5 = \frac{(4A_2 + 3\sqrt{3} \times A_1)W}{2(5A_2 + 3\sqrt{3} \times A_1)}$$

9.9. Deflection of bent members

The deflection of isostatic bent members is also obtained by the equations $X = B^T \delta$. Here δ is the vector of rotations at the ends of the members and matrix B is obtained by calculating the bending moments at each end of the members.

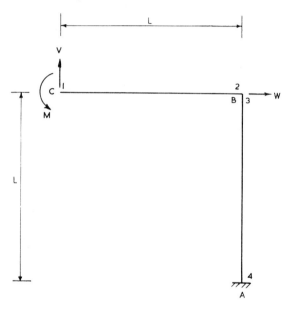

Figure 9.8

As an example, the bent member ABC, shown in *Figure 9.8*, is subject to a vertical force V and a moment M at C as shown in the figure. The vertical deflection y_c at C, the rotation θ_c and the horizontal deflection x_B at B are calculated as follows. A force W is applied at B and the bending moments at points 1, 2, 3 and 4 are calculated, thus

$$\left. \begin{array}{l} M_1 = M \\ M_2 = M - VL \\ M_3 = M - VL \\ M_4 = M - VL - WL \end{array} \right\} \tag{9.13a}$$

Thus **P = B W** becomes

$$\begin{bmatrix} M_1 \\ M_2 \\ M_3 \\ M_4 \end{bmatrix} = \begin{bmatrix} 1 & 0 & 0 \\ 1 & -L & 0 \\ 1 & -L & 0 \\ 1 & -L & -L \end{bmatrix} \begin{bmatrix} M \\ V \\ W \end{bmatrix} \qquad (9.13b)$$

The flexibility equations for the members are

$$\theta_1 = (L/6EI)(2M_1 + M_2)$$
$$\theta_2 = (L/6EI)(M_1 + 2M_2)$$
$$\theta_3 = (L/6EI)(2M_3 + M_4)$$
$$\theta_4 = (L/6EI)(M_3 + 2M_4)$$

Using equations (9.13), with $W = 0$, we obtain

$$\theta_1 = (L/6EI)(3M - VL)$$
$$\theta_2 = (L/6EI)(3M - 2VL)$$
$$\theta_3 = (L/6EI)(3M - 3VL)$$
$$\theta_4 = (L/6EI)(3M - 3VL)$$

These values of δ and the transpose of matrix **B** given by equations (9.13), gives

$$\begin{bmatrix} \theta_C \\ y_C \\ x_B \end{bmatrix} = \begin{bmatrix} 1 & 1 & 1 & 1 \\ 0 & -L & -L & -L \\ 0 & 0 & 0 & -L \end{bmatrix} (L/6EI) \begin{bmatrix} 3M - VL \\ 3M - 2VL \\ 3M - 3VL \\ 3M - 3VL \end{bmatrix}$$

∴ $\theta_C = (L/6EI)(3M - VL + 3M - 2VL + 3M - 3VL + 3M - 3VL)$

$\theta_C = L(2M - 1.5VL)/(EI)$
$y_C = (-L^2/6EI)(3M - 2VL + 3M - 3VL + 3M - 3VL)$
$y_C = (8VL^3 - 9ML^2)/(6EI)$

and $x_B = (-L^2/6EI)(3M - 3VL) = -[ML^2/(2EI)] + [VL^3/(2EI)]$

9.10. Examples

Example 1. Calculate the vertical and the horizontal deflections at joint A for the bent member ABCD shown in *Figure 9.9a*. What are these deflections when $H = -10$ kN?

Answer: The bending moments at points 1 to 6 are

$$M_1 = 0$$
$$M_2 = HL - VL$$
$$M_3 = HL - VL$$
$$M_4 = HL - 2VL$$
$$M_5 = HL - 2VL$$
$$M_6 = -2VL$$

It is noticed that the substitution $V = 2H$ is not made. The columns corresponding to H and V in matrix **B** are thus kept separate. In this manner matrix $\mathbf{B}^T$ will have two rows corresponding to x_A and y_A. The equations $\mathbf{P} = \mathbf{B}\,\mathbf{W}$ then become

$$\begin{bmatrix} M_1 \\ M_2 \\ M_3 \\ M_4 \\ M_5 \\ M_6 \end{bmatrix} = \begin{bmatrix} 0 & 0 \\ L & -L \\ L & -L \\ L & -2L \\ L & -2L \\ 0 & -2L \end{bmatrix} \begin{bmatrix} H \\ V \end{bmatrix}$$

Using these the flexibility equations become

$$\theta_1 = (\sqrt{2} \times L/6EI)(2M_1 + M_2) = (\sqrt{2} \times L/6EI)(2 \times 0 + HL - VL)$$
$$= (L/6EI)(\sqrt{2} \times LH - \sqrt{2} \times LV)$$
$$\theta_2 = (\sqrt{2} \times L/6EI)(2M_2 + M_1) = (\sqrt{2} \times L/6EI)(2HL - 2VL)$$
$$= (L/6EI)(2\sqrt{2} \times LH - 2\sqrt{2} \times LV)$$
$$\theta_3 = (L/6EI)(2M_3 + M_4) = (L/6EI)(3LH - 4LV)$$
$$\theta_4 = (L/6EI)(2M_4 + M_3) = (L/6EI)(3LH - 5LV)$$
$$\theta_5 = (L/6EI)(2M_5 + M_6) = (L/6EI)(2LH - 6LV)$$
$$\theta_6 = (L/6EI)(2M_6 + M_5) = (L/6EI)(LH - 6LV)$$

Thus $\mathbf{X} = \mathbf{B}^T \boldsymbol{\delta}$ becomes

$$\begin{bmatrix} x_A \\ y_A \end{bmatrix} = \begin{bmatrix} 0 & L & L & L & L & 0 \\ 0 & -L & -L & -2L & -2L & -2L \end{bmatrix} (L/6EI) \begin{bmatrix} \sqrt{2} \times LH - \sqrt{2} \times LV \\ 2\sqrt{2} \times LH - 2\sqrt{2} \times LV \\ 3LH - 4LV \\ 3LH - 5LV \\ 2LH - 6LV \\ LH - 6LV \end{bmatrix}$$

(9.14)

$$\therefore x_A = (L/6EI)(2\sqrt{2} \times L^2H - 2\sqrt{2} \times L^2V + 3L^2H - 4L^2V + 3L^2H - 5L^2V + 2L^2H - 6L^2V)$$

$$x_A = (L/6EI)[(8 + 2\sqrt{2})L^2H - (15 + 2\sqrt{2})L^2V]$$

and

$$y_A = (L/6EI)(-2\sqrt{2}L^2H + 2\sqrt{2}L^2V - 3L^2H + 4L^2V - 6L^2H + 10L^2V - 4L^2H + 12L^2V - 2L^2H + 12L^2V)$$

$$y_A = (L/6EI)[(38 + 2\sqrt{2})L^2V - (15 + 2\sqrt{2})L^2H]$$

With $V = 2H$

and

$$x_A = -2(11 + \sqrt{2})HL^3/6EI$$

$$y_A = (61 + 2\sqrt{2})HL^3/6EI$$

and when $H = -10$, $V = -20$; i.e. H is to the left and V is downwards

$$x_A = 20(11 + \sqrt{2})L^3/6EI \text{ (to the right)}$$

and

$$y_A = -10(61 + 2\sqrt{2})L^3/6EI \text{ (downwards)}$$

Example 2. Calculate the rotation of joint C for the bent member shown in Figure 9.9.

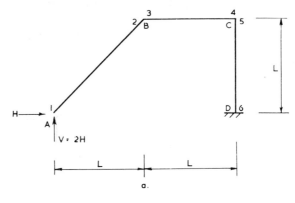

Figure 9.9

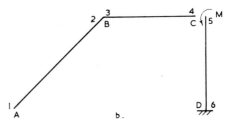

Answer: Apply a moment M at C as shown in *Figure 9.9b*. The bending moments at points 1 to 6 due to M are

$$M_1 = M_2 = M_3 = M_4 = 0, \quad M_5 = M_6 = M$$

An extra equation is added to equations (9.14) which is of the form

$$\theta_C = [0 \quad 0 \quad 0 \quad 0 \quad 1 \quad 1](L/6EI)\begin{bmatrix} 2\sqrt{2} \times LH - \sqrt{2} \times LV \\ 2\sqrt{2} \times LH - 2\sqrt{2} \times LV \\ 3LH - 4LV \\ 3LH - 5LV \\ 2LH - 6LV \\ LH - 6LV \end{bmatrix}$$

$$\therefore \quad \theta_C = (L/6EI)(2LH - 6LV + LH - 6LV)$$
$$\theta_C = (L/6EI)(3LH - 12LV)$$

with $V = 2H$; i.e. V is upwards while H is to the right.

$$\theta_C = (L/6EI)(3LH - 24LH)$$
$$\theta_C = -7HL^2/2EI \text{ (clockwise)}$$

Example 3. The portal frame shown in *Figure 9.10* is pinned at A and supported on rollers at D. Calculate the horizontal deflection of point D due to a force H_D acting there.

Answer: The bending moments at points 1 to 6 are

$$M_1 = M_6 = 0, \quad M_2 = M_3 = M_4 = M_5 = hH_D$$

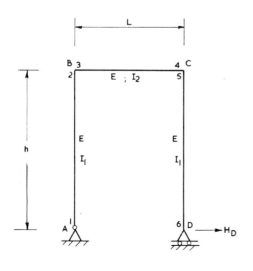

Figure 9.10

Thus **P** = **B W** becomes

$$\begin{bmatrix} M_1 \\ M_2 \\ M_3 \\ M_4 \\ M_5 \\ M_6 \end{bmatrix} = \begin{bmatrix} 0 \\ h \\ h \\ h \\ h \\ 0 \end{bmatrix} \begin{bmatrix} H_D \end{bmatrix} \qquad (9.15)$$

and

$$\mathbf{B}^T = \begin{bmatrix} 0 & h & h & h & h & 0 \end{bmatrix}$$

The flexibility equations **θ** = **f M** for the members are

$$\begin{bmatrix} \theta_1 \\ \theta_2 \\ \theta_3 \\ \theta_4 \\ \theta_5 \\ \theta_6 \end{bmatrix} = \begin{bmatrix} \dfrac{h}{3EI_1} & \dfrac{h}{6EI_1} & 0 & 0 & 0 & 0 \\ \dfrac{h}{6EI_1} & \dfrac{h}{3EI_1} & 0 & 0 & 0 & 0 \\ 0 & 0 & \dfrac{L}{3EI_2} & \dfrac{L}{6EI_2} & 0 & 0 \\ 0 & 0 & \dfrac{L}{6EI_2} & \dfrac{L}{3EI_2} & 0 & 0 \\ 0 & 0 & 0 & 0 & \dfrac{h}{3EI_1} & \dfrac{h}{6EI_1} \\ 0 & 0 & 0 & 0 & \dfrac{h}{6EI_1} & \dfrac{h}{3EI_1} \end{bmatrix} \begin{bmatrix} M_1 \\ M_2 \\ M_3 \\ M_4 \\ M_5 \\ M_6 \end{bmatrix}$$

(9.16)

Thus using equations (9.15)

$$\begin{bmatrix} \theta_1 \\ \theta_2 \\ \theta_3 \\ \theta_4 \\ \theta_5 \\ \theta_6 \end{bmatrix} = \begin{bmatrix} h^2/6EI_1 \\ h^2/3EI_1 \\ hL/2EI_2 \\ hL/2EI_2 \\ h^2/3EI_1 \\ h^2/6EI_1 \end{bmatrix} \begin{bmatrix} H_D \end{bmatrix}$$

The equations $X = B^T\delta$ therefore become

$$x_D = [0 \quad h \quad h \quad h \quad h \quad 0] \begin{bmatrix} h^2/6EI_1 \\ h^2/3EI_1 \\ hL/2EI_2 \\ hL/2EI_2 \\ h^2/3EI_1 \\ h^2/6EI_1 \end{bmatrix} [H_D]$$

Thus

$$x_D = H_D \left(\frac{h^3}{3EI_1} + \frac{h^2 L}{2EI_2} + \frac{h^2 L}{2EI_2} + \frac{h^3}{3EI_1} \right)$$

$$x_D = H_D \left(\frac{2h^3}{3EI_1} + \frac{h^2 L}{EI_2} \right) \qquad (9.17)$$

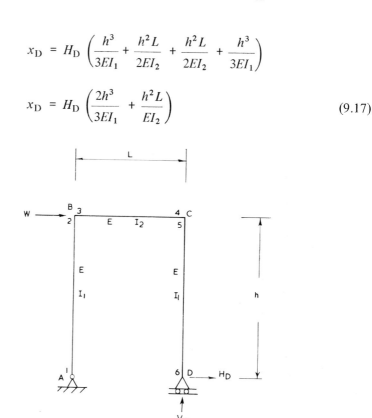

Figure 9.11

Example 4. The frame in the last example is subject to a horizontal force W at B as shown in *Figure 9.11*. Calculate the horizontal deflection at D.
Answer: Apply a horizontal force H_D at D. The vertical reaction V_D at D is obtained by taking moments about A, thus

$$V_D \times L - Wh = 0$$
$$\therefore \quad V_D = Wh/L$$

Analysis of Structures by the Force Method

The bending moments at points 1 to 6 are

$$M_1 = M_6 = 0$$
$$M_2 = M_3 = H_D h + V_D L = H_D h + WhL/L = H_D h + Wh$$
$$M_4 = M_5 = H_D h$$

Thus **P = B W** becomes

$$\begin{bmatrix} M_1 \\ M_2 \\ M_3 \\ M_4 \\ M_5 \\ M_6 \end{bmatrix} = \begin{bmatrix} 0 & 0 \\ h & h \\ h & h \\ 0 & h \\ 0 & h \\ 0 & 0 \end{bmatrix} \begin{bmatrix} W \\ H_D \end{bmatrix} \qquad (9.18)$$

and

$$\mathbf{B}^T = \begin{bmatrix} 0 & h & h & 0 & 0 & 0 \\ 0 & h & h & h & h & 0 \end{bmatrix}$$

The flexibility equations are given by equations (9.16). Using these together with equations (9.18), we obtain

$$\begin{bmatrix} \theta_1 \\ \theta_2 \\ \theta_3 \\ \theta_4 \\ \theta_5 \\ \theta_6 \end{bmatrix} = \begin{bmatrix} h^2/6EI_1 & h^2/6EI_1 \\ h^2/3EI_1 & h^2/3EI_1 \\ Lh/3EI_2 & Lh/2EI_2 \\ Lh/6EI_2 & Lh/2EI_2 \\ 0 & h^2/3EI_1 \\ 0 & h^2/6EI_1 \end{bmatrix} \begin{bmatrix} W \\ H_D \end{bmatrix}$$

The equations $\mathbf{X} = \mathbf{B}^T \boldsymbol{\delta}$ thus become

$$\begin{bmatrix} x_B \\ x_D \end{bmatrix} = \begin{bmatrix} 0 & h & h & 0 & 0 & 0 \\ 0 & h & h & h & h & 0 \end{bmatrix} \begin{bmatrix} h^2/6EI_1 & h^2/6EI_1 \\ h^2/3EI_1 & h^2/3EI_1 \\ Lh/3EI_2 & Lh/2EI_2 \\ Lh/6EI_2 & Lh/2EI_2 \\ 0 & h^2/3EI_1 \\ 0 & h^2/6EI_1 \end{bmatrix} \begin{bmatrix} W \\ H_D \end{bmatrix}$$

With $H_D = 0$

$$x_D = [0 \quad h \quad h \quad h \quad h \quad 0] \begin{bmatrix} h^2/6EI_1 \\ h^2/3EI_1 \\ Lh/3EI_2 \\ Lh/6EI_2 \\ 0 \\ 0 \end{bmatrix} [W]$$

i.e.

$$x_D = (h^3/3EI_1 + Lh^2/3EI_2 + Lh^2/6EI_2)W$$
$$x_D = W(h^3/3EI_1 + Lh^2/2EI_2) \tag{9.19}$$

Example 5. Calculate the reactions H_D and H_A for the pinned base portal shown in Figure 9.12.

Answer: In the last two examples the values of x_D were calculated when the support at D was on rollers. When the rollers are prevented from moving horizontally, support D acts as a pin and x_D becomes equal to zero. The horizontal reaction H_D for the pinned portal can therefore be calculated by equating the sum of the values of x_D obtained in the last two examples, to zero. Thus using equations (9.17) and (9.19)

$$H_D \times (2h^3/3EI_1 + h^2L/EI_2) + W(h^3/3EI_1 + Lh^2/2EI_2) = 0$$

which gives $H_D = -W/2$

Resolving horizontally

$$W + H_A + H_D = 0$$

$$\therefore \qquad H_A = -W - H_D = -W + W/2 = -W/2$$

The bending moment at C is $H_D h$ and that at B is $H_A h$, both causing tension to the right of the columns.

Thus, using the force method, a rigidly jointed hyperstatic frame is analysed as follows: first the redundant forces are selected and released and a basic isostatic frame is obtained. This new frame is analysed under two systems of loads. The first system consists of the external loads alone and the second system consists of the unknown redundant forces. The deflection of the isostatic frame in the direction of each redundant is thus calculated twice. In the actual hyperstatic frame these two deflections add up to zero and thus as many equations as the number of redundant forces are derived and solved for the unknown values of these redundants.

9.11. Assembly and thermal forces in hyperstatic frames

In the previous sections the analysis of hyperstatic frames were carried out by first removing the redundant members or redundant supports and treating them as if they were external forces. The resulting isostatic frames were then analysed on the assumption that when the redundant members or supports are replaced they fit into

Analysis of Structures by the Force Method 231

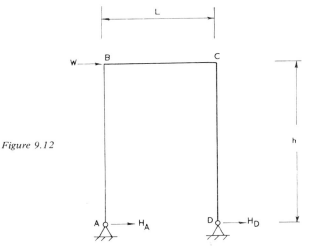

Figure 9.12

their positions exactly. In reality, some small imperfection is always introduced into every structure during fabrication or construction. Members are often slightly too short or too long, supports may be erected below or above their expected position. It is also possible that these supports settle because they are made out of elastic materials. All these introduce into the hyperstatic structure what is known as assembly or lack of fit forces.

In Chapter 2 it was stated that if the temperature of one or more of the members of a hyperstatic frame is increased, then member forces develop in these frames. Such forces, resulting from changes in the temperature or due to lack of fit must be superimposed on the member forces caused by the external loads. In this section a number of examples are solved to cover these topics.

Example 1. During the construction of the frame shown in *Figure 9.7a* (example 2, section 9.8), it was found that member CD is λ units of length too short.

(a) Calculate the assembly forces developed in the members when CD is forced into its position before the application of the external vertical load W. What is the extension in CD?
(b) If the external load W is also applied on this imperfect frame, calculate the total force in CD.
(c) What is the procedure to calculate the force in CD if this member is initially λ units too long?

Answer: (a) Member CD has to be extended before being pinned to the joints at C and D. When this operation is completed the member tends to contract back to its original length. This relieves some of the force in the member but causes forces to develop in the other members. Thus compressive forces develop in members 1 to 4 and a tensile force develops in member 6.

The tensile force applied to CD must be sufficient to cause a relative movement $y_{Dp} = \lambda$ between joints C and D. Now since W is not yet applied to the frame, y_{DW} is zero and from the results obtained in example 2 of section 9.8

$$y_{Dp} = \frac{p_5 L(5A_2 + 3\sqrt{3}A_1)}{A_1 A_2 E} = \lambda$$

∴ $p_5 = \lambda A_1 A_2 E / [L(5A_2 + 3\sqrt{3}A_1)]$

Due to this 'lack of fit' in member 5, the assembly forces in the members are calculated as in page 221, thus

$$\begin{bmatrix} p_1 \\ p_2 \\ p_3 \\ p_4 \\ p_5 \\ p_6 \end{bmatrix} = \frac{\lambda A_1 A_2 E}{L(5A_2 + 3\sqrt{3}A_1)} \begin{bmatrix} -1 \\ -1 \\ -1 \\ -1 \\ 1 \\ \sqrt{3} \end{bmatrix}$$

The final extension in member CD is

$$\delta_5 = \frac{p_5 L}{EA_1} = \frac{\lambda A_1 A_2 E}{L(5A_2 + 3\sqrt{3}A_1)} \times \frac{L}{EA_1} = \frac{\lambda A_2}{5A_2 + 3\sqrt{3}A_1}$$

(b)   If the external force W is also acting then the value of the force in member 5 is obtained from

$$y_{DW} + y_{Dp} = \lambda$$

and using the results of section 9.8 for y_{DW} and y_{Dp} (see page 221)

$$-\frac{WL(4A_2 + 3\sqrt{3}A_1)}{2A_1 A_2 E} + \frac{p_5 L(5A_2 + 3\sqrt{3}A_1)}{A_1 A_2 E} = \lambda$$

Hence

$$p_5 = [2A_1 A_2 E\lambda + WL(4A_2 + 3\sqrt{3}A_1)] / [2L(5A_2 + 3\sqrt{3}A_1)]$$

(c)   If member CD is initially λ units too long then p_5 is calculated from the equation

$$y_{DW} + y_{Dp} = -\lambda$$

The negative sign given to λ indicates that member 5 has to contract before it can be connected to the joints at C and D.

Example 2. In the pin jointed frame shown in *Figure 9.6a* (example 1, section 9.8), calculate the support reaction under the following conditions

(a)   The support at D was constructed λ units below the horizontal line AD.
(b)   The support at D was initially constructed λ units above the line AD.
(c)   The support at D is replaced by a vertical pin ended elastic column of length L', modulus of elasticity E' and area A'.

Answer: (a) In example 1 in section 9.8, the deflection y_{DV} was calculated by applying a vertical force V_D which was acting downwards as shown in *Figure 9.6c*. Here the support at D is λ units below the line AD. If the deflection of point D is y_{DW} due to the applied load W_1 and y_{DV} due to the force V_D (*Figure 9.6c*), then the sum of these two downward deflections cannot exceed λ.

i.e.
$$y_{DW} + y_{DV} = \lambda$$

$$\therefore \quad \frac{5W_1L}{3EA} + \frac{11V_DL}{3EA} = \lambda$$

This gives

$$V_D = (3\lambda EA - 5W_1L)/(11L)$$

(b) If the support at D is initially constructed λ units above the line AD, then joint D must move upwards by an amount λ before it can be connected to the support. Thus the two downward deflections y_{DW} and y_{DV} must add up to $-\lambda$. The negative sign indicates that the resulting deflection is measured in the opposite direction to the applied forces W_2 and V_D shown in *Figures 9.6b* and *c* respectively.

$$\therefore \quad y_{DW} + y_{DV} = -\lambda$$

i.e.
$$\frac{5W_1L}{3EA} + \frac{11V_DL}{3EA} = -\lambda$$

giving

$$V_D = -(3\lambda EA + 5W_1L)/(11L)$$

i.e. upwards. In both cases (a) and (b) the assembly forces developed in the frame are due to the lack of fit at support D.

(c) When the support at D is replaced by a vertical pin ended column, joint D remains free to move horizontally. Thus no horizontal reaction is developed at D and the new system is hyperstatic, with just one redundant force, this being the reactive axial force V_D developed in the column. The application of the load W_1 to the frame tends to push joint D down by an amount $y_{DW} + y_{DV}$. As a result settlement takes place at D. The amount of this settlement is λ which is equal to the contraction of the column at D, i.e. $V_DL'/(E'A')$

Hence

$$\frac{5W_1L}{3EA} + \frac{11V_DL}{3EA} = \lambda = \frac{V_DL'}{E'A'}$$

which gives

$$V_D = -5W_1LE'A'/(11E'A'L - 3EAL')$$

with $E = E'$, $A = A'$ and $L = L'$, for instance, $V_D = -5W_1/8$ which is compressive.

The force in each of the other members can now be calculated in the usual manner from $\mathbf{P} = \mathbf{BW}$.

Example 3. The force W acting on the frame shown in *Figure 9.7a* (example 2, section 9.8) is removed and then the temperature of the frame is increased by $t°C$. Calculate the resulting member forces. The coefficient of linear thermal expansion is θ for members 1, 2, 3 and 4 and ϕ for members 5 and 6.

Answer: When the temperature of the frame is increased a force p_5 is developed in the redundant member 5. Due to this, forces will develop in all the other members. From example 2, section 9.8, these forces are

$$\mathbf{P} = \{p_1 \quad p_2 \quad p_3 \quad p_4 \quad p_5 \quad p_6\}$$
$$= \{-1 \quad -1 \quad -1 \quad -1 \quad 1 \quad \sqrt{3}\} p_5$$

The corresponding member deformations are

$$\boldsymbol{\delta} = (p_5 L/E) \{-1/A_1 \quad -1/A_1 \quad -1/A_1 \quad -1/A_1 \quad 1/A_1 \quad 3/A_2\}$$

On the other hand the deformations of these members due to the temperature rise are

$$\boldsymbol{\delta}_t = \{\theta Lt \quad \theta Lt \quad \theta Lt \quad \theta Lt \quad \phi Lt \quad \phi Lt\sqrt{3}\}$$

There is no external force acting on the frame and as the members expand and the frame deflects, no external work is done. The work done by the internal member forces is

$$U = 0.5(\mathbf{P}^T\boldsymbol{\delta} + \mathbf{P}^T\boldsymbol{\delta}_t) = 0.5\mathbf{P}^T[\boldsymbol{\delta} + \boldsymbol{\delta}_t]$$

The virtual work equations (7.44b) become

$$\mathbf{P}^T[\boldsymbol{\delta} + \boldsymbol{\delta}_t] = 0$$

where $\boldsymbol{\delta}$ and $\boldsymbol{\delta}_t$ are column vectors. Thus

$$[-1 \quad -1 \quad -1 \quad -1 \quad 1 \quad \sqrt{3}]p_5 \times$$
$$\times (p_5 L/E)\{-1/A_1 \quad -1/A_1 \quad -1/A_1 \quad -1/A_1 \quad 1/A_1 \quad 3/A_2\} +$$
$$+ [-1 \quad -1 \quad -1 \quad -1 \quad 1 \quad \sqrt{3}]p_5 \times$$
$$\times \{\theta Lt \quad \theta Lt \quad \theta Lt \quad \theta Lt \quad \phi Lt \quad \phi Lt\sqrt{3}\} = 0$$

$$\therefore \quad p_5 = 4EA_1A_2 t(\theta - \phi)/(5A_2 + 3\sqrt{3}A_1)$$

The member forces now become

$$\{p_1 \ p_2 \ p_3 \ p_4 \ p_5 \ p_6\} = \frac{4EA_1A_2t(\theta - \phi)}{5A_2 + 3\sqrt{3}A_1} \begin{bmatrix} -1 \\ -1 \\ -1 \\ -1 \\ 1 \\ \sqrt{3} \end{bmatrix}$$

Example 4. In the two pinned portal shown in *Figure 9.12* (example 5, section 9.10) the support at D was initially constructed λ units to the right of end D of the frame. This end was then forced into its position before the load W was applied. Calculate the horizontal reaction at D after the application of W.

Answer: When calculating the horizontal movement x_D of point D in equations (9.17) and (9.19), it was assumed that the force at D was acting to the right. The final movement of end D due to H_D and W must now add up to $+\lambda$ which is a positive movement as it is also to the right. Using the results given by equations (9.17) and (9.19) we therefore establish

$$H_D \left(\frac{2h^3}{3EI_1} + \frac{h^2 L}{EI_2} \right) + W \left(\frac{h^3}{3EI_1} + \frac{Lh^2}{2EI_2} \right) = \lambda$$

$$\therefore \quad H_D = \left[\lambda - W \left(\frac{h^3}{3EI_1} + \frac{Lh^2}{2EI_2} \right) \right] \bigg/ \left(\frac{2h^3}{3EI_1} + \frac{h^2 L}{EI_2} \right)$$

The horizontal reaction at A is again calculated from the horizontal equilibrium of the whole frame, thus

$$H_A + H_D + W = 0$$

$$\therefore \quad H_A = -W - H_D = -(W + H_D)$$

Exercises on Chapter 9

1. The cross sectional areas of the members of the pin jointed frame shown in *Figure 9.13* are all equal to 100 mm^2. The length L is 2 m and $E = 200$ kN/mm^2. A vertical load of 10 kN is applied at C. Calculate the value of a vertical load that should be applied at E so that the vertical deflection of joint C under the combined effect of both loads is zero. What is the horizontal deflection of joint C?
Ans. 24.1 kN, 0.41 mm.

2. The members of the pin jointed frame shown in *Figure 9.14* all have an area of 100 mm^2. Calculate the vertical deflection of joint H. Take $E = 200$ kN/mm^2.
Ans. 69.67 mm.

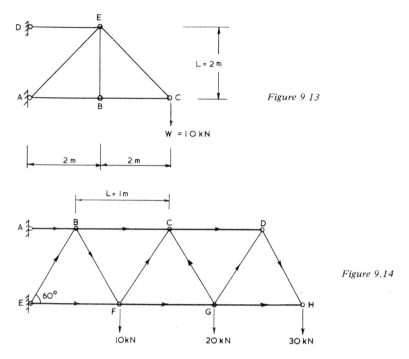

Figure 9.13

Figure 9.14

3. The bent member shown in *Figure 9.15* is made from a steel circular bar of radius 50mm and $E = 200$ kN/mm^2. Calculate the horizontal deflection at D.
Ans. 54.2 mm.

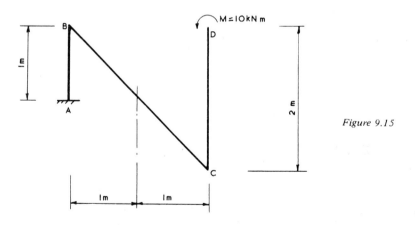

Figure 9.15

4. In *Figure 9.16* members AB and CD are welded together at B. The support at A is fixed. Calculate the vertical deflection of joint D. The second moment of area of the horizontal member is 10^6 mm^4 and that of the vertical member is 2×10^6 mm^4. Take $E = 200$ kN/mm^2.
Ans. +12.5 mm.

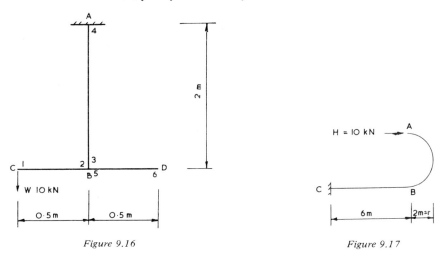

Figure 9.16   Figure 9.17

5. Calculate the vertical deflection of point B in the bent member shown in *Figure 9.17*. The member is of constant cross section with $I = 10^8 \text{ mm}^4$ and $E = 200 \text{ kN/mm}^2$.
Ans. 36 mm.

6. The ratio L/EA for all the members of the two pinned arch shown in *Figure 9.18* is constant. Calculate the horizontal reactions at A and B.
Ans. $H_A = -H_B = 118.3$ kN.

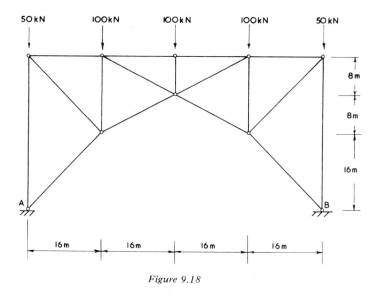

Figure 9.18

7. Calculate the forces in members AB and CD in the pin jointed frame shown in *Figure 9.21*. EA is constant.
Ans. $p_{AB} = -1.11$ kN, $p_{DC} = 1.11$ kN.

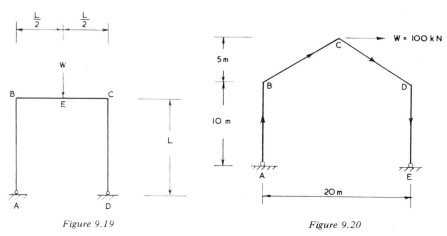

Figure 9.19   Figure 9.20

8. Calculate the reactions at A and D and the bending moments at B, E and C for the portal shown in *Figure 9.19*. Draw the bending moment diagram. EI is constant.
Ans. $V_A = V_D = W/2$

$H_A = -H_D = 3W/40$

$M_B = M_C = 3WL/40$

$M_E = -7WL/40$

9. Calculate the bending moments at B, C and D in the frame shown in *Figure 9.20*. EI is constant.
Ans. $M_B = M_D = -500$ kN m, $M_C = 0$.

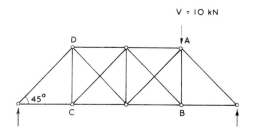

Figure 9.21

10. For the beam of Exercise 12, Chapter 5, make use of the fact that there is no deflection at B to calculate the support reaction R_B.

11. A two pinned semicircular arch of radius R is subject to a vertical load W at its crown. Calculate the horizontal reaction at the supports. EI is constant.
 If the right support is, by mistake, constructed λ units nearer the left support, calculate the bending moment at the crown due to W and the assembly stresses.
Ans. W/π, $0.5[WR^3(\pi - 2) - 4EI\lambda]/\pi R^2$.

10

Analysis of columns

10.1. Introduction

A column is a slender straight member subject to an axial force, bending moments and shearing forces. So far, we have considered two types of structures consisting of straight members. The first type had members under axial loads connected together by frictionless pins. Real structures are not manufactured in this manner and the rigidity of the joints cause bending moments and shearing forces. Thus these members are really columns. A second type of structure considered consisted of beams subject to bending moments and shearing forces. In this case, too axial forces may develop in the members. Structures consisting of bent members demonstrate this.

For a member under axial compressive force, bending and shear, two distinct aspects are of interest. These are:

(1) Increasing the axial load causes the member to buckle. The buckling load is known as the 'critical' or the 'Euler critical' load. Buckling takes place with elastic members with or without the application of bending moments.

(2) In a real member, the applied load is much less than the elastic critical load. As this load is increased, imperfection in the member deflects it transversely and failure takes place by plastic yielding below the critical load. At any point in a member, the bending moment is aggravated by the axial force. *Figure 7.5* of Chapter 7 demonstrated this. The actual bending moment must be calculated before the member can be designed safely. In this chapter both these aspects are studied. Two methods will be given both of which have strong connections with the modern approach to structural analysis.

10.2. The deformed shape of a column

In *Figure 10.1*, a pin ended column of length L is subject to an axial compressive load p and a bending moment M at end 1. Because the column becomes curved, end

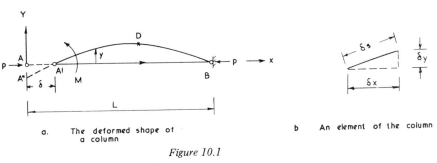

a. The deformed shape of a column

b. An element of the column

Figure 10.1

A moves by a small amount δ to A'. An element δs of the column is shown in *Figure 10.1b*, where

$$\delta s^2 = \delta x^2 + \delta y^2$$

i.e. $\qquad \delta s = [1 + (dy/dx)^2]^{0.5}$ \hfill (10.1)

Assuming that the actual length L does not change, the curved length $A'DB$ is equal to L and is given by

$$A'DB = L = \int_\delta^L ds = \int_\delta^L [1 + (dy/dx)^2]^{0.5} dx$$

Expanding $[1 + (dy/dx)^2]^{0.5}$ as a power series

$$[1 + (dy/dx)^2]^{0.5} = 1 + 0.5 (dy/dx)^2 + \text{terms involving higher orders of } dy/dx.$$

$$\therefore \qquad L = \int_\delta^L [1 + 0.5 (dy/dx)^2] dx = \int_\delta^L dx + 0.5 \int_\delta^L (dy/dx)^2 dx$$

But $\int_\delta^L dx =$ the straight length $A'B$. Hence

$$L = A'B + 0.5 \int_\delta^L (dy/dx)^2 dx$$

and since $AB = L = A'B + \delta$, it follows that

$$\delta = 0.5 \int_\delta^L (dy/dx)^2 dx$$

The origin of the coordinate axes is at A and the limits of the integration can be changed so that the integration is carried out from zero to L. Thus

$$\delta = 0.5 \int_\delta^L (dy/dx)^2 dx = 0.5 \left[\int_0^L (dy/dx)^2 dx - \int_0^\delta (dy/dx)^2 dx \right]$$

Now the length δ is small and $(dy/dx)^2$ is also small

thus $0.5 \int_0^\delta (dy/dx)^2 dx \simeq 0$

$$\therefore \qquad \delta = 0.5 \int_0^L (dy/dx)^2 dx \hfill (10.2)$$

This expression is now used to calculate the elastic critical load of a column by a method which is also the basis of the powerful finite element technique.

10.3. Method 1: The use of energy and displacement functions

Consider a pin ended perfect and ideal column AB of length L and subject to an axial compressive force p. The column is assumed to be made out of a linear elastic material and will remain so indefinitely. The load p makes the column contract by a

Analysis of Columns

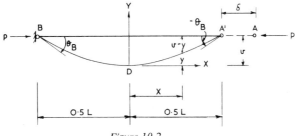

Figure 10.2

small amount which is neglected here. However, as the load increases and reaches the critical value p_{cr}, the column suddenly fails elastically and bends to the shape $A'DB$ shown in Figure 10.2. Taking the origin of the coordinate axes at D, the strain energy in bending is given by

$$U = 2 \int_0^{0.5L} \frac{M^2 \, dx}{2EI}$$

From the figure the anticlockwise bending moment at a distance x from the origin is given by

$$M = p_{cr}(v - y) \tag{10.3}$$

Substituting this in the expression for the strain energy

$$U = 2 \int_0^{0.5L} \frac{p_{cr}^2 (v - y)^2}{2EI} \, dx \tag{10.4a}$$

If the column has non-uniform sectional properties then the quantity EI is kept inside the integration. For a column with a constant cross sectional property throughout, equation (10.4a) becomes

$$U = \frac{p_{cr}^2}{EI} \int_0^{0.5L} (v - y)^2 \, dx \tag{10.4b}$$

At the critical load, the work done by the external load in moving pin A suddenly by an amount δ to A' is $p_{cr}\delta$ and using equation (10.2), this work is given as

$$w = 0.5 \, p_{cr} \times 2 \int_0^{0.5L} \left(\frac{dy}{dx}\right)^2 dx \tag{10.5}$$

The work w is converted to the strain energy in bending and $U = w$, hence

$$p_{cr} \int_0^{0.5L} (dy/dx)^2 \, dx = \frac{p_{cr}^2}{EI} \int_0^{0.5L} (v - y)^2 \, dx$$

i.e.

$$EI \int_0^{0.5L} (dy/dx)^2 \, dx = p_{cr} \int_0^{0.5L} (v - y)^2 \, dx \tag{10.6}$$

Analysis of Columns

The shape of the deflected form of the column is not known yet. It is known, however, that at incipient failure the deflection v is small. In that case any equation, i.e. any displacement function, should give a good result for p_{cr}. As a trial the displacement function

$$y = ax^2 + bx + c \tag{10.7}$$

is selected. Thus it is assumed, wrongly, that the shape of the deflected column is parabolic. The arbitrary constants a, b and c are found from the boundary conditions of the column. These are

(1) At $x = 0$; the value of y is zero and in equation (10.7), $c = 0$.
(2) At $x = 0$; $dy/dx = 0$: now $dy/dx = 2ax + b$ and with $x = 0$, $2a \times 0 + b = 0$, thus $b = 0$.
Equation (10.7) therefore becomes $y = ax^2$
(3) At $x = 0.5L$; $y = v$, thus $v = 0.25\,aL^2$ and $a = 4v/L^2$. The shape of the function is therefore

$$y = 4vx^2/L^2 \tag{10.8}$$
$$dy/dx = 8vx/L^2 \tag{10.9}$$
and
$$d^2y/dx^2 = 8v/L^2 \tag{10.10}$$

Equation (10.6), when solved, gives the value of the elastic critical load p_{cr} which defines the state when the column is about to buckle at a stage when it is in neutral equilibrium and with its stiffness equal to zero. These points will be studied in further detail later. Using equations (10.8) and (10.9), equation (10.6) becomes

$$EI \int_0^{0.5L} (8vx/L^2)^2 \, dx = p_{cr} \int_0^{0.5L} (v - 4vx^2/L^2)^2 \, dx$$

which gives

$$p_{cr} = 10\,EI/L^2 \tag{10.11}$$

Consider now, as another trial, that the displacement function is

$$y = a\cos(\pi x/L) + bx + c \tag{10.12}$$

The boundary conditions are used as usual to find a, b and c

(1) When $x = 0$; $y = 0$ and thus equation (10.12) gives $c = -a$.
(2) When $x = 0$, $dy/dx = (-a\pi/L)\sin(\pi x/L) = 0 \therefore b = 0$
and
(3) When $x = 0.5L$, $y = v \therefore a = -v$.

Equation (10.12) therefore becomes

$$y = -v\cos(\pi x/L) + v \tag{10.13}$$

Using this equation and its derivative in equation (10.6) we obtain

$$P_E = P_{cr} = \pi^2 EI/L^2 \tag{10.14}$$

With $\pi = 3.1416$, $\pi^2 = 9.87$ the difference between the two values of the elastic critical load given by equations (10.11) and (10.14) is in fact 1.3%. Nevertheless the accuracy of the parabolic displacement functions (10.7) and the trigonometric function (10.12) are investigated further.

Substituting for y from equations (10.8) or (10.13) with $x = \pm 0.5L$, into equation (10.3) it is found that both these equations give the value of the bending moments at the pins at A' and B as zero. This of course is valid because a pin cannot sustain a bending moment and the force p passes through these points.

However, for the parabolic displacement function, equation (10.10) gives the value of d^2y/dx^2 as $8v/L^2$ which is constant. Since $EI d^2y/dx^2$ is the expression for the bending moment, it follows that the parabolic displacement function (10.7) or (10.8) makes the implicit wrong assumption that $M = -EI d^2y/dx^2$ has a finite value of $8EIv/L^2$ which is different from zero at A' and B. The trigonometric function (10.10) or (10.13) does not suffer from this weakness. This is therefore the exact displacement function and the value of the elastic critical load is given by equation (10.14).

The parabolic function (10.7) gives satisfactory results in every other respect and it is a good function. Because it does not satisfy one requirement, such a function is called a 'non-conforming' function. The trigonometric function, on the other hand, is a 'conforming' function. Often non-conforming functions are used in the finite element analysis and give good results while not satisfying every requirement of the structural behaviour.

10.4. Method 2: The use of stability functions

This method gives the behaviour of a column subject to axial forces and bending moments at any stage of loading up to and including the elastic critical load.

Consider that the propped cantilever of *Figure 7.1*, section 7.2, is subjected to an axial load as well as the external moment M_{AB} as shown in *Figure 10.3*.

It was explained in example 3 of section 7.5, with the aid of *Figure 7.5*, that applying an axial force to a member aggravates its end rotations and thus reduces the member stiffness. The stiffness coefficient of the member in *Figure 10.3*, is no longer given by equations (7.4), (7.5) or (7.6). These equations have to be altered to some other form such as

$$M_A = M_{AB} = (4EI/L) i\, \theta_A = 4 k i\, \theta_A \tag{10.15a}$$

$$M_B = M_{BA} = (2EI/L) t\, \theta_A = 2 k t\, \theta_A \tag{10.16a}$$

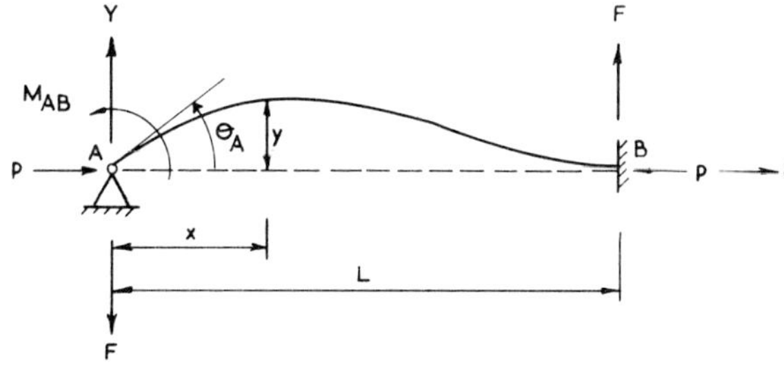

Figure 10.3

and

$$F = F_{AB} = -(6EI/L^2) j \theta_A = -(6k/L) j \theta_A \qquad (10.17a)$$

where M_A is the external moment applied at A, M_B is the bending moment developed at B, F is the shearing force in the member and $k = EI/L$.

The factors i, t and j are unknown except, it is evident, that when the value of the axial load is zero, $i = t = j = 1$ so that equations (10.15), (10.16) and (10.17) revert to the original equations (7.4), (7.5) and (7.6) respectively.

At a distance x from A, we have

$$M_x = -EI (d^2 y/dx^2) = (M_A + Fx + py) \qquad (10.18)$$

Defining a constant

$$\rho = p/P_E = p/(\pi^2 EI/L^2) \qquad (10.19)$$

and thus

$$p = \rho \pi^2 k/L$$

which means that ρ is the ratio between the axial force in the propped cantilever and the critical load of a pin ended column, and using equations (10.15) and (10.17), equation (10.18) gives

$$d^2 y/dx^2 = -(1/EI) [4 k i \theta_A - (6kj \theta_A x/L) + (\rho \pi^2 k/L) y]$$

The solution of this differential equation is

$$y = A \sin a x + B \cos a x + \{2 \theta_A [(3/L) jx - 2 i] / (a^2 L)\} \qquad (10.20)$$

where $a^2 = \pi^2 \rho/L^2$ and A and B are the arbitrary constants obtainable from the boundary conditions. These are: (1) when $x = 0$, $y = 0$ and (2) when $x = L$, $y = 0$, thus

$$A = -(L/4\alpha^2 k) (M_A \cot 2\alpha + M_B \csc 2\alpha)$$

and

$$B = (L/4\alpha^2 k) M_A$$

where $\alpha = 0.5\pi \sqrt{\rho}$ \qquad (10.21)

Now when $x = L$, $dy/dx = 0$, because end B is fixed. Differentiating equation (10.20) and substituting $x = L$ and $dy/dx = 0$ we obtain

$$j = \alpha^2/(3 - 3\phi_1) = \phi_2 \qquad (10.22)$$

where $\phi_1 = \alpha \cot \alpha$ \qquad (10.23)

Furthermore, when $x = 0$, $(dy/dx) = \theta_A$, hence we obtain, using equation (10.20)

$$i = 0.25 (3 \phi_2 + \phi_1) = \phi_3 \tag{10.24}$$

where $\phi_2 = j$ and is given by equation (10.22).
Considering the equilibrium of the member and taking moments about point B

$$M_{AB} + M_{BA} + FL = 0$$

Using equations (10.15) through (10.17) we obtain

$$t = \phi_4 = 3j - 2i = 3\phi_2 - 2\phi_3 \tag{10.25}$$

The intricacy of the mathematics need not be remembered.

10.5. The stability functions

Attention should be paid only to ϕ_1, ϕ_2, ϕ_3 and ϕ_4. These are the well known 'Livesley's stability functions'. They decide, as will soon be seen, whether any member in a structure is stable or has lost its stiffness and failed. Because they are used extensively, various computer programs have been written to calculate them.

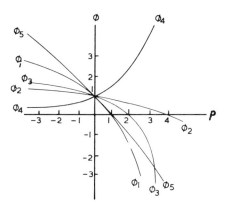

Figure 10.4. Livesley's functions

Their values are tabulated in Appendix 1 at the end of this book. Thus for a given value of p and hence p, the ϕ functions are read off directly, without any calculation. In *Figure 10.4* the graphs of these functions are plotted against $\rho = p/P_E$ and it is noticed that when the axial load in a member is zero, i.e. when p is zero, $\phi_1 = \phi_2 = \phi_3 = \phi_4 = 1$.
Equations (10.15) to (10.17) now become

$$M_{AB} = 4k \phi_3 \theta_{AB} \tag{10.15b}$$

$$M_{BA} = 2k \phi_4 \theta_{AB} \tag{10.16b}$$

$$F = -6k \phi_2 \theta_{AB}/L \tag{10.17b}$$

246   Analysis of Columns

and when $\phi_2 = \phi_3 = \phi_4 = 1$ equations (7.4), (7.5) and (7.6) are obtained. Except for the ϕ functions, the two sets of equations are identical.

10.6. The critical load of a propped cantilever

Figure 10.4 and Appendix 1 shows that at the value of $\rho = p/P_E = 2.046$, the value of ϕ_3 becomes zero. This stiffness of the propped cantilever becomes

$$M_{AB}/\theta_{AB} = 4k\phi_3 = 4k \times 0 = 0$$

Hence at $\rho = 2.046$ the propped cantilever loses its stiffness completely and becomes unstable. The member, at this instant, is in neutral equilibrium. The elastic critical load for a propped cantilever is therefore obtained from

$$\rho = p_{cr}/P_E = 2.046$$
$$\therefore \quad p_{cr} = 2.046\pi^2 \, EI/L^2 \tag{10.26}$$

In other words the elastic critical load for a propped cantilever is 2.046 times that for a pin ended member.

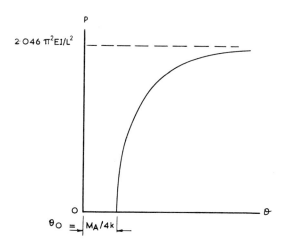

Figure 10.5

Below the critical load, the value of ρ is calculated from $\rho = p/P_E$ and the ϕ functions are read off from the tables in Appendix 1. In this manner the bending moments are given directly by equations (10.15b) and (10.16b).

10.7. The force-rotation curve for a propped cantilever

From equation (10.15b) we have

$$M_A = 4k\phi_3 \theta_A$$
$$\therefore \quad \theta_A = M_A/4k\phi_3 \tag{10.27}$$

First consider the case when M_A is constant but has a value other than zero and the axial force p is zero, $\phi_3 = 1$ and

$$\theta_{AB} = M_A/4k = M_A L/4EI$$

When p is increased, ϕ_3 decreases and for various values of p, the value of ρ is calculated from $\rho = pL^2/\pi^2 EI$ and from the tables for the stability functions, the corresponding values of ϕ_3 are obtained. Equation (10.27) is then used to calculate the value of θ_A. The graph of θ_A against p takes the form shown in *Figure 10.5*. It is noticed that although the material of the column is assumed to be linear elastic, the force rotation curve is non-linear. When $p = p_{cr} = 2.046 \, \pi^2 EI/L^2$, the value of ϕ_3 becomes zero and equation (10.27) gives $\theta_A = \infty$. The graph is therefore asymptotic to the line $p = p_{cr}$.

Referring to the equation $M_A = 4k\phi_3\theta_A$ again, consider now the case where $\theta_A \neq 0$ and $k \neq 0$ but M_A is changing because of changes in ϕ_3. This can be achieved by altering the value of the axial force p in the member.

Now, $M_A = 4k\,\phi_3\theta_A$, it is clear that, irrespective of the values of k and θ_A, if the axial load is increased to p_{cr} then ϕ_3 reduces to zero and $M_A = 0$. This means that at $p = p_{cr}$ no external moment at A is required to produce a rotation there. Hence at the critical load, the member can buckle without the application of an external moment at A. This also means that the critical loads for two members, one with a moment at A and the other without, are equal.

10.8. Example

Calculate the bending moment at the fixed end of a propped cantilever when it is subject to a moment of 1 kN mm at its pinned end.

(a) When no axial force is applied to the column and (b) when an axial force of 0.5 p_{cr} is applied.

Answer: (a) When the axial force is zero, $\phi_3 = \phi_4 = 1$, and equations (10.15b) and (10.16b) give $M_B = 0.5 M_A = 0.5$ kN mm where B is the fixed end and A is the pin end.

(b)
$$\rho = p/(\pi^2 EI/L^2)$$
$$\therefore \quad \rho = 0.5 \times 2.046 \,(\pi^2 EI/L^2) \times L^2/(\pi^2 EI) = 1.023$$

With $\rho = 1.023$, it is found from the tables for stability functions that $\phi_3 = 0.6078$ and $\phi_4 = 1.2403$. Equation (10.15b) gives

$$\theta_A = M_A/(4k\phi_3) = 1/(4k\phi_3)$$

Therefore equation (10.16b) gives

$$M_B = 2k\phi_4/(4k\phi_3) = 0.5\phi_4/\phi_3$$

$$\therefore \quad M_B = 0.5 \times 1.2403/0.6078 = 1.02 \text{ kN mm}$$

Thus the axial load has reduced the stiffness of the member so much that M_B is now larger than M_A. In a practical column, designers prevent this situation by subjecting it to a much lower axial force so that ρ is considerably less than 1.023.

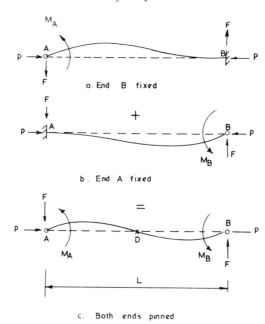

a. End B fixed

b. End A fixed

c. Both ends pinned

Figure 10.6

10.9. A column in double curvature

The column in *Figure 10.6a* has its end A pinned and end B fixed. Equations (10.15b) through (10.17b) give the bending moments and the shearing force in this column. If, on the other hand a column has its end A fixed and its end B pinned as shown in *Figure 10.6b*, then equations (10.15b) to (10.17b) change to

$$M_B = 4k\phi_3\theta_B \tag{10.28}$$

$$M_A = 2k\phi_4\theta_B \tag{10.29}$$

and

$$F = -6k\phi_2\theta_B/L \tag{10.30}$$

A column in double curvature, shown in *Figure 10.6c*, is obtained by the direct addition of the above two cases, thus

$$M_A = 4k\phi_3\theta_A + 2k\phi_4\theta_B \tag{10.31}$$

$$M_B = 2k\phi_4\theta_A + 4k\phi_3\theta_B \tag{10.32}$$

and

$$F = (-6k\phi_2\theta_A - 6k\phi_2\theta_B)/L \tag{10.33}$$

This column has both its ends pinned and it is under pure rotation without any sway. For a given value of the axial load p, in this column, the value of ρ is calculated

Analysis of Columns

from $\rho = p/(\pi^2 EI/L^2)$ and equations (10.31) to (10.33) are used to calculate the end rotations and the shearing force. Or if the end rotations are known, the bending moments are calculated.

If in this column $M_A = M_B$ then equations (10.31) and (10.32) give $\theta_A = \theta_B$ and equation (10.31) becomes

$$M_A = 4k\phi_3\theta_A + 2k\phi_4\theta_A \tag{10.34}$$

The stiffness K of this member is

$$K = M_A/\theta_A = 4k\phi_3 + 2k\phi_4 = 6k\phi_2 \tag{10.35}$$

The fact that $4k\phi_3 + 2k\phi_4 = 6k\phi_2$ is obtained from the definition of ϕ_4 given by equation (10.25).

Thus at the elastic critical load, when this column loses its stiffness, $K = 0$ and hence

$$6k\phi_2 = 0 \tag{10.36}$$

This condition is satisfied when $\phi_2 = 0$ at $\rho = 4$, as shown in *Figure 10.4* and in the tables for the stability functions. The elastic critical load is obtained from

$$\rho = 4 = p_{cr}/(\pi^2 EI/L^2)$$

i.e.
$$p_{cr} = 4\pi^2 EI/L^2 \tag{10.37}$$

In other words the critical load of a column in double curvature is four times that for a pin ended column in single curvature. It is noticed from equation (10.35) that ϕ_2 measures the stiffness of this column and as ρ increases ϕ_2 decreases and so does K. On the other hand, equation (10.15) shows that ϕ_3 measures the stiffness of a propped cantilever.

10.10. A column in pure sway

Consider a member with a *small* inclination ϕ which is pinned at one end and fixed at the other. Apply a moment m_{AB} at end A as shown in *Figure 10.7a* so that the

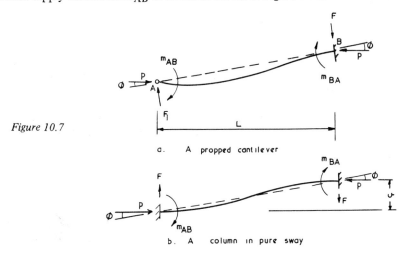

Figure 10.7

a. A propped cantilever

b. A column in pure sway

tangent at A to the member becomes horizontal and the rotation at that end is $-\phi = -v/L$, where v is the sway in the member as shown in the figure. Equations (10.15b) and (10.16b) give

$$m_{AB} = -4\, k\phi_3 v/L$$

and

$$m_{BA} = -2\, k\phi_4 v/L$$

The application of another moment m'_{AB} at B in order to make that end also horizontal, while keeping A fixed, results in

$$m'_{BA} = -4\, k\phi_3 v/L$$

and

$$m'_{AB} = -2\, k\phi_4 v/L$$

Both ends of the member are now horizontal with one of them subsided relative to the other as shown in *Figure 10.7b*. This member is now in pure sway without any end rotation. The final moments at the ends of the member are obtained by adding m and m', thus

$$M_A = -4\, k\phi_3 v/L - 2\, k\phi_4 v/L = -6\, k\phi_2 v/L \qquad (10.38)$$

$$M_B = -2\, k\phi_4 v/L - 4\, k\phi_3 v/L = -6\, k\phi_2 v/L \qquad (10.39)$$

The shear force F in the member is obtained by taking moments about one end, say B, giving

$$-FL - M_{AB} - M_{BA} + pv = 0$$

Using the above two equations and also (10.21), (10.22) and (10.23) we obtain

$$F = 12\, k\phi_5 v/L^2$$

where $\phi_5 = \phi_1 \phi_2$ is also a stability function. This function is also shown in *Figure 10.4* and tabulated together with the other stability functions in Appendix 1. Once again when the axial load p is zero, $\phi_5 = 1$.

Neither M_A nor M_B can be increased without making the ends rotate. The only way to make a member in pure sway unstable and buckle is by influencing the shear force F. This can alter if the axial force is altered. The quantity $K = F/v = 12\, k\phi_5/L^2$ is the stiffness of the member against lateral translation and at the elastic critical load, $K = 0$ giving

$$12\, k\phi_5/L^2 = 0 \qquad (10.40)$$

This condition is satisfied when $\phi_5 = 0$. Now $\phi_5 = \phi_1 \phi_2$ and it follows that ϕ_5 becomes zero as soon as either ϕ_1 or ϕ_2 is zero. It is noticed in *Figure 10.4* that ϕ_1

Analysis of Columns

becomes zero at $\rho = 1$ long before ϕ_2. Hence elastic instability takes place when $\phi_s = \phi_1 = 0$ and the elastic critical load is obtained from

$$\rho = 1 = p_{cr}/(\pi^2 \, EI/L^2)$$

i.e. $\qquad p_{cr} = \pi^2 \, EI/L^2 \qquad (10.41)$

Thus the critical load of a fixed ended member in pure sway is the same as that of a pin ended column. Here again, it is noticed that the value of ϕ_s is a measure of the stiffness of a column in pure sway.

It will be shown later that ϕ_1 measures the stiffness of a pin ended member. It is evident therefore that each column in the table for the stability functions is a direct record of the stiffness of a column with a certain end condition. This indeed stresses the significance of the stability function in the analysis and the design of columns. There are other stability functions but none has the same *physical* significance just demonstrated for the ϕ functions.

10.11. The general column

Sometimes columns are constructed with one end or the other either pinned or fixed. Most columns in a structure, however, are attached to other members or to the foundations. The ends of such columns are neither fully fixed nor pinned. They may be rigidly connected at their ends and thus are subject, at both ends, to axial loads, shearing forces and bending moments. A column of this type is known as a general

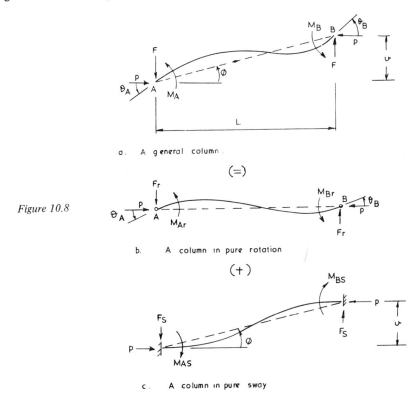

Figure 10.8

a. A general column

b. A column in pure rotation

c. A column in pure sway

column. Its ends rotate and sway with respect to each other as shown in *Figure 10.8a*. A general column is therefore obtained by adding two columns together; one in pure rotation (*Figure 10.8b*) and the other in pure sway (*Figure 10.8c*). The bending moments and the shear force in the general column are obtained by adding the two cases together. Thus

$$M_A = M_{Ar} + M_{As} = -(6EI/L^2)\phi_2 v + (4EI/L)\phi_3 \theta_A + (2EI/L)\phi_4 \theta_B \quad (10.42)$$

$$M_B = M_{Br} + M_{Bs} = -(6EI/L^2)\phi_2 v + (2EI/L)\phi_4 \theta_A + (4EI/L)\phi_3 \theta_B \quad (10.43)$$

$$F = F_r + F_s = (12EI/L^3)\phi_5 v - (6EI/L^2)\phi_2 \theta_A - (6EI/L^2)\phi_2 \theta_B \quad (10.44)$$

It is immediately evident that these equations are very similar to the well known slope deflection equations (5.14) and (7.3). In fact, they are the general slope deflection equations used in the non-linear analysis of structures. The load deflection diagrams obtained by using these equations are non-linear because as the axial load in a member increases, the stability functions decrease and thus the member loses some of its stiffness and deflects more. It has already been stated that when the axial load in a member is zero, the ϕ functions become unity and equations (10.42) to (10.44) revert to the ordinary slope deflection equations used in linear analysis. This indicates how, fundamentally, the non-linear analysis is a natural extension of the linear one.

10.12. Examples

Example 1. From the general column, derive the behaviour and the elastic critical load of a pin ended member in single curvature.
Answer: For a column in single curvature, the sway $v = 0$ and for $\theta_A = -\theta_B$, (*Figure 10.2*) equation (10.42) for a general column becomes

$$M_A = 4k\phi_3 \theta_A - 2k\phi_4 \theta_A$$

This defines the relations between M_A, θ_A and the stability functions. The stiffness K is given by

$$K = M_A/\theta_A = 2(2\phi_3 - \phi_4)k$$

and when $K = 0$ at the elastic critical load, $2(2\phi_3 - \phi_4)k = 0$. This means that when the member is unstable, zero moment is sufficient to rotate the ends. Now $\phi_3 = 0.25(3\phi_2 + \phi_1)$, $\phi_4 = 0.5(3\phi_2 - \phi_1)$, hence $2(2\phi_3 - \phi_4)k = 2k\phi_1$ and at the critical load

$$K = 2k\phi_1 = 0 \quad (10.45)$$

i.e. the column becomes unstable when $\phi_1 = 0$ and from *Figure 10.4* or the tables for ϕ functions this takes place when

$$\rho = 1 = P_{cr}/P_E$$

Thus at $\quad \rho = 1: p_{cr} = \pi^2 EI/L^2$

which is the elastic critical load for a pin ended column and was obtained earlier in equation (10.14) using the energy approach. From $K = 2 k\phi_1$ it is evident that ϕ_1 is a measure of the stability of the pin-ended column.

Example 2. Derive the value of the elastic critical load for a vertical fixed ended column which has no sway.

Answer: When the ends A and B of a column are fixed so that they do not rotate, the middle half of the column DC at the elastic critical load deforms as if it were a separate pin ended column, with points of contraflexure, i.e. points with zero moments, at the quarter points as shown in *Figure 10.9a*. The buckling load for a fixed ended column is thus the same as that of a pin ended column of half the length.

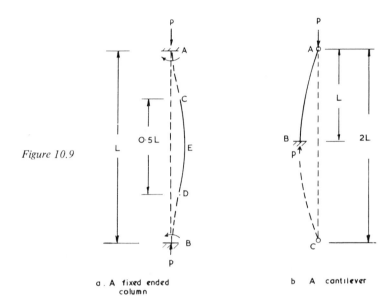

Figure 10.9

a. A fixed ended column

b. A cantilever

$$\therefore \quad p_{cr} = \pi^2 EI/(L/2)^2 = 4\pi^2 EI/L^2 \qquad (10.46)$$

Notice also, that in *Figure 10.9a*, for the case of a fixed ended column points A, B and E have zero rotations, that is each half of the column is in pure sway. If $\ell = 0.5L$, then from *Figure 10.9a*, the critical load for AE in pure sway is $4\pi^2 EI/(2\ell)^2 = \pi^2 EI/\ell^2$, as was obtained in equation (10.41).

Example 3. What is the buckling load for a vertical cantilever under an axial load?

Answer: An inspection of *Figure 10.9b* indicates that when the cantilever AB with length L buckles, it takes the form of half a pin ended column which has a height of $2L$, thus

$$p_{cr} = \pi^2 EI/(2L)^2 = 0.25 \, \pi^2 EI/L^2 \qquad (10.47)$$

Example 4. End A of a column in a tall building is subject to a moment of 200 kN m and a vertical load of 12 500 kN as shown in *Figure 10.10*. End B is fixed to the

Analysis of Columns

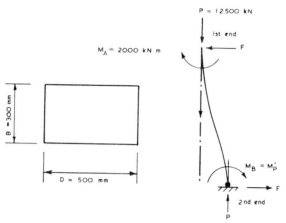

Figure 10.10

foundation. Calculate the value of the shear force F that makes the end of the column at B yield completely. The column is 3 mm long and has a rectangular cross section with width $B = 300$ mm and depth $d = 500$ mm. The yield stress of the material is 0.2 kN/mm^2 and E is 200 kN/mm^2.

Answer: From section (8.1) the reduced plastic hinge moment of the cross section is

$$M'_p = 0.25\sigma_y [BD^2 - (p^2/B\sigma_y^2)]$$

$$\therefore \quad M'_p = 0.25 \times 0.2 [300 \times 500^2 - 12\,500^2/(300 \times 0.04)]$$

$$= 3\,098\,958 \text{ kN mm}$$

There is no rotation at B before the formation of the plastic hinge and $\theta_B = 0$. The general slope deflection equations (10.42) and (10.43) become

$$M_{AB} = 4k\phi_3\theta_A - 6k\phi_2 v/L = -2\,000\,000$$

$$M_{BA} = 2k\phi_4\theta_A - 6k\phi_2 v/L = -3\,098\,958$$

subtracting the second equation from the first

$$4k\phi_3\theta_A - 2k\phi_4\theta_A = 1\,098\,958$$

$$\therefore \quad k\theta_A = 0.5 \times 1\,098\,958/(2\phi_3 - \phi_4)$$

with $p = 12\,500$

$$\rho = \frac{pL^2}{\pi^2 EI} = \frac{12\,500 \times 3\,000^2 \times 12}{\pi^2 \times 200 \times 300 \times (500)^3} = 0.018\,25$$

Analysis of Columns

From Appendix 1, by interpolation: $\phi_2 = 0.997$, $\phi_3 = 0.994$, $\phi_4 = 1.0003$,
$\phi_5 = 0.982$ and $k\theta_A = 0.5 \times 1\,098\,958/(2 \times 0.994 - 1.000\,3) = 556\,321.758$ with
$k = EI/L = 200 \times 300 \times 500^3/(3\,000 \times 12) = 208.3 \times 10^6$
The first slope deflection equation gives

$$6kv\phi_2/L = 2\,000\,000 + 4k\phi_3\theta_A$$

Hence $\quad v = \dfrac{3\,000(2\,000\,000 + 4 \times 0.994 \times 556\,321.758)}{6 \times 208.3 \times 10^6 \times 0.997} = 10.14 \text{ mm}$

Equation (10.44) gives the shear force as

$$F = (12k\phi_5 v/L^2) - 6k\phi_2\theta_A/L$$

$$\therefore \quad F = \dfrac{12 \times 208.3 \times 10^6 \times 0.982 \times 10.14}{3\,000 \times 3\,000} - \dfrac{6 \times 556\,321.758 \times 0.997}{3\,000}$$

$$F = 1\,656.22 \text{ kN.}$$

10.13. Deflection of an eccentrically loaded column

In practice, all columns are imperfect. When an axial load is applied to the ends of an imperfect column, it acts eccentrically to the longitudinal axis of the column as shown in *Figure 10.11a*. As soon as the load is applied, the column deflects sideways. At a distance y from one end of the column, the deflection is $e + x$ where e is the eccentricity of the load. The maximum deflection at mid-height D is $e + v$ as shown in *Figure 10.11b*. The load produces a variable bending moment along the column. This has the values of pe at A, $p(e + v)$ at D and $-pe$ at B as shown in *Figure 10.11c*.

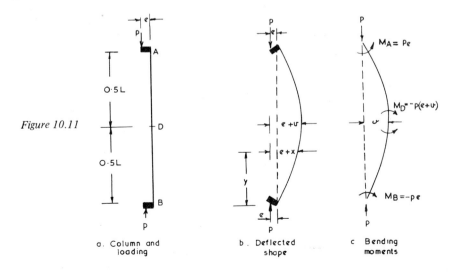

Figure 10.11

a. Column and loading
b. Deflected shape
c. Bending moments

Analysis of Columns

The deflection v is obtained by considering the top half AD of the column with length $0.5L$ and by using the slope deflection equations, thus

$$M_{AD} = 4k\phi_3\theta_A - 12k\phi_2 v/L = pe \qquad (10.48)$$

$$M_{DA} = 2k\phi_4\theta_A - 12k\phi_2 v/L = -p(e+v) \qquad (10.49)$$

where $k = EI/(0.5L) = 2EI/L$. The rotation θ_D at D is zero because of symmetry. Equation (10.48) gives

$$2k\theta_A = (0.5pe + 6k\phi_2 v/L)/\phi_3$$

and substituting this in equation (10.49), we obtain

$$12k\phi_2(\phi_4 - 2\phi_3)v + 2pLv\phi_3 = -peL(\phi_4 + 2\phi_3)$$

Now as $\phi_4 - 2\phi_3 = -\phi_1$ and $\phi_4 + 2\phi_3 = 3\phi_2$, it follows that

$$v = 3peL\phi_2/(12k\phi_5 - 2pL\phi_3) \qquad (10.50)$$

Noting that, for half the column, $p = 2\pi^2 \rho k/L$, equation (10.50) becomes

$$v = 3e\phi_2\pi^2\rho/(6\phi_5 - 2\pi^2\rho\phi_3) \qquad (10.51)$$

When $p = 0$, $\rho = 0$ and therefore $v = 0$.
On the other hand as p increases, ρ also increases and so does v. Equation (10.51) relates the central deflection v in a non-linear manner to ρ and therefore p. At $\rho = 0.25$, $p = \pi^2 EI/L^2$, the value of ϕ_3 is, from the tables in Appendix 1, equal to 0.91494 and $\phi_5 = 0.7525$. Hence at $\rho = 0.25$

$$6\phi_5 - 2\pi^2\rho\phi_3 = 6 \times 0.7525 - 2\pi^2 \times 0.25 \times 0.91494 = 0$$

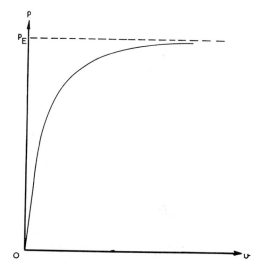

Figure 10.12

Analysis of Columns

Thus at the critical load, equation (10.51) gives the deflection $v = \infty$. The load deflection diagram for the column is obtained by selecting various values of p, calculating the corresponding values of ρ, obtaining the values of ϕ_2, ϕ_3 and ϕ_5 from the tables and using equation (10.51) to calculate v. The diagram of v against p is shown in Figure 10.12 where it is noticed that the curve is asymptotic to the elastic critical load p_E.

10.14. The stresses in an eccentrically loaded column

In the above column, the largest deflection and bending moment occur at mid-height D. The numerical value of the moment at D is $p(e + v)$ and using equation (10.50), we obtain

$$M_D = p\left(e + \frac{3peL\,\phi_2}{12k\phi_5 - 2pL\,\phi_3}\right)$$

Noting that $3\phi_2 - 2\phi_3 = \phi_4$, it follows that

$$M_D = \frac{12\,k\,e\,p\,\phi_5 + p^2 L\,e\,\phi_4}{12\,k\,\phi_5 - 2\,p\,L\,\phi_3} \tag{10.51}$$

Let s be the distance between the centroidal axis of the column and the extreme fibres. The longitudinal bending stress due to M is

$$\sigma_b = \frac{Ms}{I} = \frac{(12\,k\,e\,p\,\phi_5 + p^2 L\,e\,\phi_4)s}{(12\,k\,\phi_5 - 2\,p\,L\,\phi_3)I} \tag{10.52}$$

The axial stress due to p is

$$\sigma_d = p/A \tag{10.53}$$

where A is the area of the cross section of the column. Thus the total maximum compressive stress at mid-height is

$$\sigma_{max} = \sigma_d + \sigma_b = \frac{p}{A} + \frac{(12\,k\,e\,p\,\phi_5 + p^2 L\,e\phi_4)s}{(12\,k\,\phi_5 - 2p\,L\,\phi_3)I} \tag{10.54}$$

Notice that in this expression the stability functions are calculated for half the column with $\rho = pL/(2\pi^2 k)$ and with $k = 2EI/L$, where L is the length of the column. Equation (10.54) is equivalent to the so-called secant formula.

To obtain this maximum stress for a given value of p, the value of ρ is first calculated and the corresponding values of ϕ_3, ϕ_4 and ϕ_5 are obtained from Appendix 1.

10.15. Example

Calculate the maximum stress in a pin-ended column subject to a load of 320 kN applied eccentrically at a distance of 5 mm from the longitudinal axis of the column. Take $E = 200$ kN/mm², $L = 1$ m, $I = 10^6$ mm⁴, $A = 200$ mm² and $s = 10$ mm.

Answer:

$$\rho = p/P_E = \frac{320 \times 10^6}{4\pi^2 \times 200 \times 10^6} = 0.0406$$

From Appendix 1, by interpolation

$$\phi_3 = 0.9865, \quad \phi_4 = 1.0068, \quad \phi_5 = 0.9599$$

$$k = 2EI/L = 0.4 \times 10^6$$

$$\therefore \sigma_b = \frac{(12 \times 0.4 \times 10^6 \times 5 \times 320 \times 0.9599 + 320 \times 320 \times 10^3 \times 5 \times 1.0068)\,10}{(12 \times 0.4 \times 10^6 \times 0.9599 - 2 \times 320 \times 10^3 \times 0.9865) \times 10^6}$$

$$\sigma_b = 0.02 \text{ kN/mm}^2$$

$$\sigma_d = \frac{320}{200} = 1.6 \text{ kN/mm}^2$$

and $\sigma_{max} = 0.02 + 1.6 = 1.62$ kN/mm²

It is noticed that if the column is made out of steel, the value of the maximum stress developed will be higher than the limit of proportionality. The direct stress alone is 1.6 kN/mm² and the column will yield under direct compression.

10.16. Initially curved pin-ended column

Real columns are not perfectly straight. The initial curvature of a column can be taken into consideration as follows.

Suppose that the mid-point of a pin-ended column is initially at a distance a from the vertical line joining the two ends as shown in *Figure 10.13*. In this figure v is the additional deflection due to the end forces p. The bending moment at A is zero because of the pin at A and p passes through A, while the bending moment at D is $-p(a + v)$. With $k = EI/0.5L = 2EI/L$, the slope deflection equations for the top half AD of the column are

$$M_{AD} = 4k\phi_3\,\theta_A - 12k\phi_2\,v/L = 0 \tag{10.55}$$

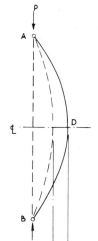

Figure 10.13

$$M_{DA} = 2k\phi_4 \theta_A - 12 k \phi_2 v/L = -p(a + v) \tag{10.56}$$

From equation (10.55), we obtain

$$2k\theta_A = 6k\phi_2 v/(L\phi_3)$$

and hence equation (10.56) gives

$$6k\phi_2 (2\phi_3 - \phi_4) v - pL\phi_3 v = paL\phi_3$$

But $\quad 2\phi_3 - \phi_4 = \phi_1$ and $\phi_1 \phi_2 = \phi_5$

$\therefore \quad v = pa\phi_3 L/(6k\phi_5 - pL\phi_3)$

i.e. $\quad v = \dfrac{a}{\dfrac{6k\phi_5}{pL\phi_3} - 1}$

and since $k = 2EI/L$, it follows that

$$v = \dfrac{a}{\dfrac{12 EI \phi_5}{pL^2 \phi_3} - 1} \tag{10.57}$$

where v is the additional lateral displacement of the column at D as shown in Figure 10.13. The stability functions ϕ_3 and ϕ_5 are for the top half AD of the column and as the value of p_{AD} increases from zero to 0.25, that is, as the column approaches the critical stage, the value of $12\phi_5/\phi_3$ reduces from 12 to π^2. Nearer buckling, equation (10.57) can be written as

$$v \simeq \frac{a}{\dfrac{\pi^2 EI}{pL^2} - 1} \approx \frac{a}{\dfrac{p_E}{p} - 1} \qquad (10.58)$$

where p is the actual axial load in the column and $p_E = \pi^2 EI/L^2$ for the whole column. This equation is hyperbolic with its asymptote being $p = p_E = \pi^2 EI/L^2$.

10.17. The maximum stress in curved pin-ended column

The maximum stress in the above column occurs at the mid-point D. The numerically maximum bending moment is

$$M = p(a + v) = pa \left(1 + \frac{1}{\dfrac{p_E}{p} - 1} \right)$$

i.e.
$$M = pa \left(\frac{p_E}{p_E - p} \right) \qquad (10.59)$$

The maximum compressive stress at the extreme fibres has the value

$$\sigma_{max} = \frac{p}{A} + \frac{pap_E}{p_E - p} \times \frac{s}{I}$$

where A is the area of the cross section and y is the distance from the centroidal axis to the extreme fibres. Here p/A is the direct stress due to the axial force; if this is equal to σ_d, then

$$\sigma_{max} = \sigma_d \left(1 + \frac{\sigma_E}{\sigma_E - \sigma_d} \times \frac{as}{r^2} \right) \qquad (10.60)$$

where

$$\sigma_E = \frac{p_E}{A} = \pi^2 E \left(\frac{r}{L} \right)^2 \qquad (10.61)$$

and r is the radius of gyration of the cross section. The quantity L/r is called the slenderness ratio of the column.

Let $\beta = as/r^2$

then

$$\sigma_{max} = \sigma_d \left(1 + \frac{\beta \sigma_E}{\sigma_E - \sigma_d} \right)$$

Analysis of Columns

Solving this equation for σ_d and considering the negative root only, since this is the smallest, we obtain

$$\sigma_d = 0.5\,[\sigma_{max} + (1 + \beta)\sigma_E] - \sqrt{\{0.25\,[\sigma_{max} + (1 + \beta)\sigma_E]^2 - \sigma_{max}\sigma_E\}}$$

This equation gives the value of the direct compressive stress σ_d at which a maximum compressive, bending and direct, stress σ_{max} would develop for a given value of β.
When first yield occurs in the column, $\sigma_{max} = \sigma_y$ and hence

$$\sigma_d = 0.5\,[\sigma_y + (1 + \beta)\sigma_E] - \sqrt{\{0.25\,[\sigma_y + (1 + \beta)\sigma_E]^2 - \sigma_y\sigma_E\}} \qquad (10.62)$$

This is known as Perry's formula.

10.18. Design of pin-ended column

From a large number of tests on initially curved pin-ended columns, Robertson concluded that a suitable value of β for mild steel columns is given by

$$\beta = 0.003\,L/r \qquad (10.63)$$

Equation (10.62) therefore becomes

$$\sigma_d = 0.5\,[\sigma_y + (1 + 0.003\,L/r)\sigma_E] -$$
$$- \sqrt{\{0.25\,[\sigma_y + (1 + 0.003\,L/r)\sigma_E]^2 - \sigma_y\sigma_E\}} \qquad (10.64)$$

This formula is known as the Perry-Robertson's formula which is sometimes used to design columns.

10.19. Laterally loaded columns

When a column is subject to several lateral point loads, the same reasoning of section 5.12 is used to treat such a column. That is to say the column is treated as if it consists of several members connected to one another at joints under the point loads.
The value of k in portion i of the column is equal to EI/L_i where L_i is the length of the portion. The value of ρ_i is calculated from

$$\rho_i = p_i/(\pi^2 EI/L_i^2) = (p_i L_i^2/\pi^2 EI)$$

where p_i is the axial force in portion i. For instance, in the column shown in Figure 10.14, the values of k and ρ for AB, BC and CD are

$$k_1 = EI/400$$
$$\rho_1 = 400^2 p/(\pi^2 EI)$$
$$k_2 = EI/500$$

262   Analysis of Columns

$$\rho_2 = 500^2 p/(\pi^2 EI)$$

$$k_3 = EI/200$$

and

$$\rho_3 = 200^2 p/(\pi^2 EI)$$

The general slope deflection equations (10.42), (10.43) and (10.44) are then applied to each portion in the same manner as in section 5.12.

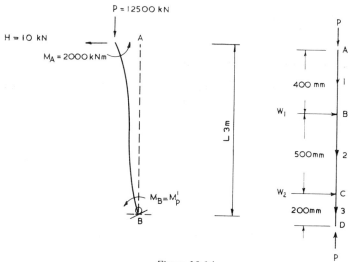

Figure 10.14

In the case of a column subject to a uniformly distributed lateral load, along either its whole length or part of it, the same reasoning of section 5.16 is used to analyse the column. Thus it is evident that no extra work is involved in dealing with laterally loaded columns. The only difference between such a column and an ordinary member subject to lateral loads without an axial force is that the general non-linear slope deflection equations are used with the column. A procedure similar to that given in section 10.4 can be used to show that the fixed end moments are now $\pm wL^2/(12\phi_2)$.

10.20. Example

Calculate the deflection at D in the beam shown in *Figure 5.10* and analysed in example 1 of section 5.13 if it is subject to the loads given as well as an axial force of 20 kN.

Answer: Numbering portion AD as member 1 and DB as member 2 and applying the general slope deflection equations to AD and DB

$$M_{AD} = -(6k_1/a)\phi_{21} v + 4k_1\phi_{31}\theta_A + 2k_1\phi_{41}\theta_D = 0$$

Referring to *Figure 5.10c*

$$M_{DA} = -(6k_1/a)\phi_{21} v + 2k_1 \phi_{41}\theta_A + 4k_1 \phi_{31}\theta_D = Ma/L$$
$$M_{BD} = (6k_2/b)\phi_{22} v + 4k_2 \phi_{32}\theta_B + 2k_2 \phi_{42}\theta_D = 0$$
$$M_{DB} = (6k_2/b)\phi_{22} v + 2k_2 \phi_{42}\theta_B + 4k_2 \phi_{32}\theta_D = Mb/L$$

where the second suffix of the stability functions refers to the member number. From the first equation we obtain

$$2k_1 \theta_A = (3k_1 \phi_{21}/a\phi_{31})v - k_1 (\phi_{41}/\phi_{31})\theta_D$$

and the second equation therefore gives

$$\theta_D = [Ma \phi_{31}/(3Lk_1\phi_{51})] + (v/a) \qquad (10.65)$$

Similarly the last two slope deflection equations give

$$\theta_D = [Mb \phi_{32}/(3Lk_2 \phi_{52})] - (v/b) \qquad (10.66)$$

Equating the right hand side of equations (10.65) and (10.66) we obtain

$$v = \frac{Mab}{3 EI L^2}\left(\frac{b^2\phi_{32}}{\phi_{52}} - \frac{a^2\phi_{31}}{\phi_{51}}\right)$$

$$\rho_1 = \frac{20 a^2}{\pi^2 EI} = \frac{20 \times 4000^2}{\pi^2 \times 200 \times 10^6} = 0.164$$

$$\rho_2 = \frac{20 b^2}{\pi^2 EI} = \frac{20 \times 6000^2}{\pi^2 \times 200 \times 10^6} = 0.365$$

∴ $\phi_{31} = 0.9448$, $\phi_{51} = 0.8378$, $\phi_{32} = 0.8738$

and $\phi_{52} = 0.6381$

Thus

$$v = \frac{10^4 \times 4000 \times 6000}{3 \times 200 \times 10^6 \times 10^8}\left(\frac{6000^2 \times 0.8738}{0.6381} - \frac{4000^2 \times 0.9448}{0.8378}\right)$$

∴ $v = 125.24$ mm

Thus the axial load has increased the deflection at D from 80 mm to 125.24 mm which is an increase of 53%.

Exercises on Chapter 10

1. In *Figure 10.15* the column AB is fixed at A and pinned at B while the column DC is pin-ended. The block BC is rigid and moves down only horizontally by the load W. The column AB is circular in cross section and has 25.4 mm diameter. The column DC has a rectangular cross section of width 50.8 mm and depth 25.4 mm. Calculate the critical value of the load W when x = 457.2 mm. What value of x makes the critical load a maximum?
Take E as 200 kN/mm².
Ans. W_{cr} = 14 kN, x = 820 mm, $W_{cr\,max}$ = 67.921 kN.

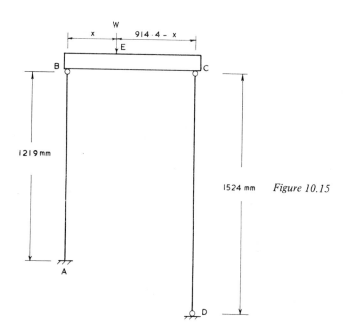

Figure 10.15

2. Design the pin-ended column shown in *Figure 10.16* by selecting an I section for its cross section. The column is pin-ended with a length of 7.620 m. It carries the load W_1 = 444.8 kN centrally and the load W_2 = 222.4 kN eccentrically at a distance of 254 mm from the centre line of the column as shown in the figure. The partial safety factor for load is 1.6 and that against stress is 1.1. E = 200 kN/mm² and the yield stress = 0.3 kN/mm².

3. A pin ended column of length 2 m is subject to an axial load of 2 kN and a lateral central load of 1 kN. Calculate the maximum bending moment in the column. E = 200 kN/mm² and I = 10^6 mm⁴.
Ans. 502 kN mm.

4. An initially curved pin-ended column of length 2 m has an initial maximum deviation of 5 mm at its mid-height from the vertical line joining its ends. Calculate the total departure of the mid-height when a load of $0.9 p_E$ is applied axially to the column. Take E = 200 kN/mm², and I = 10^6 mm⁴.
Ans. 50 mm.

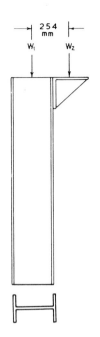

Figure 10.16

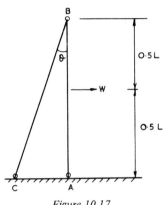

Figure 10.17

5. In *Figure 10.17*, the pin-ended column AB is connected to the tie CB at B. Calculate the deflection of the mid-height when a horizontal force W acts there. Plot the $W - \delta$ graph

Ans.
$$\delta \approx \frac{WL^3}{48EI\left(p - \frac{\pi^2 EI}{L^2}\right)}$$

where p is the compressive force in the column.

6. In the frame shown in *Figure 10.18*, the columns AB and DE are fixed at A and D and pinned at B and E. Prove that the buckling load W_{cr} is given by the formula

$$\tan L \sqrt{\frac{W_{cr}}{EI}} = L\sqrt{\frac{W_{cr}}{EI}} - \frac{L^3}{3}\left(\frac{W_{cr}}{EI}\right)^{3/2}$$

where L is the height of the columns and EI is their flexural rigidity.

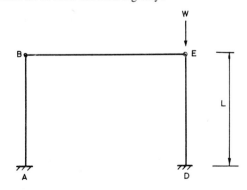

Figure 10.18

11

Pin jointed space structures

11.1. Introduction

It was stated at the beginning of this book that real structures are all three dimensional. This chapter is concerned with such structures but is restricted to those in which the members are subject to axial forces only. A system of members, joined together at their ends to form a rigid space structure, is often called a space frame or truss. Transmission towers and cranes are examples of such constructions. These structures are manufactured so that the members are either welded together at their joints or connected by means of bolts. The rigidity of the joints therefore gives rise to the development of shearing forces and bending moments in the members. However, it is assumed that these affect the stresses in the members in a secondary manner compared to the primary stresses due to the direct axial forces. Thus it is assumed that the joints of these space structures are ideal spherical hinges that can only transmit direct axial forces to the members.

11.2. Concurrent forces in space

Several forces, in a three dimensional space, that act at a point can be reduced to a single resultant by the principle of the parallelogram of forces. This principle can be extended to deal with several forces acting at a point. They can all be reduced to a single resultant which acts at the point of intersection of all the forces.

The algebraic sum of the projections of several forces intersecting at a point on any axis is equal to the projection of their resultant on the same axis. It follows that if a system of intersecting forces is in equilibrium then the algebraic sums of their projections on the cartesian X-Y-Z axes are equal to zero. Thus the equations of equilibrium in the three dimensional form are

$$\left. \begin{array}{l} \sum_{1}^{N} p_{xi} = 0 \\ \sum_{1}^{N} p_{yi} = 0 \\ \sum_{1}^{N} p_{zi} = 0 \end{array} \right\} \quad (11.1)$$

where p_{xi} is the projection of a typical force p_i on the X axis and N is the total number of forces acting at a point.

11.3. Equilibrium of isostatic space structures

A general system of forces acting on a space structure can be reduced to a single resultant force and a single resultant couple. For instance, the force p_1 which is acting at point A of a structure and has a general orientation in the three dimensional space X-Y-Z as shown in *Figure 11.1*, can be replaced by a force p_1 acting at the origin O and a couple p_1 AB. The lever arm of this couple is the perpendicular distance AB. This replacement is achieved by applying two equal and opposite forces p_1 and $-p_1$ at the origin and in the same plane as the original force acting at A. Since the forces p_1 and $-p_1$ at O are in equilibrium, they do not in any way alter the effect of the original force p_1 acting at A. Similarly the force p_2 can be replaced by a force p_2 acting at the origin together with a couple whose lever arm is the perpendicular distance between the original force p_2 and the two equal and opposite forces acting at O.

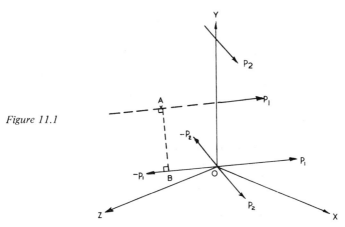

Figure 11.1

Two such systems of forces can only be in equilibrium if the resultant force and the resultant couple both vanish. For the resultant force to be zero, the algebraic sums of the projections of the given forces on the cartesian X-Y-Z axes must vanish and for the resultant couple to be zero, the algebraic sums of the moments of the forces about the cartesian axes must also vanish. Thus

$$\left. \begin{array}{l} \sum_{1}^{N} p_{xi} = 0, \quad \sum_{1}^{N} p_{yi} = 0, \quad \sum_{1}^{N} p_{zi} = 0 \\ \\ \sum_{1}^{N} M_{xi} = 0, \quad \sum_{1}^{N} M_{yi} = 0, \quad \sum_{1}^{N} M_{zi} = 0 \end{array} \right\} \quad (11.2)$$

where M_{xi} is the moment of a typical force p_i about the X axis. These equations are sufficient to evaluate six unknown member forces or reactions in a pin jointed space structure.

11.4. Stability of pin jointed space structures

The simplest stable space structure can be formed by starting with a rigid foundation to which a joint is connected by means of three members not in one plane. In

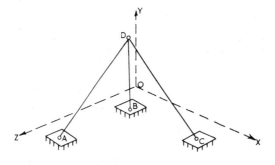

a. The simplest space structure

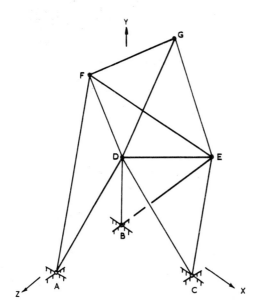

b. Building up joints D-E-F-G

Figure 11.2

Figure 11.2a such a simple structure is shown with joint D connected to the supports at A, B and C by three members. Once this structure is constructed, a further joint such as E (*Figure 11.2b*) can be connected to it by means of three members EB, EC and ED. The resulting structure ABCDE is also stable. It is noticed that two of the members EB and EC are connected to the foundations at B and C. The third member could also have been connected to the support at A but the stability of the structure is obtained with equal efficiency by joining the third member to D.

Further joints such as F and G can be added successively to the existing structure provided that each extra joint is connected to it by means of three extra members. On constructing joint F next, it is found that while it could be connected by two members to D and E, the third member has to be connected to a support such as A in *Figure 11.2b*. Thus any stable structure with at least three supports but with more than two joints is found to require to be connected to its supports by at least six members. In fact these six members are also sufficient to produce a structure connected to its support in a stable manner.

Building up a stable space structure in this manner indicates that each joint requires three members, thus for m members and j joints, we find that

$$m = 3j \qquad (11.3)$$

Here the supports are not counted as joints.

11.5. Direction cosines of a space member

In *Figure 11.3* the member AB has a general orientation. As usual an arrow is placed on the member the head of which is pointing to the second end B and the positive direction of the P axis of the member is from the first end A to B as shown in the figure. The dotted lines AC, AD and AE are drawn from the first end A parallel to the cartesian axes X, Y and Z respectively. The angles between the member and these three lines are α, β and γ respectively.

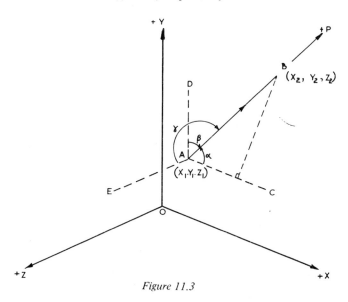

Figure 11.3

Let the coordinates of the first and the second ends of the member be (X_1, Y_1, Z_1) and (X_2, Y_2, Z_2). The length L of the member is then given by

$$L = \sqrt{[(X_2 - X_1)^2 + (Y_2 - Y_1)^2 + (Z_2 - Z_1)^2]} \qquad (11.4)$$

The direction cosines of the P axis of this member are $\cos \alpha$, $\cos \beta$ and $\cos \gamma$ given by

$$\left.\begin{array}{l} \cos \alpha = (X_2 - X_1)/L = \ell_P \\[4pt] \cos \beta = (Y_2 - Y_1)/L = m_P \\[4pt] \cos \gamma = (Z_2 - Z_1)/L = n_P \end{array}\right\} \qquad (11.5)$$

These are the cosines of the angles between the positive directions of the coordinate axes X, Y and Z and the positive direction of the P axis of the member. The positive direction of each angle is measured from one of the coordinate axes to the member as shown in the figure by the arrows allocated to the angles α, β and γ. In equations (11.4) and (11.5) it is noticed that the coordinates X_2, Y_2 and Z_2 of the second end B of the member always appear first. From these the coordinates X_1, Y_1 and Z_1 of the first end are subtracted. This procedure is strictly followed throughout. If the value of one of the coordinates, such as Y_2 is negative, then its minus sign is also included in the equations.

11.6. Example

Calculate the lengths and the direction cosines of the members of the space frame ABCD shown in *Figure 11.4*. The members are numbered and the positive P axis for each of them is specified by an arrow.

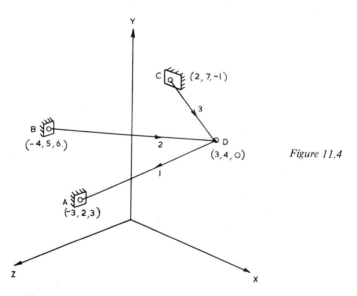

Figure 11.4

Answer: The arrow on member 1 indicates that D is the first end and A is the second.

$L_1 = \sqrt{[(-3-3)^2 + (2-4)^2 + (3-0)^2]} = 7$ units of length

$\ell_{P_1} = (-3-3)/7 = -6/7 = -0.857$

$m_{P_1} = (2-4)/7 = -2/7 = -0.286$

$n_{P_1} = (3-0)/7 = 3/7 = 0.429$

For member 2, the arrow indicates that B is the first end and D is the second.

$L_2 = \sqrt{\{[3-(-4)]^2 + (4-5)^2 + (0-6)^2\}} = 9.274$

$\ell_{P_2} = [3-(-4)]/9.274 = 7/9.274 = 0.756$

$m_{P_2} = (4 - 5)/9.274 = -1/9.274 = -0.1078$

$n_{P_2} = (0 - 6)/9.274 = -6/9.274 = -0.647$

For member 3, C is the first end and D is the second.

$L_3 = \sqrt{\{(3 - 2)^2 + (4 - 7)^2 + [0 - (-1)]^2\}} = 3.317$

$\ell_{P_3} = (3 - 2)/3.317 = 1/3.317 = 0.302$

$m_{P_3} = (4 - 7)/3.317 = -3/3.317 = -0.906$

$n_{P_3} = [0 - (-1)]/3.317 = 1/3.317 = 0.302$

11.7. The cartesian components of a force

In *Figure 11.5* the vector AB represents a force of magnitude p. The vector is drawn in a three dimensional space with its first end A having the coordinates (X_1, Y_1, Z_1) and its second end having coordinates (X_2, Y_2, Z_2). The length AB = L, the magnitude of the force is given by

$$L = \sqrt{[(X_2 - X_1)^2 + (Y_2 - Y_1)^2 + (Z_2 - Z_1)^2]}$$

The component of p in the X direction is equal to $p_x = Ac = p \cos \alpha = p\ell_p$ where α is the angle between the force and the X axis, thus

$$p_x = p\ell_p = (X_2 - X_1)p/L \tag{11.6}$$

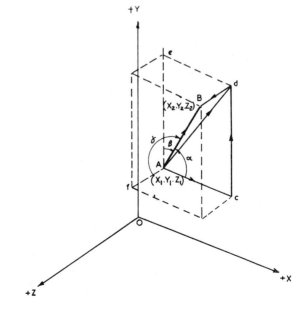

Figure 11.5

Similarly the component of the force p in the Y direction is equal to $p_y = Ae = p \cos \beta = pm_P$, thus

$$p_y = pm_P = (Y_2 - Y_1)p/L \tag{11.7}$$

Finally the component of p in the Z direction is $p_z = Af$ given by

$$p_z = pn_P = (Z_2 - Z_1)p/L \tag{11.8}$$

In *Figure 11.5*, cd = Ae = p_y and dB = Af = p_z and it is noticed, in the plane ABd, the force p is the resultant of the vectors Ad and dB where the vector Ad is in the plane parallel to the X-Y plane. This vector, itself, is the resultant of the vectors p_x = AC and p_y = cd.

11.8. Analysis of isostatic space frames

Equations (11.1) define the state of equilibrium of a point subject to a number of forces concurrent at that point. These equations state that the algebraic sums of the components of the forces in the X, Y and Z directions are all equal to zero. This fact is used in the analysis of isostatic pin-jointed space frames. Since at each joint of the frame only three such equations of equilibrium can be derived, the procedure is to start with a joint which has only three members so that the unknown forces in all three can be calculated. Once this is done, the next joint is selected which is also subjected to three unknown forces. This may be either a joint with three members or a joint with more than three members but with the forces in the 4th, 5th etc. already calculated by considering the equilibrium of previous joints.

Initially a member is assumed to be subject, at each joint, to a force *acting in the same direction as its positive P axis* and when the result of the analysis shows that the force in the member is negative, it indicates that it is in the opposite direction to that assumed originally. The component of an external force acting at a joint is considered to be positive if it acts in the same direction as the positive X, Y or Z axis.

Consider a joint j with members i, k, t etc. and subject to external loads W_f, W_g, W_h. These forces are first resolved to their components W_{fx}, W_{fy}, W_{fz}, W_{gx}, W_{gy} etc. in the manner explained in section 11.7. The components of these forces in the direction of one of the Cartesian axis can then be added up, thus in the X direction

$$W_x = W_{fx} + W_{gx} + W_{hx} + \text{etc.}$$

In the Y direction

$$W_y = W_{fy} + W_{gy} + W_{hy} + \text{etc.} \tag{11.9}$$

Finally in the Z direction

$$W_z = W_{fz} + W_{gz} + W_{hz} + \text{etc.}$$

Similarly each unknown member force p_i, p_t etc. is resolved into its components and each component is added to the relevant equation in (11.9). Equating the result to zero yields three equations of equilibrium for joint j, thus at j

$$W_{fx} + W_{gx} + W_{hx} + \text{etc.} + p_{xi} + p_{xk} + p_{xt} + \text{etc.} = 0$$
$$W_{fy} + W_{gy} + W_{hy} + \text{etc.} + p_{yi} + p_{yk} + p_{yt} + \text{etc.} = 0 \quad\quad (11.10)$$
$$W_{fz} + W_{gz} + W_{hz} + \text{etc.} + p_{zi} + p_{zk} + p_{zt} + \text{etc.} = 0$$

Considering all the joints and the supports in this manner yields a sufficient number of equations to calculate all the unknown member forces and reactions.

11.9. Examples

Example 1. The pin-jointed space structure shown in *Figure 11.4* is subject at D to a vertical force of -10 kN, i.e. acting downwards. Calculate the member forces if the coordinates are measured in metres.

Answer: Using the values obtained in the example of section 11.6, the three equations of equilibrium for joint D become

In the X direction
$$-0.857\, p_1 + 0.756 \times -p_2 + 0.302 \times -p_3 + 0 = 0$$

In the Y direction
$$-0.286\, p_1 - 0.1078 \times -p_2 - 0.906 \times -p_3 - 10 = 0$$

In the Z direction
$$0.429\, p_1 - 0.647 \times -p_2 + 0.302 \times -p_3 + 0 = 0$$

Joint D is at the first end of member 1 and a positive force $+p_1$ is assumed in this member and inserted into the equations. On the other hand this joint is at the second end of members 2 and 3 and the forces in these members are assumed to be $-p_2$ and $-p_3$. They are inserted with their negative sign into the equations.

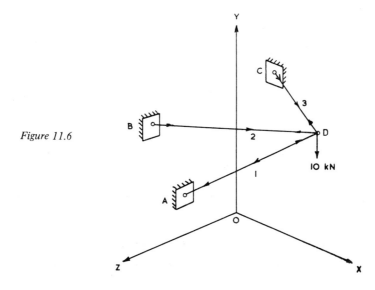

Figure 11.6

Solving these three equations, we obtain

$$p_1 = -11.45 \text{ kN}$$
$$p_2 = 10.47 \text{ kN}$$
$$p_3 = 6.16 \text{ kN}$$

Thus at D the signs of all the three forces turn out to be opposite to those assumed. The arrows that identify the sign of the forces in a member are in the opposite direction to the arrows for identifying their ends. The two sets of arrows are shown in *Figure 11.6* and it is clear from this figure that p_1 is a compressive force while both p_2 and p_3 are tensile. In fact tensile forces turn out to be positive and compressive forces negative.

Example 2. Calculate the forces in the members of the space frame shown in *Figure 11.7*.

Answer: Joint H is connected to four members of which three are in one plane. There is no external force acting at H and since any three concurrent forces can be in equilibrium only if they are coplanar, it follows that member FH is inactive because it has no force in it. Any force in HF acting transversely to the plane HKGJ cannot be balanced by the forces in members GH, HJ and HK which are acting in the plane. Similar reasoning shows that members HK, KM and MD are also inactive.

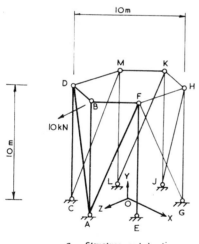

a. Structure and loading

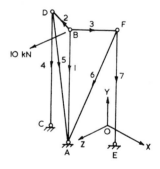

b. Active members

Figure 11.7

Once these inactive members are removed from the structure it becomes evident that all the other members represented by thin lines are also inactive. The active members of the structure are shown in *Figure 11.7b*.

From the dimensions of the structure, it is found that the coordinates of the joints are

$$A = (0, 0, 5); \; B = (0, 10, 5); \; C = (-2.5\sqrt{3}, 0, 2.5),$$
$$D = (-2.5\sqrt{3}, 10, 2.5); \; E = (2.5\sqrt{3}, 0, 2.5) \text{ and } F = (2.5\sqrt{3}, 10, 2.5).$$

Pin Jointed Space Structures

Numbering the members as shown in *Figure 11.7b* where the arrow shown on each member defines its positive P axis, the direction cosines of the members are calculated as

$$L_1 = 10$$
$$\ell_{P_1} = (0 - 0)/10 = 0, \quad m_{P_1} = (0 - 10)/10 = -1,$$
$$n_{P_1} = (5 - 5)/10 = 0$$
$$L_2 = 5$$
$$\ell_{P_2} = (-2.5\sqrt{3} - 0)/5 = -\tfrac{1}{2}\sqrt{3}, \quad m_{P_2} = (10 - 10)/5 = 0,$$
$$n_{P_2} = (2.5 - 5)/5 = -0.5$$
$$L_3 = 5$$
$$\ell_{P_3} = (2.5\sqrt{3} - 0)/5 = \tfrac{1}{2}\sqrt{3}, \quad m_{P_3} = (10 - 10)/5 = 0,$$
$$n_{P_3} = (2.5 - 5)/5 = -0.5$$

Similarly $\ell_{P_4} = 0, m_{P_4} = -1, n_{P_4} = 0, \ell_{P_7} = 0, m_{P_7} = -1$ and $n_{P_7} = 0$.

$$L_5 = 11.2$$
$$\ell_{P_5} = [0 - (-2.5\sqrt{3})]/11.2 = 0.387 \quad m_{P_5} = (0 - 10)/11.2 = -0.894$$
$$n_{P_5} = (5 - 2.5)/11.2 = 0.223$$

For the equilibrium of joint B

In the X direction
$$0 \times p_1 - 0.5\sqrt{3}\, p_2 + 0.5\sqrt{3}\, p_3 + 0 = 0$$
In the Y direction
$$-1 \times p_1 + 0 \times p_2 + 0 \times p_3 + 0 = 0$$
In the Z direction
$$0 \times p_1 - 0.5 p_2 - 0.5 p_3 + 10 = 0$$
$$\therefore \quad p_1 = 0, \; p_2 = p_3 = 10 \text{ kN (tensile)}$$

For the equilibrium of joint D

In the X direction
$$-0.5\sqrt{3} \times -p_2 + 0\, p_4 + 0.387\, p_5 + 0 = 0$$
In the Y direction
$$0 \times -p_2 - 1\, p_4 - 0.894\, p_5 + 0 = 0$$

In the Z direction

$$-0.5 \times -p_2 + 0 \times p_4 + 0.223\, p_5 = 0$$

$$\therefore \quad p_4 = 20 \text{ kN (tensile)}$$

$$p_5 = -22.4 \text{ kN (compressive)}$$

From symmetry

and

$$p_6 = p_5 = -22.4 \text{ kN (compressive)}$$

$$p_7 = p_4 = 20 \text{ kN (tensile)}$$

11.10. Deflections of isostatic space frames

As in Chapter 9, the deflection of a space frame is calculated from the equations $X = B^T \delta$. The procedure is to calculate the member forces in terms of the external loads, thus construct matrix B. Transposing this matrix and post multiplying it by the member deformations δ yields the joint deflections X.

11.11. Example

The table below shows the direction cosines of the members for the space frame shown in *Figure 11.8*. Calculate the deflections in the X, Y and Z directions of joint 1. Take $E = 200 \text{ kN/mm}^2$ and $A = 100 \text{ mm}^2$ for all the members.

Member	l_P	m_P	n_P
1	1	0	0
2	0.8	0	−0.6
3	0.8	−0.6	0
4	1	0	0
5	0	0.707	−0.707
6	0.8	0.6	0

Answer:

Considering the equilibrium of joint 1

In the X direction

$$-0.8\, p_3 - p_4 + 0 \times p_5 + H = 0$$

In the Y direction

$$-0.6 \times -p_3 + 0 \times -p_4 + 0.707\, p_5 - V = 0$$

In the Z direction

$$0 \times -p_3 + 0 \times -p_4 - 0.707\, p_5 + L = 0$$

$$\therefore \quad p_3 = 1.667\, (V - L)$$

$$p_4 = H - 1.33\, (V - L)$$

$$p_5 = 1.414 L$$

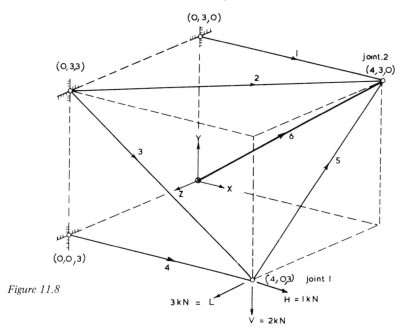

Figure 11.8

Considering the equilibrium of joint 2

In the X direction

$$1 \times -p_1 + 0.8 \times -p_2 + 0 \times -p_5 + 0.8 \times -p_6 + 0 = 0$$

In the Y direction

$$0 \times -p_1 + 0 \times -p_2 + 0.707 \times -p_5 + 0.6 \times -p_6 + 0 = 0$$

In the Z direction

$$0 \times -p_1 - 0.6 \times -p_2 - 0.707 \times -p_5 + 0 \times -p_6 + 0 = 0$$

$\therefore \quad p_1 = 2.667L$

$p_2 = -1.667L$

$p_6 = -1.667L$

Thus **P = B W** becomes

$$\begin{bmatrix} p_1 \\ p_2 \\ p_3 \\ p_4 \\ p_5 \\ p_6 \end{bmatrix} = \begin{bmatrix} 0 & 0 & 2.667 \\ 0 & 0 & -1.667 \\ 0 & 1.667 & -1.667 \\ 1 & -1.333 & 1.333 \\ 0 & 0 & 1.414 \\ 0 & 0 & -1.667 \end{bmatrix} \begin{bmatrix} H \\ V \\ L \end{bmatrix}$$

and $\mathbf{B}^T = \begin{bmatrix} 0 & 0 & 0 & 1 & 0 & 0 \\ 0 & 0 & 1.667 & -1.333 & 0 & 0 \\ 2.667 & -1.667 & -1.667 & 1.333 & 1.414 & -1.667 \end{bmatrix}$

With $H = 1$ kN, $V = 2$ kN, $L = 3$ kN, we find that the actual member forces are

$$p_1 = 8 \text{ kN}, \ p_2 = -5 \text{ kN}, \ p_3 = -1.667 \text{ kN}, \ p_4 = 2.333 \text{ kN},$$
$$p_5 = 4.24 \text{ kN and } p_6 = -5 \text{ kN}.$$

Thus: $\delta_1 = [4000/(200 \times 100)] \times 8 = 1.60$ mm

Similarly $\delta_2 = -1.25$ mm, $\delta_3 = -0.42$ mm, $\delta_4 = 0.47$ mm,

$$\delta_5 = 0.90 \text{ mm and } \delta_6 = -1.25 \text{ mm}.$$

Therefore the deflections u, v and w of joint 1 in the X, Y and Z directions are obtained from

$$\begin{bmatrix} u \\ v \\ w \end{bmatrix} = \begin{bmatrix} 0 & 0 & 0 & 1 & 0 & 0 \\ 0 & 0 & 1.667 & -1.333 & 0 & 0 \\ 2.667 & -1.667 & -1.667 & 1.333 & 1.414 & -1.667 \end{bmatrix} \begin{bmatrix} 1.60 \\ -1.25 \\ -0.42 \\ 0.47 \\ 0.90 \\ -1.25 \end{bmatrix}$$

$$\therefore \quad u = 1 \times 0.47 = 0.47 \text{ mm}$$
$$v = 1.667 \times -0.42 - 1.333 \times 0.47 = -1.33 \text{ mm}$$

and

$$w = 11.0 \text{ mm}.$$

11.12. Relationship between member deformations and joint deflections

The member AB shown in *Figure 11.9* is connected to joint A at its first end and joint B at its second end. The coordinates of these joints are (X_1, Y_1, Z_1) and (X_2, Y_2, Z_2). Under the influence of some external loads this member deforms and moves to its new position $A'B'$. The movement AA' of the first end is u_1 while that at the second end is $BB' = u_2$. The net deformation of the member is u given by

$$u = u_2 - u_1 \tag{11.11}$$

where u_1, u_2 and u are parallel to the original direction AB of the member. Thus u is an extension parallel to the P axis of the member.

Let the displacements of joint A be x_1, y_1 and z_1 parallel to the X, Y and Z axes respectively and those of joint B be x_2, y_2 and z_2. The component of x_2 in the direction AB is BG which is equal to $x_2 \cos\alpha = x_2 \ell_p$, the component of y_2 in the same direction is BH $= y_2 \cos\beta = y_2 m_p$ and that of z_2 is BF $= z_2 \cos\gamma = z_2 n_p$.

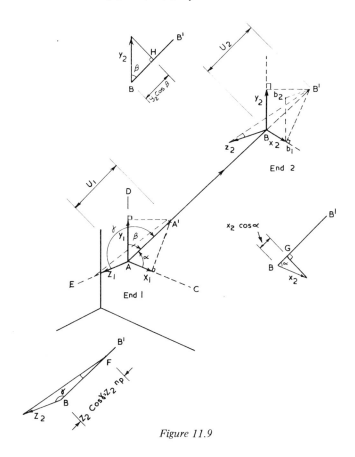

Figure 11.9

These components add up to u_2, thus

$$u_2 = x_2 \ell_p + y_2 m_p + z_2 n_p \qquad (11.12)$$

Similarly at the first end

$$u_1 = x_1 \ell_p + y_1 m_p + z_1 n_p \qquad (11.13)$$

Thus the net extension u, in equation (11.11) becomes

$$u = x_2 \ell_p + y_2 m_p + z_2 n_p - (x_1 \ell_p + y_1 m_p + z_1 n_p) \quad (11.14)$$

or in matrix form

$$u = \begin{bmatrix} -\ell_p & -m_p & -n_p & \ell_p & m_p & n_p \end{bmatrix} \begin{bmatrix} x_1 \\ y_1 \\ z_1 \\ x_2 \\ y_2 \\ z_2 \end{bmatrix} \quad (11.14a)$$

or just

$$u = D\Delta \quad (11.14b)$$

where the row vector $D = [-\ell_p \ -m_p \ -n_p \ \ell_p \ m_p \ n_p]$ is known as the displacement transformation vector and the column vector Δ lists the deflections at joints A and B to which the member is connected. A similar relationship between the deformation u_i in member i and the joint displacements Δ_{jk} can be written for any member i and the displacements of the joints j and k to which member i is connected. For the whole space frame the equations relating the member deformations and the joint displacements become

$$\delta = D\Delta \quad (11.15)$$

where $\delta = \{u_1 \ u_2 \ldots u_i \ldots u_M\}$ and is the column vector of member deformations. The column vector $\Delta = \{x_1 \ y_1 \ z_1 \ x_2 \ y_2 \ z_2 \ldots \ldots x_j \ y_j \ z_j \ldots x_J \ y_J \ z_J\}$ consists of the joint deflections of the space frame and matrix D is the displacement transformation matrix of the frame. Each row of this matrix belongs to one member and contains the direction cosines of the member. These direction cosines are negative at the first end of the member and positive at the second end.

11.13. The joint equilibrium equations

The stiffness equation $p = EAu/L$ relates the axial force p in a pin ended member to its deformation u and makes use of Hooke's law. On the other hand, equation (11.14) relates the member deformation u to the joint displacements. Thus the force p in a member can be expressed in terms of the displacements of the joints to which it is connected as

$$p = (EA/L)u = (EA/L)(-x_1\ell_p - y_1 m_p - z_1 n_p + x_2\ell_p + y_2 m_p + z_2 n_p) \quad (11.16)$$

Now if a number of members e, f, g etc. are connected to a joint j, then to preserve the equilibrium of that joint, the components of p_e, p_f, p_g etc. in the X, Y and Z directions must add up to the external forces W_{Xj}, W_{Yj} and W_{Zj} applied to the joint. Here W_{Xj} is the external force acting at j in the X direction. Similarly for W_{Yj} and W_{Zj}. Thus at each joint three equations of equilibrium can be formulated.

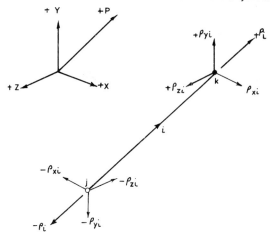

Figure 11.10

The components of a force p in X, Y and Z directions are $p\ell_P$, pm_P and pn_P as given by equations (11.6), (11.7) and (11.8). In *Figure 11.10* the member i with force p_i is shown to be connected to joint j at its first end and joint k at its second end. The member is in tension so the force p_i at the second end k will act in the direction of the positive P axis of the member, i.e. from j to k. At the first end j this same force will be acting in the opposite direction, i.e. from k to j and thus considered to be negative.

The contribution of member i to the three equations of equilibrium for joint k are thus $p_{xi} = p\ell_{Pi}$, $p_{yi} = pm_{pi}$ and $p_{zi} = pn_{pi}$. Its contribution to the equilibrium equations for joint j, on the other hand, are $-p_{xi}$, $-p_{yi}$ and $-p_{zi}$. With $a = EA/L$, using equations (11.16), the six contributions that member i makes to the equilibrium of the whole space frame are

To joint j at the first end of the member

$$-p_{xi} = -\ell_P a \left(-x_j \ell_P - y_j m_P - z_j n_P + x_k \ell_P + y_k m_P + z_k n_P \right)$$

$$= -\ell_P a u_i$$

$$-p_{yi} = -m_P a u_i \qquad (11.17a)$$

$$-p_{zi} = -n_P a u_i$$

where $u_i = -x_j \ell_P - y_j m_P - z_j n_P + x_k \ell_P + y_k m_P + z_k n_P$

To joint k at the second end of the member

$$\left. \begin{array}{l} p_{xi} = \ell_P a u_i \\ p_{yi} = m_P a u_i \\ p_{zi} = n_P a u_i \end{array} \right\} \qquad (11.17b)$$

The set of equations for the joint equilibrium prepared, for all the joints, are square, i.e. there are as many equations as there are unknown joint displacements. They are also symmetrical, that is the coefficient k_{st} for the unknown x_{xt}, which is the unknown number t in the equation number s, is equal to k_{ts} for the unknown x_{st}.

The solution of these equations gives the unknown joint displacements. Once this is done, the member forces are calculated using equation (11.16).

In matrix form the right hand sides of equations (11.17) can be written as

$$K = \begin{array}{c|ccc|ccc|}
& \multicolumn{3}{c|}{\text{At joint } j} & \multicolumn{3}{c|}{\text{At joint } k} \\
& x_j & y_j & z_j & x_k & y_k & z_k \\
\hline
\text{At} & a\ell_p^2 & & & & & \\
\text{joint} & a\ell_p m_p & am_p^2 & & & & \\
j & a\ell_p n_p & an_p m_p & an_p^2 & & \text{Symmetrical} & \\
\hline
\text{At} & -a\ell_p^2 & -a\ell_p m_p & -a\ell_p n_p & a\ell_p^2 & & \\
\text{joint} & -a\ell_p m_p & -am_p^2 & -am_p n_p & a\ell_p m_p & am_p^2 & \\
k & -a\ell_p n_p & -an_p m_p & -an_p^2 & a\ell_p n_p & an_p m_p & an_p^2 \\
\end{array} \begin{array}{c} W_{xj} \\ W_{yj} \\ W_{zj} \\ W_{xk} \\ W_{yk} \\ W_{zk} \end{array}$$

(11.18a)

where **K** is known as the overall stiffness matrix of the space frame and

$$\mathbf{W} = \mathbf{K}\mathbf{\Delta} \tag{11.18b}$$

where **W** is the vector of the external loads applied to the joints of the space frame.

The method just described can be used to analyse isostatic as well as hyperstatic space frames with equal efficiency. It can also be used for the analysis of isostatic and hyperstatic pin jointed plane frames, since in this case the equations of equilibrium in the Z direction are disregarded.

This method is known as the 'displacement method' because it calculates all the joint displacements first. Once this is done, the member forces can be calculated.

11.14. Examples

Example 1. The symmetrical pin-jointed space frame shown in *Figure 11.11* is subject at A to a horizontal force $W_x = 2000$ kN in the X-Y plane. Calculate the member forces. Members 1 and 2 have area $A_1 = 5 \times 10^3$ mm^2 while the area of member 3 is 2.5×10^3 mm^2. $E = 200$ kN/mm^2.

Answer: This space frame is hyperstatic (why?). The frame is symmetrical about the X-Y plane and is loaded symmetrically in the same plane. It can therefore be cut into two halves one of which is shown in *Figure 11.11b*. Because of symmetry the deflection of joint A in the Z direction is zero and the direction cosine n_p for the members need not be calculated. There are only two unknown joint displacements x_A and y_A.

Pin Jointed Space Structures

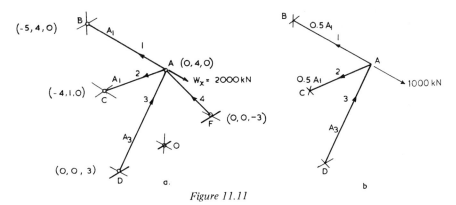

Figure 11.11

The arrows on each member indicate that A is at the first end of members 1 and 2 and at the second end of member 3.

$$L_1 = L_2 = L_3 = 5 \text{ m}$$

$$\ell_{P1} = (-5 - 0)/5 = -1, \quad m_{P1} = (4 - 4)/5 = 0$$

$$\ell_{P2} = (-4 - 0)/5 = -0.8, \quad m_{P2} = (1 - 4)/5 = -0.6$$

$$\ell_{P3} = (0 - 0)/5 = 0, \quad m_{P3} = (4 - 0)/5 = 0.8$$

$$a_1 = a_2 = (E \times 0.5A_1)/L_1 = 200 \times 2.5 \times 10^3/(5 \times 10^3)$$
$$= 100 \text{ kN/mm}$$

$$a_3 = EA_3/L_3 = 200 \times 2.5 \times 10^3/(5 \times 10^3) = 100 \text{ kN/mm}.$$

Joints B, C and D are supports and do not deflect.

The contributions of members 1 and 2 to the equilibrium equations of joint A: A is at the first end and replaces j in equations (11.17a) while terms with suffix k are disregarded.

(1) In the X direction

$$-P_{x1} = 1 \times 100 (1 \times x_A - 0 \times y_A) = 100 x_A$$

$$-P_{x2} = 0.8 \times 100 (0.8 x_A + 0.6 y_A) = 64 x_A + 48 y_A$$

(2) In the Y direction

$$m_{P1} = 0, \text{ thus } -P_{y1} = 0$$

$$-P_{y2} = 0.6 \times 100 (0.8 x_A + 0.6 y_A) = 48 x_A + 36 y_A$$

The contributions of member 3 to the equilibrium of joint A: A is at the second end and replaces k in equations (11.17b) while terms with suffix j are disregarded.

(1) In the X direction

$$\ell_{P3} = 0, \text{ thus } p_{x3} = 0$$

(2) In the Y direction

$$p_{y3} = 0.8 \times 100 \,(0 \times x_A + 0.8 y_A) = 64 y_A$$

The equations of equilibrium for joint A are

$$\Sigma p_x = 1000$$

∴

$$(100 x_A) + (64 x_A + 48 y_A) = 1000$$

i.e. $164 x_A + 48 y_A = 1000$

$$\Sigma p_y = 0$$

∴

$$(48 x_A + 36 y_A) + (64 y_A) = 0$$

i.e. $48 x_A + 100 y_A = 0$

These equations can be written in matrix form $K\Delta = W$ as

$$\begin{bmatrix} 164 & 48 \\ 48 & 100 \end{bmatrix} \begin{bmatrix} x_A \\ y_A \end{bmatrix} = \begin{bmatrix} 1000 \\ 0 \end{bmatrix}$$

which can be obtained directly using equations (11.18). This matrix is noticed to be symmetrical.
Solving these for x_A and y_A

$$x_A = 7.094 \text{ mm}$$
$$y_A = -3.405 \text{ mm}$$

Using equations (11.16), the member forces become

$$p_1 = \frac{200 \times 5 \times 10^3}{5 \times 10^3} (1 \times 7.094) = 1418.8 \text{ kN (tensile)}$$

$$p_2 = \frac{200 \times 5 \times 10^3}{5 \times 10^3} (0.8 \times 7.094 + 0.6 \times -3.405)$$
$$= 726.44 \text{ kN (tensile)}$$

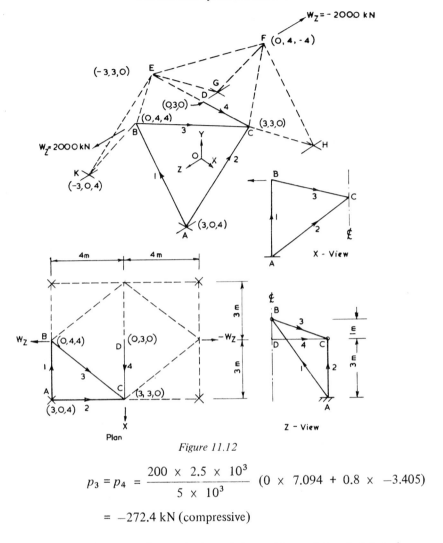

Figure 11.12

$$p_3 = p_4 = \frac{200 \times 2.5 \times 10^3}{5 \times 10^3}(0 \times 7.094 + 0.8 \times -3.405)$$

$$= -272.4 \text{ kN (compressive)}$$

Example 2. The table shows the particulars of the members of the pin-jointed space frame shown in *Figure 11.12*. The frame is symmetrical about both X and Z axes. It is supported in the Z-X plane at AGH and K and carries two horizontal loads $W_Z = \pm 2000$ kN at B and F in the direction of the Z axis. Joints C and E are connected by a horizontal member which intersects the vertical axes of symmetry at D. Calculate the deflections at B, C and D. Take E as 200 kN/mm².

Member	length m	area mm²	l_p	m_p	n_p
1	5	2.5×10^3	-0.6	0.8	0
2	5	2.5×10^3	0	0.6	-0.8
3	5.1	2.55×10^3	$3/5.1$	$-1/5.1$	$-4/5.1$
4	3	3×10^3	1	0	0

Answer: Because of symmetry only a quarter of the frame is analysed. This is shown in full lines in the isometric drawing. Joint B moves in Z and Y directions, joint C moves in X and Y directions while joint D moves in Y direction only. However, the vertical deflection of joints D and C are equal. Thus there are four unknown deflections. These are y_B, z_B, x_C and y_C.

$$a_1 = 200 \times 2.5 \times 10^3/(5 \times 10^3) = 100 = a_2 = a_3$$
$$a_4 = 200 \times 0.5 \times 3 \times 10^3/(3 \times 10^3) = 100$$

The columns of the 4 × 4 overall stiffness matrix **K** for the whole structure correspond to y_B, z_B, x_C and y_C and the rows correspond to the loads W_{yB}, W_{zB}, W_{xC} and W_{yC}.

The first member is connected to joint B at its second end. Thus from equations (11.18) we select the elements in rows W_{yk}, W_{zk} and columns y_k and z_k, for this member and contribute them to the rows of W_{yB}, W_{zB} and the columns of y_B and z_B. These elements are

$$\begin{array}{c} \\ W_{yB} \\ W_{zB} \end{array} \begin{array}{cc} y_B & z_B \\ \left[\begin{array}{c|c} a_1 m_{P1}^2 & a_1 n_{P1} m_{P1} \\ \hline a_1 n_{P1} m_{P1} & a_1 n_{P1}^2 \end{array} \right] \end{array} = \left[\begin{array}{c|c} 64 & 0 \\ \hline 0 & 0 \end{array} \right]$$

The second member is connected to joint C at its second end. From matrix **K** of equations (11.18) we select the elements in rows W_{xk}, W_{yk} and in columns x_k, y_k and contribute them to the rows of W_{xC}, W_{yC} and columns x_C and y_C. These are

$$\begin{array}{c} \\ W_{xC} \\ W_{yC} \end{array} \begin{array}{cc} x_C & y_C \\ \left[\begin{array}{cc} a_2 \ell_{P2}^2 & a_2 \ell_{P2} m_{P2} \\ a_2 \ell_{P2} m_{P2} & a_2 m_{P2}^2 \end{array} \right] \end{array} = \left[\begin{array}{cc} 0 & 0 \\ 0 & 36 \end{array} \right]$$

Member 3 is connected to joint B at its first end and joint C at its second end. From matrix **K** of equation (11.18) we select the elements in rows $W_{yj}, W_{zj}, W_{xk}, W_{yk}$ and columns y_j, z_j, x_k and y_k and contribute them to rows of W_{yB}, W_{zB}, W_{xC} and W_{yC} and columns y_B, z_B, x_C and y_C. These are shown in equations (11.19a).

Member 4 is connected to joint D at its first end and C at its second. From equations (11.18) we select elements on rows W_{yj}, W_{xk}, W_{yk} and columns y_j, x_k and y_k and contribute them to rows of W_{yD}, W_{xC}, W_{yC} and columns y_D, x_C and y_C and in that order. These are

$$\begin{array}{c} \\ W_{xC} \\ W_{yC} \\ W_{yD} \end{array} \begin{array}{ccc} x_C & y_C & y_D \\ \left[\begin{array}{c|c|c} a_4 \ell_{P4}^2 & a_4 \ell_{P4} m_{P4} & -a_4 \ell_P m_{P4} \\ a_4 \ell_{P4} m_{P4} & a_4 m_{P4}^2 & -a_4 m_{P4}^2 \\ -a_4 \ell_{P4} m_{P4} & -a_4 m_{P4}^2 & a_4 m_{P4}^2 \end{array} \right] \end{array}$$

Pin Jointed Space Structures 287

$$
= \begin{array}{c} W_{xC} \\ W_{yC} \\ W_{yD} \end{array} \begin{bmatrix} x_C & y_C & y_D \\ 100 & 0 \leftarrow -0 \\ 0 & 0 \leftarrow -0 \\ -0 & -0 & 0 \end{bmatrix}
$$

But $y_D = y_C$. Thus the contributions of member 4 to row W_{yD} and column y_D go to row W_{yC} and column y_C as shown by the arrows. The final contributions of member 4 then become

$$
\begin{array}{c} W_{xC} \\ W_{yC} \end{array} \begin{bmatrix} x_C & y_C \\ 100 & 0-0 \\ 0-0 & 0-0-0+0 \end{bmatrix} = \begin{bmatrix} 100 & 0 \\ 0 & 0 \end{bmatrix}
$$

The matrix equations $W = K \Delta$ for the frame is thus

$$
\begin{bmatrix} W_{yB} \\ W_{zB} \\ W_{xC} \\ W_{yC} \end{bmatrix} = \begin{bmatrix} a_1 m_{P1}^2 + a_3 m_{P3}^2 & a_1 n_{P1} m_{P1} + a_3 n_{P3} m_{P3} & -a_3 \ell_{P3} m_{P3} & -a_3 m_{P3}^2 \\ a_1 n_{P1} m_{P1} + a_3 n_{P3} m_{P3} & a_1 n_{P1}^2 + a_3 n_{P3}^2 & -a_3 \ell_{P3} n_{P3} & -a_3 m_{P3} n_{P3} \\ -a_3 \ell_{P3} m_{P3} & -a_3 \ell_{P3} n_{P3} & a_2 \ell_{P2}^2 + a_3 \ell_{P3}^2 + a_4 \ell_{P4}^2 & a_2 \ell_{P2} m_{P2} + a_3 \ell_{P3} m_{P3} \\ -a_3 m_{P3}^2 & -a_3 m_{P3} n_{P3} & a_2 \ell_{P2} m_{P2} + a_3 \ell_{P3} m_{P3} & a_2 m_{P2}^2 + a_3 m_{P3}^2 \end{bmatrix} \begin{bmatrix} y_B \\ z_B \\ x_C \\ y_C \end{bmatrix}
$$

(11.19a)

288   Pin Jointed Space Structures

i.e.

$$\begin{bmatrix} 0 \\ 1000 \\ 0 \\ 0 \end{bmatrix} = \begin{bmatrix} 64 + 3.85 & 0 + 15.38 & 11.54 & -3.85 \\ 0 + 15.38 & 0 + 61.54 & +46.15 & -15.38 \\ 11.54 & 46.15 & 100 + 34.62 & 0 - 11.54 \\ -3.85 & -15.38 & 0 - 11.54 & 36 + 3.85 \end{bmatrix} \begin{bmatrix} y_B \\ z_B \\ x_C \\ y_C \end{bmatrix}$$

(11.19b)

Solving these equations gives

$$y_B = -3.91 \text{ mm}, \quad z_B = 24.62 \text{ mm}$$

$$x_C = -7.50 \text{ mm}, \quad y_C = 7.04 \text{ mm} = y_D$$

Example 3. A and B are two joints of a pin jointed space frame. Joint A is restrained in the Z direction while joint B is restrained in the Y direction. It is known that the deflection z_B is half the deflection x_A. What are the contributions of member AB to the overall stiffness of the frame
  (a) When A is the first end of the member as shown in *Figure 11.13a*, and
  (b) When A is the second end of the member.

$$EA/L = 100 \text{ kN/mm}.$$

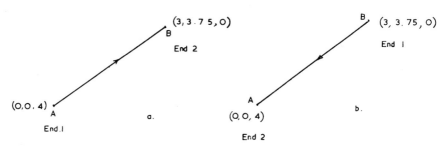

Figure 11.13

Answer:

(a) $$L = \sqrt{[(3 - 0)^2 + (3.75 - 0)^2 + (0 - 4)^2]} = 6.25$$

$$\ell_p = (3 - 0)/6.25 = 0.48, \quad m_p = (3.75 - 0)/6.25 = 0.6,$$

$$n_p = (0 - 4)/6.25 = -0.64$$

In equations (11.18) j corresponds to A and k to B.
Because $z_A = y_B = 0$, the elements in columns of z_j and y_k and rows W_{zj} and W_{yk}

are disregarded. The rest of the elements in equations (11.18) contribute to the stiffness of the frame. These are

$$
\begin{array}{c}
 & \begin{array}{cccc} x_A & y_A & x_B & z_B \end{array} \\
\begin{array}{c} W_{xA} \\ W_{yA} \\ W_{xB} \\ W_{zB} \end{array} &
\left[\begin{array}{cccc}
a\ell_p^2 & a\ell_p m_p & -a\ell_p^2 & -a\ell_p n_p \\
a\ell_p m_p & a m_p^2 & -a\ell_p m_p & -a n_p m_p \\
-a\ell_p^2 & -a\ell_p m_p & a\ell_p^2 & a\ell_p n_p \\
-a\ell_p n_p & -a n_p m_p & a\ell_p n_p & a n_p^2
\end{array} \right]
\end{array}
$$

Because $z_B = \tfrac{1}{2} x_A$, the elements in column z_B are divided by two and added to the elements in column x_A. The elements in row W_{zB} are also halved and added to row W_{xA}. Column z_B and row W_{zB} are then disregarded. In this manner the number of unknowns are reduced from four to three. The above matrix then becomes

$$
K_{AB} = \left[\begin{array}{c|c|c}
\begin{array}{c} a\ell_p^2 - (a/2)\ell_p n_p \\ -a\ell_p n_p/2 + a n_p^2/4 \end{array} & a\ell_p m_p & -a\ell_p^2 \\
 & -a n_p m_p/2 & +a\ell_p n_p/2 \\
\hline
a\ell_p m_p - a n_p m_p/2 & a m_p^2 & -a\ell_p m_p \\
\hline
-a\ell_p^2 + a\ell_p n_p/2 & -a\ell_p m_p & a\ell_p^2
\end{array} \right] \quad (11.20)
$$

Thus the contributions of the member become

$$
\begin{bmatrix}
64 & 48 & -38.4 \\
48 & 36 & -28.8 \\
-38.4 & -28.8 & 23.04
\end{bmatrix}
$$

It should be remembered that the load W_{zB} should also be halved and added to W_{xA} before solving the stiffness equations.

(b) Equations (11.20) do not change but the direction cosines do. These are now

$$\ell_p = (0 - 3)/6.25 = -0.48, \quad m_p = (0 - 3.75)/6.25 = -0.6 \text{ and}$$

$$n_p = (4 - 0)/6.25 = 0.64$$

Nevertheless, the contributions of the member do not alter.

Exercises on Chapter 11

1. A space frame consists of three members AD, BD and CD. It is supported at A(0, 0, 0), B(8, 0, 6) and C(12, 0, 2) and the three members are connected together at D(16, 12, 0) where a vertical upward force of 2 kN is acting. Calculate the member forces. The coordinates are in metres.
Ans. $p_{AD} = +2.03$ kN, $p_{BD} = -0.95$ kN, $p_{CD} = -2.33$ kN.

2. Calculate the vertical support reactions at 0, B and C of the pin jointed frame shown in *Figure 11.14* when it carries a vertical load of 50 kN at D.
Ans. $R_o = 166.7$ kN, $R_B = -66.7$ kN, $R_C = -50$ kN.

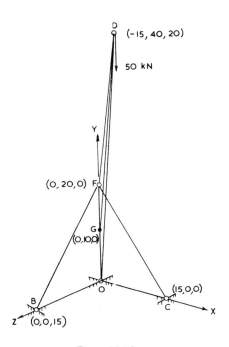

Figure 11.14

3. The pin jointed space structure shown in *Figure 11.15* is subject at D to a horizontal force of $W = 10$ kN which is at 30° to the X axis. Calculate the deflection at D in the direction of the X axis. Take $E = 200$ kN/mm². The coordinates are in metres.
Ans. 1.85 mm.

4. Calculate the forces in the members of the cubic space structure shown in *Figure 11.16*. The forces W are acting along the diagonal AG.
Ans. $p_{AB} = p_{AD} = p_{HG} = p_{GF} = p_{AE} = p_{GC} = p_{EF} = p_{DC} = W/\sqrt{3}$

$p_{BE} = p_{BD} = p_{FH} = p_{CH} = -\sqrt{2}\,W/\sqrt{3}$, $p_{BH} = W$

Others = 0

Pin Jointed Space Structures   291

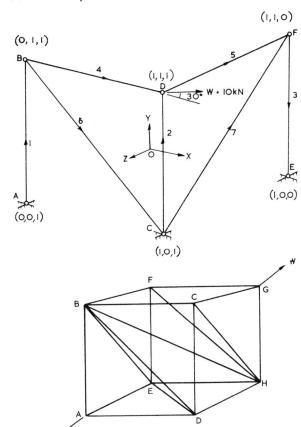

Figure 11.15

Figure 11.16

5. In the space structure shown in *Figure 11.17*, ABC is an equilateral triangle of side $2L$ while DEF is another equilateral triangle of side L. The vertical distance between these two horizontal triangles is L. Calculate the member forces.

Ans. $p_{AD} = p_{BE} = p_{CF} = 2W/\sqrt{3}$, $p_{DE} = p_{EF} = p_{DF} = W/\sqrt{3}$
$p_{AE} = p_{AF} = p_{BF} = 0$.

Figure 11.17

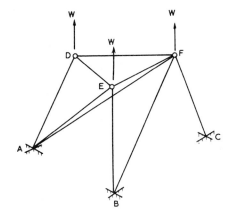

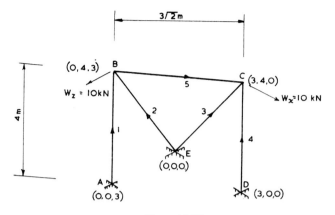

Figure 11.18

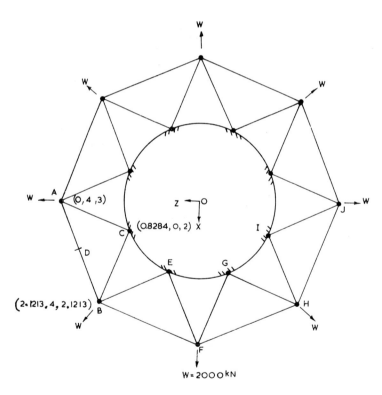

Figure 11.19

5. Construct matrix **K** for the space frame shown in *Figure 11.18*. Take $E = 200$ N/mm^2 and the area of each member as 10^3 mm^2.

Ans.

$$\begin{bmatrix} 23.57 & & & & & & \\ 0 & 75.6 & & \text{Symmetrical} & & & \\ -23.57 & 19.2 & 37.97 & & & & \\ -23.57 & 0 & 23.57 & 37.94 & & & \\ 0 & 0 & 0 & 19.2 & 65.6 & & \\ 23.57 & 0 & -23.57 & -23.57 & 0 & 23.57 \end{bmatrix}$$

7. The symmetrical pin jointed space frame shown in *Figure 11.19* is supported along CEGI etc. and carries radial loads $W = 2000$ kN at A, B, F, H etc. Calculate the joint deflections; (a) by analysing the portion ACB and (b) by analysing the portion ACD. Take E as 200 kN/mm² and $A = 10^3$ mm² for all the members. The coordinates shown are in metres.

Ans. $y_A = y_B = y_F = -4.9$ mm, radial displacement $z_A = 19.6$ mm.

12

The stiffness method for the analysis of rigidly jointed plane frames

12.1. Introduction

A complete analysis of a rigidly jointed plane frame involves the calculation of all the axial forces, the shearing forces and the bending moments in the members, as well as the deflections at some or all the joints. The linear elastic method, introduced in Chapters 5, 7 and 9, assumes that the stress-strain relationship for the material of the frame is linear elastic and obeys Hooke's law. This method also assumes that the effect of the axial forces in the members on the bending moments and deflections is negligible. Deflections and bending moments calculated by this method are linearly related to the external loads and the principle of superposition is valid. We begin this chapter by formalising the steps required to carry out the linear analysis of rigidly jointed frames using the stiffness method.

An extension of the linear elastic method is known as the 'non-linear elastic' method. This latter method takes the effect of the axial loads in the members on the bending moments and deflections into consideration. The non-linear elastic method also assumes that the material of the frame is elastic and linear, obeying Hooke's law. However, because the effect of the member axial loads on the bending moments and deflections are taken into consideration, these latter quantities are larger than their corresponding values obtained by the linear elastic method. As the external loads acting on a frame are increased, the axial forces in the members become progressively significant. Because of this, the bending moments and deflections are non-linearly related to the externally applied loads. The principle of superposition is not valid here. The later part of this chapter is concerned with this method.

12.2. The steps required in a linear analysis

To analyse a frame, neglecting the effect of member axial forces, the following steps are necessary:

(1) For each member, the sway v and the rotations θ_1 and θ_2 at either end are expressed in terms of the displacements x, y and θ of the joints to which the member is connected.

(2) For each member, the slope deflection equations (5.14) and (7.2) or (7.3) are written.

(3) For each joint, three equations of equilibrium are prepared. These are:

(a) The sum of the horizontal components of the forces in members meeting at a joint is equal to the external horizontal force acting at that joint.

Stiffness Method for Analysis 295

(b) The sum of the vertical components of the forces in members meeting at a joint is equal to the external vertical force acting at that joint.
(c) The sum of the moments developed at the ends of the members meeting at a joint is equal to the external moment acting at that joint.

(4) The joint equilibrium equations are solved for the unknown values of joint displacements x, y and θ.

(5) The shearing force, the bending moments and the axial force in each member are calculated using the slope deflection equations and Hooke's law. This completes the linear elastic analysis of the frame.

12.3. Relationships between member deformations and joint displacements

Member AB shown in *Figure 12.1* is connected to joint j at its first end and joint k at its second end. Under the influence of some external loads let joint j move to

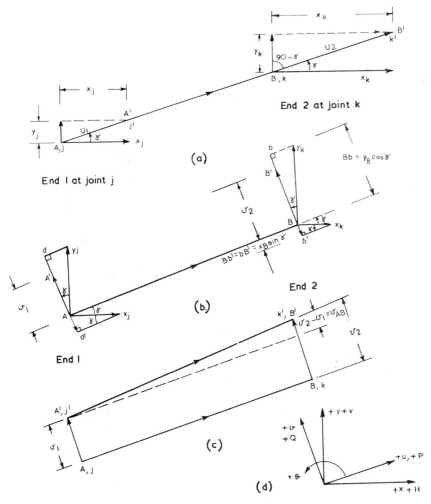

Figure 12.1

Stiffness Method for Analysis

j′ and joint k to k′ as shown in *Figure 12.1a*. Let the displacements of j be x_j horizontally and y_j vertically while those of joint k are x_k and y_k.

The movement of end 1 of the member parallel to the direction AB is u_1. Now the components of x_j in the direction of AB is $x_j \cos \gamma$ while that of y_j is $y_j \sin \gamma$ where γ is the positive, anticlockwise, angle of inclination of the member. These components (see *Figure 12.1a*) vectorially add up to u_1, i.e.

$$u_1 = x_j \cos \gamma + y_j \sin \gamma$$

Similarly, the components $x_k \cos \gamma$ and $y_k \sin \gamma$ of the displacements x_k and y_k of joint k add up to u_2 which is the movement of joint k to k′ parallel to AB, thus

$$u_2 = x_k \cos \gamma + y_k \sin \gamma$$

The net extension of member AB is u_{AB}, given by

$$u_{AB} = u_2 - u_1 = (x_k \cos \gamma + y_k \sin \gamma) - (x_j \cos \gamma + y_j \sin \gamma)$$

i.e.

$$u_{AB} = -x_j \cos \gamma - y_j \sin \gamma + x_k \cos \gamma + y_k \sin \gamma \tag{12.1}$$

Next consider the case of member AB moving perpendicularly to its original position. Let j move to j′, thus A moves to A′ by an amount v_1 and B moves to B′ by an amount v_2 as shown in *Figure 12.1b*. At k the component of the displacement y_k parallel to AA′ is $y_k \cos \gamma$ and that of x_k is $x_k \sin \gamma$. These two components vectorially add up to v_2, thus $BB' = v_2 = y_k \cos \gamma - x_k \sin \gamma$. Similarly at A

$$AA' = v_1 = y_j \cos \gamma - x_j \sin \gamma$$

The net sway v_{AB} of the member AB is shown in *Figure 12.1c* which is given by

$$v_{AB} = v_2 - v_1 = (y_k \cos \gamma - x_k \sin \gamma) - (y_j \cos \gamma - x_j \sin \gamma)$$

i.e.

$$v_{AB} = x_j \sin \gamma - y_j \cos \gamma - x_k \sin \gamma + y_k \cos \gamma \tag{12.2}$$

Because of the rigidity of joints j and k, if they rotate by amounts θ_j and θ_k, then the rotation θ_{AB} of end 1 of the member is equal to θ_j and θ_{BA} at end 2 is equal to θ_k, thus

$$\left. \begin{array}{l} \theta_{AB} = \theta_j \\ \theta_{BA} = \theta_k \end{array} \right\} \tag{12.3}$$

Stiffness Method for Analysis 297

Equations (12.1) to (12.3) are written in matrix form as

$$\begin{bmatrix} u_{AB} \\ v_{AB} \\ \theta_{AB} \\ \theta_{BA} \end{bmatrix} = \begin{bmatrix} -\cos\gamma & -\sin\gamma & 0 & | & \cos\gamma & \sin\gamma & 0 \\ \sin\gamma & -\cos\gamma & 0 & | & -\sin\gamma & \cos\gamma & 0 \\ 0 & 0 & 1 & | & 0 & 0 & 0 \\ 0 & 0 & 0 & | & 0 & 0 & 1 \end{bmatrix} \begin{bmatrix} x_j \\ y_j \\ \theta_j \\ x_k \\ y_k \\ \theta_k \end{bmatrix}$$

(12.4a)

or simply: $\boldsymbol{\delta} = \mathbf{D}\boldsymbol{\Delta}$ (12.4b)

Matrix **D** is called the displacement transformation matrix and relates the member deformations $\boldsymbol{\delta}$ to the joint displacements $\boldsymbol{\Delta}$.

12.4. Examples

Example 1. ABCD, shown in *Figure 12.2*, is part of a rigidly jointed plane frame. Define the member deformations for this part in terms of the displacements of the joints and express these in matrix form.
Answer:
For member 1: $\cos\gamma_1 = 0.6$, $\sin\gamma_1 = 0.8$

Referring to member AB in *Figure 12.2b*

$$u_{AB} = -\cos\gamma_1 x_A - \sin\gamma_1 y_A + \cos\gamma_1 x_B + \sin\gamma_1 y_B$$

$$= -0.6\, x_A - 0.8\, y_A + 0.6\, x_B + 0.8\, y_B$$

$$v_{AB} = \sin\gamma_1 x_A - \cos\gamma_1 y_A - \sin\gamma_1 x_B + \cos\gamma_1 y_B$$

$$= 0.8\, x_A - 0.6\, y_A - 0.8\, x_B + 0.6\, y_B$$

End 1 of the member is connected to joint A and end 2 is connected to joint B, thus

$$\theta_{AB} = \theta_A$$

$$\theta_{BA} = \theta_B$$

For member 2 $\gamma_2 = 315°$ or $\gamma_2 = -45°$, thus

$$u_{BC} = -x_B \cos(-45) - y_B \sin(-45) + x_C \cos(-45) + y_C \sin(-45)$$

298 *Stiffness Method for Analysis*

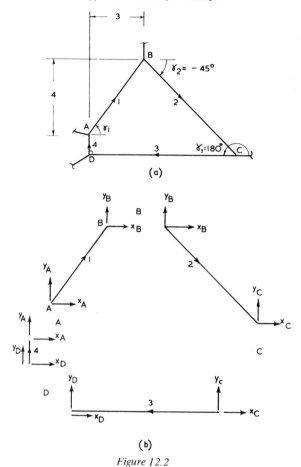

Figure 12.2

$$u_{BC} = (-x_B/\sqrt{2}) + (y_B/\sqrt{2}) + (x_C/\sqrt{2}) - (y_C/\sqrt{2})$$

$$v_{BC} = x_B \sin(-45) - y_B \cos(-45) - x_C \sin(-45) + y_C \cos(-45)$$

$$v_{BC} = (-x_B/\sqrt{2}) - (y_B/\sqrt{2}) + (x_C/\sqrt{2}) + (y_C/\sqrt{2})$$

$$\theta_{BC} = \theta_B$$

$$\theta_{CB} = \theta_C$$

For member 3: $\gamma_3 = 180°$

$$u_{CD} = -x_C \cos 180 - y_C \sin 180 + x_D \cos 180 + y_D \sin 180$$

$$\therefore \quad u_{CD} = x_C - 0 \times y_C - x_D + 0 \times y_D$$

Stiffness Method for Analysis

$$v_{CD} = x_C \sin 180 - y_C \cos 180 - x_D \sin 180 + y_D \cos 180$$

$$v_{CD} = 0 \times x_C + y_C - 0 \times x_D - y_D$$

$$\theta_{CD} = \theta_C$$

$$\theta_{DC} = \theta_D$$

For member 4: $\gamma_4 = 90°$

$$u_{DA} = -x_D \cos 90 - y_D \sin 90 + x_A \cos 90 + y_A \sin 90$$

$$= -0 \times x_D - y_D + 0 \times x_A + y_A$$

$$v_{DA} = x_D \sin 90 - y_D \cos 90 - x_A \sin 90 + y_A \cos 90$$

$$v_{DA} = x_D - 0 \times y_D - x_A + 0 \times y_A$$

$$\theta_{DA} = \theta_D$$

$$\theta_{AD} = \theta_A$$

In the matrix form these equations become

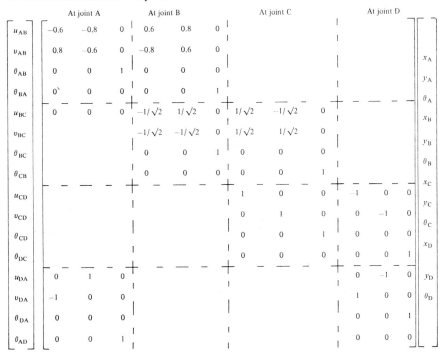

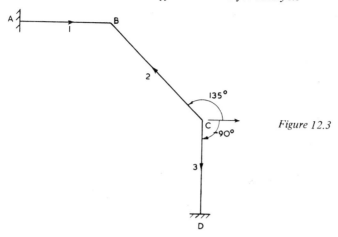

Figure 12.3

Other elements are zeros.

Example 2. Construct the displacement transformation matrix for the frame shown in *Figure 12.3*

(a) when x_B and y_C are small and

(b) when x_B and y_C are small while $x_C = -y_B$ and $\theta_C = \theta_B$.

Answer:

(a) Joints A and D are fixed supports and all their displacements are zero. Since x_B and y_C are small, they will be disregarded. The remaining joint displacements are

$$\Delta = \{y_B \quad \theta_B \quad x_C \quad \theta_C\}$$

For member 1: $\gamma_1 = 0$

$$u_{AB} = x_B \cos 0 + y_B \sin 0$$

$$= x_B = 0, \text{ since } x_B \text{ is disregarded.}$$

$$v_{AB} = -x_B \sin 0 + y_B \cos 0$$

$$= y_B$$

$$\theta_{AB} = \theta_A = 0$$

$$\theta_{BA} = \theta_B$$

For member 2: Joint C is at the first end

$$\gamma_2 = 135°, \cos 135° = -1/\sqrt{2}, \sin 135° = 1/\sqrt{2}$$

$$u_{CB} = -x_C \times (-1/\sqrt{2}) - y_C \times (1/\sqrt{2}) + x_B \times (-1/\sqrt{2}) + y_B \times (1/\sqrt{2})$$

Stiffness Method for Analysis

$$\therefore \quad u_{CB} = (y_B/\sqrt{2}) + (x_C/\sqrt{2})$$

$$v_{CB} = x_C \times (1/\sqrt{2}) - y_C \times (-1/\sqrt{2}) - x_B \times (1/\sqrt{2}) + y_B \times (-1/\sqrt{2})$$

$$v_{CB} = (-y_B/\sqrt{2}) + (x_C/\sqrt{2})$$

$$\theta_{CB} = \theta_C$$

$$\theta_{BC} = \theta_B$$

For member 3: Joint C is at the first end, $\gamma_3 = -90°$, $\cos(-90) = 0$
$\sin(-90) = -1$

$$u_{CD} = -x_C \times 0 - y_C \times -1 + x_D \times 0 + y_D \times -1$$

$\therefore \quad u_{CD} = 0$, since y_C and y_D are disregarded.

$$v_{CD} = x_C \times -1 - y_C \times 0 - x_D \times -1 + y_D \times 0$$

$$\therefore \quad v_{CD} = -x_C$$

$$\theta_{CD} = \theta_C$$

$$\theta_{DC} = \theta_D = 0$$

The equations $\pmb{\delta} = \mathbf{D}\pmb{\Delta}$ become

	At joint B		At joint C		
v_{AB}	1	0	0	0	y_B
θ_{BA}	0	1	0	0	θ_B
u_{CB}	$1/\sqrt{2}$	0	$1/\sqrt{2}$	0	x_C
v_{CB}	$-1/\sqrt{2}$	0	$1/\sqrt{2}$	0	θ_C
θ_{CB}	0	0	0	1	
θ_{BC}	0	1	0	0	
v_{CD}	0	0	-1	0	
θ_{CD}	0	0	0	1	

(b) When $x_C = -y_B$ and $\theta_C = \theta_B$

$$u_{CB} = (y_B/\sqrt{2}) + (x_C/\sqrt{2}) = (y_B/\sqrt{2}) - (y_B/\sqrt{2}) = 0$$

$$v_{CB} = -(y_B/\sqrt{2}) + (x_C/\sqrt{2}) - (y_B/\sqrt{2}) = -(y_B/\sqrt{2})$$

$$\theta_{CB} = \theta_C = \theta_B$$

$$v_{CD} = -x_C = y_B$$

$$\theta_{CD} = \theta_C = \theta_B$$

The joint displacement vector is now

$$\Delta = \{y_B \quad \theta_B\}$$

and equations $\delta = D\Delta$ become

$$\begin{bmatrix} v_{AB} \\ \theta_{BA} \\ v_{CB} \\ \theta_{CB} \\ \theta_{BC} \\ v_{CD} \\ \theta_{CD} \end{bmatrix} = \begin{bmatrix} 1 & 0 \\ 0 & 1 \\ -\sqrt{2} & 0 \\ 0 & 1 \\ 0 & 1 \\ 1 & 0 \\ 0 & 1 \end{bmatrix} \begin{bmatrix} y_B \\ \theta_B \end{bmatrix}$$

12.5. The components of the member forces

In *Figure 12.4* member AB of a frame is at an inclination γ to the horizontal X axis of the frame. The arrow on the member points to B, thus the positive P axis of the member is from A to B and the positive Q axis is from A to C. The shearing force in the member is F_{AB} which is shown positive and the axial force is p_{AB} which is also shown to be positive, i.e. acting in +P direction, at both ends. The sum of the horizontal components of these forces at A is H_A given by

$$H_A = F_{AB} \sin\gamma + p_{AB} \cos\gamma \qquad (12.5)$$

The sum of the vertical components at A is V_A given by

$$V_A = -F_{AB} \cos\gamma + p_{AB} \sin\gamma \qquad (12.6)$$

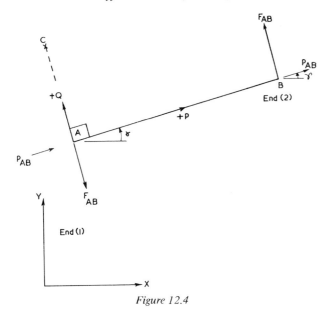

Figure 12.4

At B the sum of the horizontal and vertical components of F_{AB} and p_{AB} are

$$\left.\begin{array}{l} H_B = -F_{AB} \sin \gamma + p_{AB} \cos \gamma \\ V_B = F_{AB} \cos \gamma + p_{AB} \sin \gamma \end{array}\right\} \qquad (12.7)$$

The joint equilibrium equations at A thus contain contributions similar to equations (12.5) and (12.6), for all the members meeting at A. These contributions add up to the external horizontal and vertical loads applied at A. Similarly for joint B.

The equation of rotational equilibrium for joint A simply states that

$$\sum_{j=1}^{n} M_{Aj} = M_A \qquad (12.8)$$

where M_A is the external moment acting at joint A, while each M_{Aj} is the bending moment at end A of member j as given by the slope deflection equations. Altogether there are n members meeting at A.

The equations of joint equilibrium prepared in this manner are of the form

$$W = K\Delta = D^T k D\Delta \qquad (12.9)$$

where D^T is the transpose of the displacement transformation matrix D and matrix $K = D^T k D$ is the overall stiffness matrix of the structure and relates the unknown joint displacements Δ to the externally applied load vector W. The solution of equations (12.9) gives the values of the joint displacements.

Using matrix algebra, the joint equilibrium equations (12.9) are derived as follows: With the joint displacements $X = \Delta$, the virtual work equations (7.44b) become

$$W^T \Delta = P^T \delta \tag{a}$$

Equations (12.4b) give $\delta = D\Delta$ and thus equations (a) can be written as

$$W^T \Delta = P^T D \Delta$$

Now if this is true irrespective of the numerical values of the elements Δ, it means that

$$W^T = P^T D \tag{b}$$

Transposing both sides of equations (b)

$$W = D^T P \tag{c}$$

But equations (7.9b) state that $P = k\delta$ and hence substituting for P in equations (c) we obtain

$$W = D^T k \delta \tag{d}$$

Again, since $\delta = D\Delta$, equations (d) finally become

$$W = D^T k D \Delta$$

The equations of joint equilibrium are square, i.e. matrix K is square, and there are exactly the same number of equations as there are unknown displacements. Matrix K is also symmetrical, the element a_{ij} in row i and column j is equal to the element a_{ji} in row j and column i.

12.6. Examples

Example 1. The propped cantilever shown in *Figure 12.5* is fixed at A and supported at B by the pin ended bar BC. At B the system is subject to a horizontal force

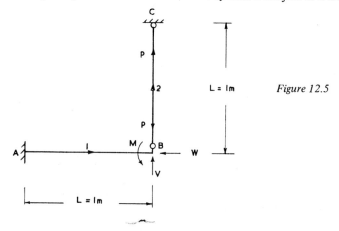

Figure 12.5

$W = -197 \times 10^3$ kN, a vertical force $V = 100$ kN and an external moment $M = 10$ kN m. Calculate the displacements at joint B and the bending moment at the fixed end A. The value of I for the beam is 10^9 mm^4 and its area is 10^5 mm^2. The area of the bar BC is 100 mm^2 and E is 200 kN/mm^2 for both members.

Answer: The horizontal displacement of joint B is calculated from the axial flexibility of member AB, thus

$$x_B = (L/EA)W = \frac{10^3 \times -197 \times 10^3}{200 \times 10^5} = -9.85 \text{ mm (to the left)}$$

Step 1: In equation (12.2), A replaces j while B replaces k. Disregarding terms associated with joint A, because $x_A = y_A = 0$, this equation becomes

$$v_{AB} = -x_B \sin \gamma + y_B \cos \gamma$$

Now $\gamma = 0$, $\sin \gamma = 0$ and $\cos \gamma = 1$, thus

$$v_{AB} = y_B$$

Because A is fixed, $\theta_{AB} = 0$, while $\theta_{BA} = \theta_B$.
For the bar BC, $\gamma_2 = 90$, $\sin \gamma_2 = 1$, $\cos \gamma_2 = 0$, equation (12.1) becomes

$$u_{BC} = -x_B \times 0 - y_B \times 1 + x_C \times 0 + y_C \times 1$$

But $y_C = 0$ because joint C is a support

$$\therefore \quad u_{BC} = -y_B$$

The equations $\boldsymbol{\delta} = \mathbf{D}\boldsymbol{\Delta}$ thus become

$$\begin{bmatrix} v_{AB} \\ \theta_{BA} \\ u_{BC} \end{bmatrix} = \begin{bmatrix} 1 & 0 \\ 0 & 1 \\ -1 & 0 \end{bmatrix} \begin{bmatrix} y_B \\ \theta_B \end{bmatrix}$$

Step 2: The slope deflection equations for member AB are

$$F_{AB} = -(6k/L)\theta_{BA} + (12k/L^2)v_{AB} = -(6k/L)\theta_B + (12k/L^2)y_B \quad (12.10)$$

$$M_{BA} = 4k\theta_{BA} - 6k\,v_{AB}/L = 4k\theta_B - 6k\,y_B/L \quad (12.11)$$

where $k = EI/L = 200 \times 10^9/10^3 = 200 \times 10^6$ kN mm.

Step 3: The vertical component of the axial force p in member BC at B is $p \sin 90 = p$ while the vertical component of the shear force in AB, at B, is $F_{AB} \cos 0 = F_{AB}$.

For the vertical equilibrium of joint B, the forces p and F_{AB} add up to the vertical external load V, thus

$$V = F_{AB} + p = 100 \qquad (12.12)$$

Let the axial stiffness of BC be

$$k_{BC} = EA/L = \alpha EI/L^3$$

where α is a constant and I is the second moment of area of the cantilever, then

$$\alpha = AL^2/I = 10^2 \times 10^6/10^9 = 0.1$$

and

$$p = (EA/L)y_B = (\alpha EI/L^3)y_B = 0.1\, ky_B/L^2$$

using this and equation (12.10), equation (12.12) becomes

$$-(6k/L)\theta_B + (12k/L^2)y_B + (0.1k/L^2)y_B = 100 \qquad (12.13a)$$

The externally applied moment at B is $M = 10^4$ kN mm and since the pin ended member CB cannot withstand a bending moment at B, it follows that the external moment M is equal to M_{BA}, thus

$$M_{BA} = 4k\theta_B - 6k\, y_B/L = 10^4 \qquad (12.13b)$$

This and equation (12.13a) are the joint equilibrium equations and can be written in matrix from $\mathbf{K}\,\boldsymbol{\Delta} = \mathbf{W}$ as

$$\begin{bmatrix} k(12 + \alpha)/L^2 & -6k/L \\ -6k/L & 4k \end{bmatrix} \begin{bmatrix} y_B \\ \theta_B \end{bmatrix} = \begin{bmatrix} 100 \\ 10^4 \end{bmatrix} \qquad (12.13c)$$

Step 4: With $k = 200 \times 10^6$ and $\alpha = 0.1$, solving equations (12.13), we obtain

$$\theta_B = 2.907 \times 10^{-4} \text{ rad}$$

$$y_B = 0.1855 \text{ mm}$$

Step 5: From the slope deflection equations

$$M_{AB} = 2k\theta_B - 6k\, v_{AB}/L = 2k\theta_B - 6k\, y_B/L$$

∴ $M_{AB} = 2 \times 200 \times 10^6 \times 2.907 \times 10^{-4} - (6 \times 200 \times 10^6$

$\times 0.1855/10^3)$

$M_{AB} = -106.320$ kN m

Equations 12.13c can also be constructed by the triple matrix multiplication $K = D^T k\ D$. The forces corresponding to v_{AB}, θ_{BA} and u_{BC} are F_{AB}, M_{BA} and p_{BC} respectively, the member stiffness equations $P = k\delta$ are

$$\begin{bmatrix} F_{AB} \\ M_{BA} \\ p_{BC} \end{bmatrix} = \begin{bmatrix} 12\ EI/L^3 & -6\ EI/L^2 & 0 \\ -6\ EI/L^2 & 4\ EI/L & 0 \\ 0 & 0 & EA/L \end{bmatrix} \begin{bmatrix} v_{AB} \\ \theta_{BA} \\ u_{BC} \end{bmatrix}$$

Thus $k\ D = \begin{bmatrix} 12\ EI/L^3 & -6\ EI/L^2 \\ -6\ EI/L^2 & 4\ EI/L \\ -EA/L & 0 \end{bmatrix}$

and $K = D^T k\ D = \begin{bmatrix} 1 & 0 & -1 \\ 0 & 1 & 0 \end{bmatrix} \begin{bmatrix} 12\ EI/L^3 & -6EI/L^2 \\ -6EI/L^2 & 4EI/L \\ -EA/L & 0 \end{bmatrix}$

Which gives matrix K of equations (12.13c).

Example 2. The fixed base portal shown in *Figure 12.6* is subject to vertical loads W at B and D and uniformly distributed wind load of w/unit height as shown. Construct the joint equilibrium equations and calculate the side sway of the frame when $W = 1.0975 \times 10^8$ kN and $w = 100$ kN/mm. Take $E = 200$ kN/mm^2, $I = 10^{12}$ mm^4, $L = 3 \times 10^3$ mm.

Answer: The uniform wind load is equivalent to those shown in *Figure 12.6b* plus fixed ended columns that have no influence on deflections. The frame shown in *Figure 12.6b* is symmetrical but loaded antisymmetrically, thus

$$x_A = y_A = \theta_A = x_C = y_C = \theta_C = 0$$

$$\theta_B = \theta_D = \theta_{BD} = \theta_{DB} = \theta_{BA}$$

Neglecting the axial contraction of the columns

$$v_{BD} = y_D - y_B = 0$$

308   *Stiffness Method for Analysis*

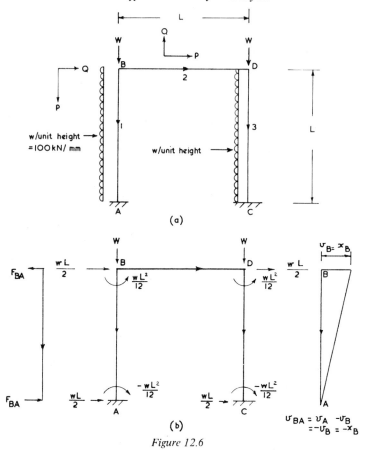

Figure 12.6

With $\gamma = -90°$, equation (12.2) gives

$$v_{BA} = v_{DC} = -x_B$$

The equation $\delta = D\Delta$ becomes

$$\begin{bmatrix} v_{BA} \\ \theta_{BA} \\ \theta_{BD} \\ \theta_{DB} \end{bmatrix} = \begin{bmatrix} -1 & 0 \\ 0 & 1 \\ 0 & 1 \\ 0 & 1 \end{bmatrix} \begin{bmatrix} x_B \\ \theta_B \end{bmatrix}$$

The slope deflection equations for members 1 and 2 are

$$\left. \begin{aligned} M_{BA} &= 4k\theta_B + 6kx_B/L \\ F_{BA} &= -6k\theta_B/L - 12kx_B/L^2 \\ M_{BD} &= 4k\theta_{BD} + 2k\theta_{DB} = 6k\theta_B \\ M_{DB} &= 2k\theta_{BD} + 4k\theta_{DB} = 6k\theta_B \end{aligned} \right\} \quad (12.14)$$

Two independent joint equilibrium equations are required to solve for θ_B and x_B. These are:

(i) For horizontal equilibrium at B

$$-F_{BA} = wL/2$$

∴
$$(6k\theta_B/L) + (12kx_B/L^2) = wL/2 \qquad (12.15a)$$

(ii) For rotational equilibrium at B

$$M_{BA} + M_{BD} = wL^2/12$$

∴
$$10k\theta_B + 6k\,x_B/L = wL^2/12 \qquad (12.15b)$$

No further independent equations can be written. Equilibrium of joint D will also yield the above two equations. In matrix form, the equations $W = K\Delta$ are

$$\begin{bmatrix} 12k/L^2 & 6k/L \\ 6k/L & 10k \end{bmatrix} \begin{bmatrix} x_B \\ \theta_B \end{bmatrix} = \begin{bmatrix} wL/2 \\ wL^2/12 \end{bmatrix} \qquad (12.15c)$$

With $w = 100$ kN/mm,

$$wL/2 = 100 \times 3 \times 10^3/2 = 150 \times 10^3$$
$$wL^2/12 = 100 \times 9 \times 10^6/12 = 75 \times 10^6$$

Solving equations (12.15) with $k = EI/L = 2 \times 10^{11}/3$ we obtain

$$\theta_B = -321.4 \times 10^{-6} \text{ rad}$$

$$x_B = 2.17 \text{ mm.}$$

Notice that because the axial contraction of the columns are neglected, joints B and D do not move vertically and the vertical loads W at these joints do not enter the analysis.

Example 3. The symmetrical fixed base pitched roof frame shown in *Figure 12.7* is loaded symmetrically at the apex and at its eaves. Neglecting the effect of axial forces, plot a graph of the load W against eaves displacement x_B. Take $E = 200$ kN/mm^2, $I = 10^{12}$ mm^4 and $L = 10$m.

Answer: The Q axis of AB is to the left, thus $+v_A$ and $+v_B$ are also to the left. The sway v_{AB} is

$$v_{AB} = v_B - v_A$$

but $v_A = 0$ because A is fixed, thus $v_{AB} = v_B$ where

$$v_B = y_B \cos 90 - x_B \sin 90.$$

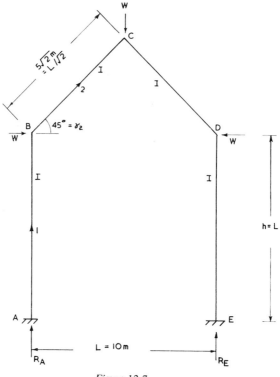

Figure 12.7

Because the axial stiffness of AB is neglected, $y_B = 0$ and

$$v_{AB} = v_B = -x_B$$

The frame is symmetrical and joint C can only move vertically by an amount y_C while $\theta_C = x_C = 0$.

If joint B moves to the right by an amount x_B, it will force joint C to move to C″ by an amount x_B (see *Figure 12.8*). However, C is restricted to move only vertically and the net result of these two factors causes member BC to sway to C′. This sway is measured perpendicular to BC. In *Figure 12.8*, the triangle CC′C″ is the displacement diagram for the half frame ABC. From this triangle

$$y_C = x_B \cot 45° = x_B$$

The sway of BC is v_{BC} given by $v_{BC} = v_2 - v_1$ as shown in *Figure 12.8*, from which

$$v_2 = y_C \cos 45 - x_C \sin 45$$

$$v_1 = y_B \cos 45 - x_B \sin 45$$

Stiffness Method for Analysis 311

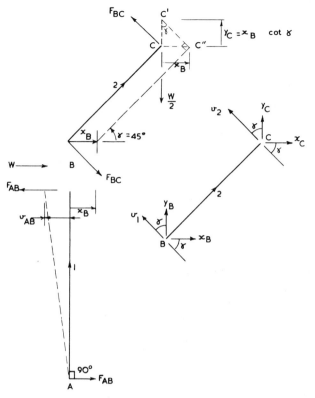

Figure 12.8

Neglecting y_B, the sway v_{BC} becomes

$$v_{BC} = (y_C/\sqrt{2}) - (x_C/\sqrt{2}) + (x_B/\sqrt{2})$$

But $y_C = x_B$ and $x_C = 0$

∴
$$v_{BC} = \sqrt{2} \times x_B$$

Joint A is fixed, thus $\theta_A = 0$.

For member 1

$$k_1 = EI/L = 200 \times 10^{12}/10^4 = 200 \times 10^8 = k$$

For member 2

$$k_2 = EI/(L/\sqrt{2}) = \sqrt{2} \times k = 282.842 \times 10^8.$$

With $e = 4EI/L = 4k$, $d = -6EI/L^2 = -6k/L$,
$b = 12EI/L^3 = 12k/L^2$, the slope deflection equations for members 1 and 2 are

$$M_{BA} = e_1\theta_B + d_1 v_{AB} = e_1\theta_B - d_1 x_B$$
$$F_{AB} = d_1\theta_B + b_1 v_{AB} = d_1\theta_B - b_1 x_B$$
$$M_{BC} = e_2\theta_B + d_2 v_{BC} = e_2\theta_B + \sqrt{2} \times d_2 x_B$$
$$F_{BC} = d_2\theta_B + b_2 v_{BC} = d_2\theta_B + \sqrt{2} \times b_2 x_B$$

For the rotational equilibrium of joint B

$$M_{BA} + M_{BC} = 0$$

$\therefore \qquad (e_1 + e_2)\theta_B + (\sqrt{2} \times d_2 - d_1) x_B = 0 \qquad (12.16a)$

For the horizontal equilibrium of joint B

$$-F_{AB} + F_{BC}/\sqrt{2} = W \qquad (12.17)$$

For the vertical equilibrium of joint C

$$F_{BC}/\sqrt{2} = -W/2 \qquad (12.18)$$

There are thus three equations of joint equilibrium but two unknowns x_B and θ_B. To reduce the number of equations to the number of unknowns, equations (12.17) and (12.18) are added together to give

$$-F_{AB} + \sqrt{2} \times F_{BC} = W/2$$

Substituting for F_{AB} and F_{BC} from the slope deflection equations, we obtain

$$-d_1\theta_B + b_1 x_B + \sqrt{2} \times d_2\theta_B + 2 b_2 x_B = W/2$$

Hence $\qquad (\sqrt{2} \times d_2 - d_1)\theta_B + (b_1 + 2 b_2) x_B = W/2 \qquad (12.16b)$

Solving the joint equilibrium equations (12.16 a and b) we obtain

$$\theta_B = -g x_B/j \qquad (12.19)$$

$$x_B = 0.5 Wj/(hj - g^2) \qquad (12.20)$$

where $j = e_1 + e_2$, $g = \sqrt{2} \times d_2 - d_1$, $h = b_1 + 2b_2$.

Stiffness Method for Analysis   313

It is noticed that both x_B and θ_B are linearly related to the applied load W. For a given value of $W = 1580408.1$ kN and with $k = 200 \times 10^8$, equations (12.20) and (12.19) give

$$x_{B \text{ Linear}} = 92.928 \text{ mm}$$

and

$$\theta_{B \text{ Linear}} = 0.01054 \text{ rads.}$$

The graph of x_B against W will be a straight line and is shown later as OC in *Figure 12.15*. For $\mathbf{D}^T\mathbf{k}\mathbf{D}$ see example 2, section 12.12.

Example 4. Using the triple multiplication $\mathbf{K} = \mathbf{D}^T\mathbf{k}\,\mathbf{D}$, obtain the joint equilibrium equations for the folded roof frame shown in *Figure 12.9*. Make the order of $\mathbf{K}$ as small as possible.
Answer: Because of symmetry half of the frame is considered. This is shown in *Figures 12.9b* and *c*. Joint C is restricted to move vertically and $\theta_C = x_C = 0$.

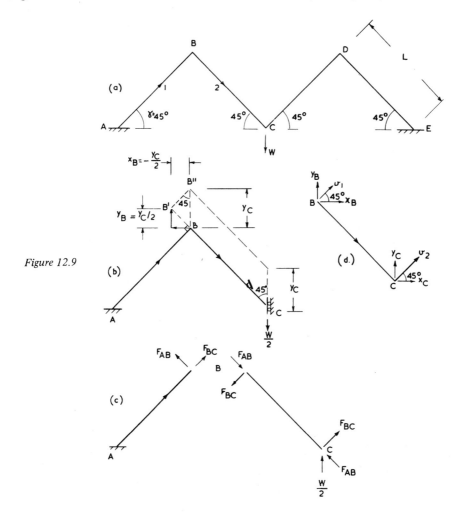

Figure 12.9

Stiffness Method for Analysis

A vertical positive displacement y_C at C tends to force joint B to move to B″, see *Figure 12.9b*. However, neglecting the axial contraction of the members, AB sways perpendicularly to its original direction and moves to B′. The triangle BB′B″ is the displacement diagram. From this it is observed that

$$y_B = y_C/2,$$
$$x_B = -y_C/2$$

The sway v_{BA} of AB is BB′ given by

$$v_{BA} = -(x_B/\sqrt{2}) + (y_B/\sqrt{2}) = (1/\sqrt{2})(y_C/2 + y_C/2) = y_C/\sqrt{2}$$

The sway v_{BC} of BC is equal to $v_2 - v_1$, shown in *Figure 12.9d* and because $x_C = 0$, $v_2 = y_C/\sqrt{2}$ while $v_1 = x_B/\sqrt{2} + y_B/\sqrt{2} = 0$

$$\therefore \qquad v_{BC} = v_2 = y_C/\sqrt{2}$$

The equations $\delta = D\Delta$ become

$$\begin{bmatrix} v_{BA} \\ \theta_{BA} \\ v_{BC} \\ \theta_{BC} \end{bmatrix} = \begin{bmatrix} 0 & 1/\sqrt{2} \\ 1 & 0 \\ 0 & 1/\sqrt{2} \\ 1 & 0 \end{bmatrix} \begin{bmatrix} \theta_B \\ y_C \end{bmatrix}$$

The slope deflection equations in matrix form $P = k\delta$ become

$$\begin{bmatrix} F_{AB} \\ M_{BA} \\ F_{BC} \\ M_{BC} \end{bmatrix} = \begin{bmatrix} b_1 & d_1 & 0 & 0 \\ d_1 & e_1 & 0 & 0 \\ 0 & 0 & b_2 & d_2 \\ 0 & 0 & d_2 & e_2 \end{bmatrix} \begin{bmatrix} v_{BA} \\ \theta_{BA} \\ v_{BC} \\ \theta_{BC} \end{bmatrix}$$

Thus $W = K\Delta = D^T k\, D\Delta$ for half the frame become

$$\{M_B \quad W_{yC}\} = \{0 \quad -W/2\} =$$

$$\begin{bmatrix} 0 & 1 & 0 & 1 \\ 1/\sqrt{2} & 0 & 1/\sqrt{2} & 0 \end{bmatrix} \begin{bmatrix} b_1 & d_1 & 0 & 0 \\ d_1 & e_1 & 0 & 0 \\ 0 & 0 & b_2 & d_2 \\ 0 & 0 & d_2 & e_2 \end{bmatrix} \begin{bmatrix} 0 & 1/\sqrt{2} \\ 1 & 0 \\ 0 & 1/\sqrt{2} \\ 1 & 0 \end{bmatrix} \begin{bmatrix} \theta_B \\ y_C \end{bmatrix}$$

i.e.

$$\begin{bmatrix} 0 \\ -W/2 \end{bmatrix} = \begin{bmatrix} e_1 + e_2 & (d_1 + d_2)/\sqrt{2} \\ (d_1 + d_2)/\sqrt{2} & (b_1 + b_2)/2 \end{bmatrix} \begin{bmatrix} \theta_B \\ y_C \end{bmatrix} \qquad (12.21)$$

Alternatively, deriving the vertical equilibrium equation for joint C, it is noticed from *Figure 12.9c*, where the shearing forces acting at the ends of the members are shown, that member BC prevents AB from moving away from joint B and thus an axial force F_{AB} develops in BC. Similarly, an axial force F_{BC} develops in AB. Joint B is thus in equilibrium under the internal forces shown. For a vertical upward force $W/2$ acting at C, the vertical equilibrium of joint C is

$$(F_{AB}/\sqrt{2}) + (F_{BC}/\sqrt{2}) = W/2$$

Thus $(1/\sqrt{2}) \times (d_1 \theta_B + b_1 v_{AB} + d_2 \theta_B + b_2 v_{BC}) = W/2$

Hence

$$[(d_1 + d_2)\theta_B/\sqrt{2}] + \tfrac{1}{2}(b_1 + b_2)y_C = W/2 \qquad (12.22)$$

In solving the joint equilibrium equations the actual value of the vertical load at C is used. Thus if this force is downwards, with a magnitude $W/2$, it will take a negative sign.

12.7. Non-linear elastic analysis of frames

To analyse a rigidly jointed plane frame, by taking the effect of axial forces in its members into consideration, the following steps are carried out.

(1) Neglecting the effect of axial forces in the members, a linear analysis of the frame is carried out. The steps for this analysis are given in section 12.2.

(2) For each member, the value of $\rho = p/(\pi^2 EI/L^2)$ is calculated from a knowledge of the axial force in the member. Using the tables in the appendix, the values of the stability functions are obtained.

(3) The equations of step 2, section 12.2, are replaced by the slope deflection equations (10.42), (10.43) and (10.44) and steps 3 to 5 of section 12.2 are repeated. The accuracy of a non-linear analysis improves if steps 2 and 3 of the present section are repeated more than once. It is noticed that the equations $\boldsymbol{\delta} = \mathbf{D}\boldsymbol{\Delta}$ are prepared only once. This is because the relationships between the joint deflections and member deformations are only functions of the topology of the frame. These are not influenced by the effect of axial forces.

12.8. Elastic instability of frames

A rigidly jointed frame becomes unstable under the sole influence of axial forces in the members of the frame. This means that, apart from external forces that act axially on the members, the frame is subject to no other external loads. As the axial forces are increased, a stage is reached when the entire frame becomes unstable and buckles. Since, the frame is subject to axial forces alone, it follows that the sum of the bending moments in the members meeting at a joint add up to zero. Similarly since the axial forces balance the external loads, it follows that the sum of the horizontal and vertical components of all the member forces meeting at a joint must add up to zero.

The equations of joint equilibrium in the deformed (i.e. buckled) frame thus become

$$\left. \begin{array}{l} \sum_{j=1}^{n} H_j = 0 \\ \\ \sum_{j=1}^{n} V_j = 0 \\ \\ \sum_{j=1}^{n} M_j = 0 \end{array} \right\} \qquad (12.23)$$

where j is a typical joint and n is the total number of joints in the frame. To demonstrate the meaning of equations (12.23) consider the frame shown in *Figure 12.10a* which is subject to a set of external loads and external moments. This frame becomes unstable and buckles to take up the deformed shape shown in *Figure 12.10b*. The frame in this figure is subject only to axial forces p_1, p_2 and p_3. In *Figure 12.10c* the external forces p_1, p_2 and p_3, shown in b, are replaced by internal member

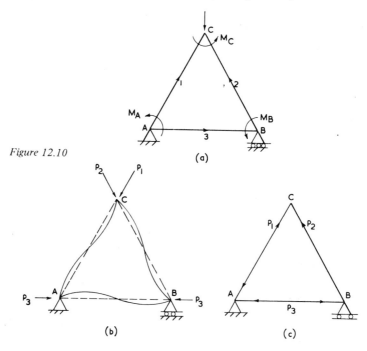

Figure 12.10

Stiffness Method for Analysis

axial forces p_1, p_2 and p_3 respectively. Now if these axial forces are increased, by any loading method whatsoever, the frame as a whole will reach a stage where it buckles. It follows that to calculate the elastic buckling load of a frame it is necessary to calculate the sum of the components of member forces and moments at all the joints as in equations (12.23). The axial forces in the members are then increased, thus reducing the stability ϕ functions, until equations 12.23 are actually satisfied. When this is achieved, the external forces that caused the buckling of the frame are obtained from the knowledge of the member axial forces.

12.9. Solution of homogeneous equations

Equations (12.23) are called homogeneous because the right hand side of each equation is equal to zero. To solve such a set of equations, consider the following two equations

$$\left. \begin{array}{l} bx + dy = 0 \\ dx + ey = 0 \end{array} \right\} \quad (12.24a)$$

or in matrix form

$$\begin{bmatrix} b & d \\ d & e \end{bmatrix} \begin{bmatrix} x \\ y \end{bmatrix} = \begin{bmatrix} 0 \\ 0 \end{bmatrix} \quad (12.24b)$$

or simply $\quad \mathbf{K}\Delta = 0 \quad (12.24c)$

where matrix **K** is square and symmetrical and thus may represent the stiffness of a frame.

One solution of these equations is $x = y = 0$ which is known as the trivial solution. On the other hand if $x \neq 0$ and $y \neq 0$ then another solution to equations (12.24a) can be obtained by rearranging them first to become

$$bx = -dy$$
$$dx = -ey$$

Dividing the first by the second

$$\frac{b}{d} = \frac{d}{e}$$

i.e.

$$be = d^2$$

Thus $\quad be - d^2 = 0$

Notice that $be - d^2$ is the determinant D of the matrix **K**. Thus a solution to equations (12.24) is obtained by calculating $D = be - d^2$ and then finding the values of b, e and d that make

$$D = be - d^2 = 0 \quad (12.25)$$

A further solution to equations (12.24) is obtained when $x = 0$ but $y \neq 0$. In that case equations (12.24) reduce to

$$dy = 0$$
$$ey = 0$$

or in matrix form

$$\begin{bmatrix} d \\ e \end{bmatrix} \begin{bmatrix} y \end{bmatrix} = \begin{bmatrix} 0 \\ 0 \end{bmatrix}$$

Now the first of these equations is redundant and has no structural significance, since the equations are no longer square. Disregarding this, therefore, the second equation gives

$$ey = 0$$

and since $y \neq 0$, it follows that the solution to equations (12.24) is given by

$$e = 0 \qquad (12.26)$$

Similarly, a further solution is obtained when $x \neq 0$ but $y = 0$, in that case, disregarding the remainder of the second of equations (12.24), the first gives

$$b = 0 \qquad (12.27)$$

A similar method can be used to solve any number of homogeneous equations. In the examples that follow, it will be shown that each of the solutions given by equations (12.25), (12.26) and (12.27) in fact gives the condition for the instability of a frame with a specific mode of deformation.

12.10. Examples

Example 1. Taking the effect of the axial force in member AB of the propped cantilever shown in *Figure 12.5*, example 1 of section 12.6, calculate the joint displacements at B and the bending moment at A. What is the buckling load of the system?
Answer: The axial force in the cantilever is known to be 197×10^3 thus its ρ value can be calculated directly as

$$\rho_{AB} = \rho = \frac{WL^2}{\pi^2 EI} = \frac{197 \times 10^3 \times 10^6}{\pi^2 \times 200 \times 10^9} = 0.1$$

From the tables for stability functions

$$\phi_2 = 0.9834, \quad \phi_3 = 0.9667, \quad \phi_4 = 1.017, \quad \phi_5 = 0.9012.$$

Stiffness Method for Analysis

The rotational equilibrium equation (12.13b) now becomes

$$M_{BA} = 4k\phi_3 \theta_B - 6k y_B \phi_2/L = 10^4 \quad (12.28a)$$

The shearing force F_{AB} is now given by

$$F_{AB} = -(6k/L)\phi_2 \theta_B + (12k\phi_5/L^2) y_B$$

and the equation (12.13a) for vertical equilibrium at joint B becomes

$$-(6k/L)\phi_2 \theta_B + (12k\phi_5/L^2) y_B + 0.1ky_B/L^2 = 100 \quad (12.28b)$$

Equations (12.13c) change to

$$\begin{bmatrix} k(12\phi_5 + \alpha)/L^2 & -6k\phi_2/L \\ -6k\phi_2/L & 4k\phi_3 \end{bmatrix} \begin{bmatrix} y_B \\ \theta_B \end{bmatrix} = \begin{bmatrix} 100 \\ 10^4 \end{bmatrix} \quad (12.28c)$$

With the known values of k and the ϕ functions given above, the solution of equations (12.28) gives

$$\theta_B = 4.7 \times 10^{-4} \text{ rad}$$

$$y_B = v = 0.3016 \text{ mm}.$$

Comparing these with the results obtained by the linear analysis in section 12.6, it is noticed that the joint displacements have increased by more than 62.5%. The bending moment M_{AB} at B is

$$M_{AB} = 2k\phi_4 \theta_B - 6k\phi_2 v/L$$

$$\therefore M_{AB} = 2 \times 200 \times 10^6 \times 1.017 \times 4.7 \times 10^{-4} - 6 \times 200 \times 10^6 \times 0.3016 \times$$
$$\times 0.984/10^3 = -163.455 \text{ kN m}.$$

an increase of 58% on the linear analysis. Notice that the percentage increase in M_{AB} is different from the percentage increase in y_B. This is due to the fact that the stability ϕ functions involved in calculating M_{AB} and y_B are influenced differently by the axial force in AB.

The system becomes unstable under the influence of the axial force $W = p_{AB}$, without the presence of M or V at B. For instability, therefore, the joint equilibrium equations become

$$\left.\begin{array}{r} 4k\phi_3 \theta_B - 6k\phi_2 y_B/L = 0 \\ (-6k\phi_2 \theta_B/L) + k(12\phi_5 + \alpha) y_B/L^2 = 0 \end{array}\right\} \quad (12.29)$$

Thus either $\theta_B = y_B = 0$ which is trivial, or equations (12.29) can be written as

$$4k\phi_3\theta_B = 6ky_B\phi_2/L$$

$$-6k\phi_2\theta_B/L = -k(12\phi_5 + \alpha)y_B/L^2$$

Dividing the first by the second

$$\frac{4k\phi_3}{-6k\phi_2/L} = \frac{6k\phi_2/L}{-k(12\phi_5 + \alpha)/L^2}$$

which gives

$$\frac{4k^2\phi_3(12\phi_5 + \alpha)}{L^2} - \frac{36k^2\phi_2^2}{L^2} = 0 \qquad (12.30)$$

Notice that the left hand side of this equation is the determinant D of the stiffness matrix

$$\mathbf{K} = \begin{bmatrix} 4k\phi_3 & -6k\phi_2/L \\ -6k\phi_2/L & k(12\phi_5 + \alpha)/L^2 \end{bmatrix}$$

i.e.

$$\frac{4k^2\phi_3(12\phi_5 + \alpha)}{L^2} - \frac{36k^2\phi_2^2}{L^2} = D = \begin{vmatrix} 4k\phi_3 & -6k\phi_2/L \\ -6k\phi_2/L & k(12\phi_5 + \alpha)/L^2 \end{vmatrix}$$

$$(12.31)$$

Solving equation (12.30) therefore means finding the value of W that makes the determinant D, of the stiffness matrix $\mathbf{K}$, equal to zero. This means that the system becomes unstable when

$$D = \text{determinant } \mathbf{K} = |\mathbf{K}| = 0 \qquad (12.32)$$

This is the condition when the system loses all its stiffness and fails. Equations (12.31) can be written as

$$d = DL^2/(4k^2) = (12\phi_5 + \alpha)\phi_3 - 9\phi_2^2 \qquad (12.33)$$

and the system becomes unstable when $d = 0$
i.e. when

$$(12\phi_5 + \alpha)\phi_3 - 9\phi_2^2 = 0 \qquad (12.34)$$

This equation is solved for the value of ρ. This is done by selecting various values of ρ, and therefore W, and calculating the corresponding values of d, and therefore D, using equation (12.33). The critical value of ρ, which causes the instability of the system, satisfies equation (12.34). In particular

(1) When $\rho = 0.24$, $\phi_2 = 0.9598$, $\phi_3 = 0.9185$, $\phi_5 = 0.7624$ and equation (12.33) gives $d = |+0.2041$.
(2) When $\rho = 0.26$, $\phi_2 = 0.9564$, $\phi_3 = 0.9114$, $\phi_5 = 0.7426$ and equation (12.33) gives $d = -0.019501$.

Interpolating between these values it is found that at $\rho = 0.2583$, $d = 0$. In *Figure 12.11* various values of ρ are plotted against the corresponding values of d and it is noticed that $d = 0$ when $\rho = 0.2583$.

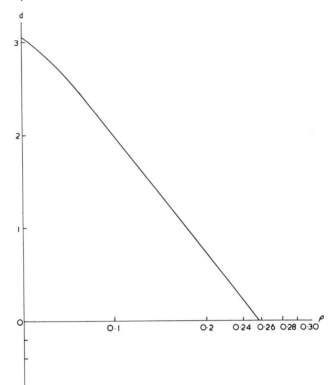

Figure 12.11

The buckling load of the system is 2.583 times greater than the load $W = 197 \times 10^3$ kN given in the first part of the question. The calculation of the critical load is valuable to safeguard against instability. However, the calculation of v, θ_B and particularly M_{AB} under a load, below the critical load, is of the utmost practical importance to the engineer since these decide the cross sectional properties of the members under the design loads. The example demonstrates how erroneous are the results obtained by the linear analysis.

Before leaving this example it is interesting to point out that there are other values of W that can make the system unstable. Some of these are:

(a) Equations (12.29) can be satisfied if $y_B = 0$ but $\theta_B \neq 0$. In that case the first of these equations gives

$$4k\phi_3\theta_B = 0$$

i.e. $\phi_3 = 0$

This condition is satisfied when $\rho = 2.046$ and takes place when the pin ended bar is infinitely stiff and the beam acts as a propped cantilever pinned to a support at B.

(b) $\theta_B = 0$ but $y_B \neq 0$. In that case the second of equations (12.29) gives

$$k(12\phi_5 + \alpha)/L^2 = 0$$

i.e. $\phi_5 = -0.1/12 = -0.0083$

which is satisfied under $\rho \simeq 1$ and takes place when end B is on vertical rollers with the beam in pure sway without end rotations.

Out of the three calculated critical values of ρ the lowest is 0.2583.

Notice too, that if $\alpha = 0$, i.e. the bar is removed then from equation (12.34) $d = 0$ $\rho = 0.25$ which is the critical value of ρ for a cantilever.

Example 2. The column ABC in *Figure 12.12* is fixed at A and C and supported laterally by the pin ended bar BD. At B the system is subject to a force $W = 210 \times 10^3$ kN acting at an angle of $45°$ to the horizontal as shown in the figure. Calculate the axial force in the bar and draw the bending moment diagram for the column. Take $E = 200$ kN/mm^2, $I = 10^9$ mm^4, and $L = 10^3$ mm for the bar and the

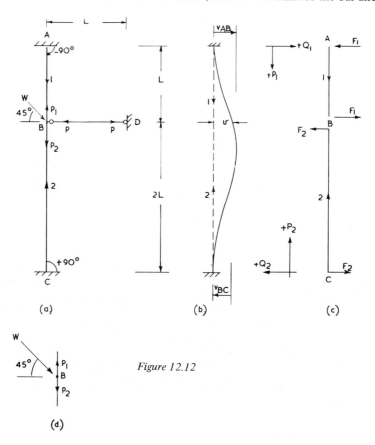

Figure 12.12

column. The area of the column is 10^4 mm^2 and that of the bar is 10^2 mm^2.
Answer: AB is numbered as member 1 and CB as member 2. The arrow on each member defines the positive P axis. Thus for AB the positive Q axis is to the right while for CB this axis is to the left as shown in *Figure 12.12c*. Because of fixity at A and C

$$x_A = y_A = \theta_A = x_C = y_C = \theta_C = 0$$

For member 1 equation (12.2) becomes

$$v_{AB} = v_2 = -x_B \sin(-90) + y_B \cos(-90) = x_B$$

For member 2 equation (12.2) gives

$$v_{CB} = v_2 = -x_B \sin 90 + y_B \cos 90 = -x_B$$

The quantity v_2 refers to the movement at the second end of a member in the +Q direction of that member.

The axial forces p_1 and p_2 in members 1 and 2 are calculated as follows
Resolving the forces at B vertically

$$p_1 - p_2 - W/\sqrt{2} = 0$$

Using the axial flexibility relation $\delta = fP$ and equating the sum of the extensions of the two members to zero, since distance AC is fixed

$$(L/EA) p_1 + (2L/EA) p_2 = 0$$

Solving these two equations we obtain

$$p_1 = 2W/(3\sqrt{2}) \text{ (tensile)}$$
$$p_2 = -W/(3\sqrt{2}) \text{ (compressive)}$$

Thus ρ_1 and ρ_2 can be calculated and the non-linear analysis can be carried out directly.

$$\rho_1 = p_1 L^2/(\pi^2 EI) = -2WL^2/(3\sqrt{2} \times \pi^2 EI)$$

and

$$\rho_2 = 4p_2 L^2/(\pi^2 EI) = 4WL^2/(3\sqrt{2} \times \pi^2 EI)$$

$$\therefore \quad \rho_1 = -0.5 \rho_2$$

324   *Stiffness Method for Analysis*

with $W = 210 \times 10^3$ kN

$$p_2 = \frac{4 \times 210 \times 10^3 \times 10^6}{3\sqrt{2} \times \pi^2 \times 200 \times 10^9} = 0.1$$

$$\therefore \quad p_1 = -0.05$$

$\phi_{2,1} = 1.0082$, $\phi_{3,1} = 1.01635$, $\phi_{4,1} = 0.9919$, $\phi_{5,1} = 1.04935$
$\phi_{2,2} = 0.9834$, $\phi_{3,2} = 0.9667$, $\phi_{4,2} = 1.0170$, $\phi_{5,2} = 0.9012$

Notice, in Appendix 1, the values of p for members in tension are considered to be negative.

The vertical deflection y_B of the joint B is equal to the extension of member 1, thus

$$y_B = (L/EA)\, p_1 = (L/EA) \times 2W/(3\sqrt{2})$$

$$y_B = \frac{10^3 \times 2 \times 210 \times 10^3}{200 \times 10^4 \times 3\sqrt{2}} = 70/\sqrt{2} \text{ mm (down)}$$

For AB, $k_1 = EI/L = k$ and for BC, $k_2 = EI/(2L) = 0.5k$.
The slope deflection equations for the members are

$$M_{AB} = 2k\phi_{4,1}\,\theta_B - 6k\phi_{2,1}\,x_B/L \; ; (v_{AB} = x_B)$$
$$M_{BA} = 4k\phi_{3,1}\,\theta_B - 6k\phi_{2,1}\,x_B/L = e_1\theta_B + d_1 x_B$$
$$F_{AB} = F_1 = 12k\phi_{5,1}\,x_B/L^2 - 6k\phi_{2,1}\,\theta_B/L = b_1 x_B + d_1\theta_B$$

where $e = 4k\phi_3$, $d = -6k\phi_2/L$ and $b = 12k\phi_5/L^2$;

$$M_{CB} = k\phi_{4,2}\,\theta_B + 1.5k\phi_{2,2}\,x_B/L \; ; (v_{CB} = -x_B)$$
$$M_{BC} = 2k\phi_{3,2}\,\theta_B + 1.5k\phi_{2,2}\,x_B/L = e_2\theta_B - d_2 x_B$$
$$F_2 = -1.5k\phi_{5,2}\,x_B/L^2 - 1.5k\phi_{2,2}\,\theta_B/L = -b_2 x_B + d_2\theta_B$$

The second suffix associated with the stability functions refers to the member number. For horizontal equilibrium of joint B

$$(W/\sqrt{2}) - p = F_1 - F_2 = (b_1 + b_2)x_B + (d_1 - d_2)\theta_B$$

where $p = EA_{BD}\,x_B/L$ is the axial force in member BD. Thus

$$(b_1 + b_2 + EA/L)x_B + (d_1 - d_2)\theta_B = W/\sqrt{2}$$

For rotational equilibrium of joint B

$$M_{BA} + M_{BC} = 0$$

$$\therefore \quad (d_1 - d_2)x_B + (e_1 + e_2)\theta_B = 0$$

Stiffness Method for Analysis

These can be written in matrix form as

$$\begin{bmatrix} b_1 + b_2 + EA/L & d_1 - d_2 \\ d_1 - d_2 & e_1 + e_2 \end{bmatrix} \begin{bmatrix} x_B \\ \theta_B \end{bmatrix} = \begin{bmatrix} W/\sqrt{2} \\ 0 \end{bmatrix}$$

Solving these equations with $W = 210 \times 10^3$ kN and using the stability functions listed above, we obtain

$$x_B = 70.3337 \text{ mm}$$
$$\theta_B = 0.0536 \text{ rad}$$

The axial force in member BD is

$$p = EAx_B/L = 200 \times 10^2 \times 70.3337/10^3 = 1406.67 \text{ kN}$$

The bending moments in the column are

$$M_{AB} = 10^6(400 \times 0.0536 \times 0.9919 - 1.2 \times 70.3337 \times 1.0082)$$
$$= -63.826 \times 10^3 \text{ kN m.}$$
$$M_{BA} = 10^6(800 \times 1.01635 \times 0.0536 - 1.2 \times 70.3337 \times 1.0082)$$
$$= -41.50 \times 10^3 \text{ kN m.}$$
$$M_{BC} = 10^6(400 \times 0.9667 \times 0.0536 + 1.5 \times 0.2 \times 0.9834 \times 70.3337)$$
$$= 41.50 \times 10^3 \text{ kN m.}$$
$$M_{CB} = 10^6(200 \times 1.017 \times 0.0536 + 0.3 \times 0.9834 \times 70.3337)$$
$$= 31.65 \times 10^3 \text{ kN m.}$$

The bending moment diagram is shown in *Figure 12.13*.

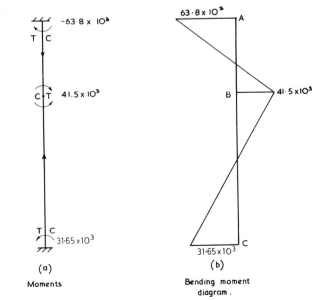

Figure 12.13

(a) Moments

(b) Bending moment diagram.

Example 3. Derive the joint equilibrium equations for the triangulated frame shown in *Figure 12.14* and obtain the condition for its elastic instability. At what value of H does the frame become unstable rotationally at B?

Answer: The vector diagrams for the deformations in the members are shown in *Figures 12.14 b, c, d* and *e*, in which x_B and y_B are the displacements of joint B. x_B is positive when it is to the right and y_B is positive when it is upwards. From these figures or by using equations (12.1) and (12.2) it is found that

The extension in member 1 = $u_1 = (x_B/\sqrt{2}) + (y_B/\sqrt{2})$

The sway in member 1 = $v_1 = (y_B/\sqrt{2}) - (x_B/\sqrt{2})$

The extension of members 2 = $u_2 = y_B$

The sway in member 2 = $v_2 = -x_B$

Let $b = 12EI\phi_5/\ell^3 \quad d = -6EI\phi_2/\ell^2 \quad a = EA/\ell$

$e = 4EI\phi_3/\ell$, where ℓ is the length of a member.

Then for member 1

$$M_{BA} = d_1 v_1 + e_1 \theta_B$$
$$F_1 = b_1 v_1 + d_1 \theta_B$$

and $\quad p_1 = a_1 u_1$

Similarly for member 2

$$M_{BC} = d_2 v_2 + e_2 \theta_B$$
$$F_2 = b_2 v_2 + d_2 \theta_B \text{ and } p_2 = a_2 u_2$$

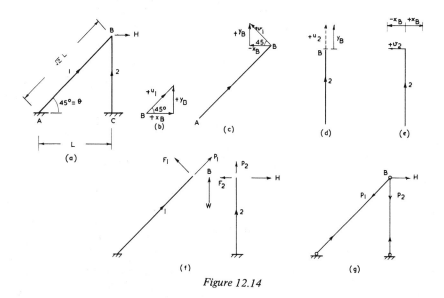

Figure 12.14

Three forces can be applied at joint B. These are a horizontal force H, a vertical force W and a moment M. In the present example W and M are set to zero. The forces acting at end B of the members are shown in *Figure 12.14f*. For the horizontal equilibrium of forces at B

$$H = -(F_1/\sqrt{2}) - F_2 + (p_1/\sqrt{2})$$

For the vertical equilibrium

$$W = 0 = (p_1/\sqrt{2}) + p_2 + (F_1/\sqrt{2})$$

and for rotation equilibrium at B

$$M = 0 = M_{BA} + M_{BC}$$

Thus using the slope deflection equations and the relationships between the member deformation and the joint displacements given above, we obtain

$$H = [(b_1/2) + b_2 + a_1/2] x_B + [(a_1/2) - (b_1/2)] y_B - [(d_1/\sqrt{2}) + d_2] \theta_B$$

$$W = 0 = [(a_1/2) - b_1/2] x_B + [(a_1/2) + (b_1/2) + a_2] y_B + (d_1/\sqrt{2}) \theta_B$$

and

$$M = 0 = -[(d_1/\sqrt{2}) + d_2] x_B + (d_1/\sqrt{2}) y_B + (e_1 + e_2) \theta_B$$

These are the joint equilibrium equations for the frame and they are written in matrix form as

$$\begin{bmatrix} (b_1/2) + b_2 + a_1/2 & (a_1/2) - (b_1/2) & -(d_2 + d_1/\sqrt{2}) \\ (a_1/2) - (b_1/2) & (a_1/2) + (b_1/2) + a_2 & d_1/\sqrt{2} \\ -(d_2 + d_1/\sqrt{2}) & d_1/\sqrt{2} & e_1 + e_2 \end{bmatrix} \begin{bmatrix} x_B \\ y_B \\ \theta_B \end{bmatrix} = \begin{bmatrix} H \\ W \\ M \end{bmatrix} = \begin{bmatrix} H \\ 0 \\ 0 \end{bmatrix}$$

(12.35)

This rigidly jointed frame is rendered isostatic by replacing the rigid joints by frictionless pins as shown in *Figure 12.14g*. The axial forces p_1 and p_2 in the rigid frame can then be calculated to a good degree of accuracy from the analysis of the isostatic pin jointed frame, thus resolving at B horizontally

$$H - p_1 \cos 45 = 0$$

$$\therefore \quad p_1 = H\sqrt{2} \text{ (tensile)}$$

resolving the forces at B vertically

$$p_1 \cos 45 + p_2 = 0$$

$$\therefore \quad p_2 = -p_1/\sqrt{2} = -H\sqrt{2}/\sqrt{2} = -H \text{ (compressive)}$$

Thus

$$p_1 = \frac{p_1 \times (L\sqrt{2})^2}{\pi^2 EI} = -\frac{H\sqrt{2} \times 2L^2}{\pi^2 EI} \quad \text{(tensile)}$$

and

$$p_2 = \frac{p_2 L^2}{\pi^2 EI} = -\frac{HL^2}{\pi^2 EI} \quad \text{(compressive)}$$

$$\therefore \quad p_1 = -2\sqrt{2} \times p_2$$

The system becomes unstable when the determinant $D = |\mathbf{K}|$ of the matrix $\mathbf{K}$ vanishes. For the case of rotational instability

$$x_B = y_B = 0 \text{ but } \theta_B \neq 0$$

In that case the last of equations (12.35) give

$$e_1 + e_2 = 0 \tag{12.36}$$

i.e.

$$\frac{4EI \, \phi_{3,1}}{L\sqrt{2}} + \frac{4EI}{L} \phi_{3,2} = 0$$

which gives

$$J = \phi_{3,1} + \sqrt{2} \times \phi_{3,2} = 0$$

At $p_2 = -1.14$ tensile, $p_1 = 2\sqrt{2} \times 1.14 = 3.22$

$$\phi_{3,2} = 1.3302, \quad \phi_{3,1} = -1.8952, \quad J = 0$$

$$\therefore \quad H = 3.22\pi^2 EI/(2\sqrt{2} \times L^2) = 1.14\pi^2 EI/L^2$$

acting to the left.

Example 4. Plot a graph of the load W against eaves displacement x_B for the pitched roof frame shown in *Figure 12.7*, example 3, section 12.6. What is the condition for the elastic instability of the frame?

Answer: From the vertical equilibrium

$$R_A = p_{AB} = W/2$$

$$\therefore \quad p_1 = \frac{W/2}{\pi^2 EI/L^2} = \frac{W \times 10^8}{2\pi^2 \times 200 \times 10^{12}} = 25.3W \times 10^{-9}$$

For $p_1 = 0.4$ say

$$W = 0.4 \times 10^9/25.3 = 15\ 804\ 081 \text{ kN}.$$

By the principle of superposition, the linear displacements of joint B, obtained in section 12.6, now become

$$x_{B\ linear} = 929.28 \text{ mm}$$

$$\theta_{B\ linear} = 0.1054 \text{ rad}$$

The linear load deflection diagram is shown in *Figure 12.15* by the straight line OC.

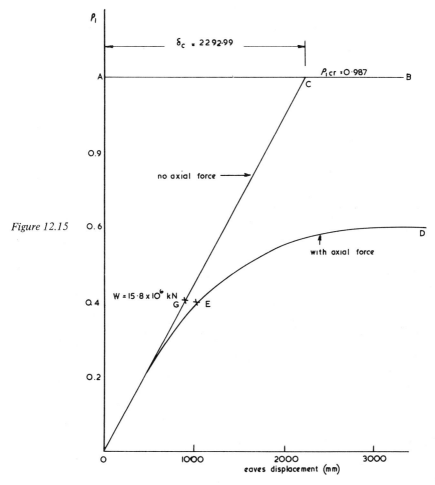

Figure 12.15

Before plotting a point on the non-linear load deflection curve, the values of the axial loads have to be calculated.

With $W = 15\,804\,081$ kN, the value of F_{AB} is $F_{AB} = -b_1 x_B + d_1 \theta_B$

$$b_1 = 12k/L^2 = 12 \times 200 \times 10^8/10^8 = 2400$$

$$d_1 = -6k/L = -6 \times 200 \times 10^8/10^4 = -1200 \times 10^4$$

∴ $\quad F_{AB} = -3495.23 \times 10^3$ kN

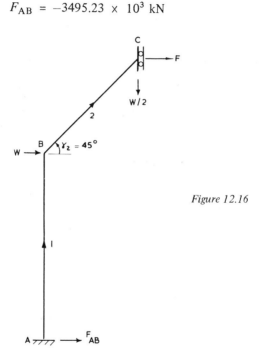

Figure 12.16

In *Figure 12.16* half of the frame is shown in which joint C is represented as a vertical roller support, with the support reaction being F. Resolving the forces acting on this half frame horizontally

$$F + W + F_{AB} = 0$$

∴ $\quad F = -W - F_{AB} = -15\,804\,081 + 3\,495\,230 = -12\,308\,851$ kN.

The negative sign indicates that F is acting to the left. The axial force p_{BC} in member BC is thus

$$p_{BC} = p_2 = (0.5W) \sin \gamma_2 - F \cos \gamma_2$$

$$p_2 = \frac{15\,804\,081}{2\sqrt{2}} + \frac{12\,308\,851}{\sqrt{2}} = 14\,291\,294 \text{ kN (compressive)}$$

$$\rho_2 = p_2/[\pi^2 EI/(L/\sqrt{2})^2] = p_2 L^2/(2\pi^2 EI)$$

$$\rho_2 = \frac{14\ 291\ 294 \times 10^8}{2\pi^2 \times 200 \times 10^{12}} = 0.3617$$

and since $\rho_1 = 0.4$

$\therefore \quad \phi_{3,1} = 0.8610, \quad \phi_{5,1} = 0.6033, \quad \phi_{2,1} = 0.9323$

$\phi_{3,2} = 0.875, \quad \phi_{5,2} = 0.6414, \quad \phi_{2,2} = 0.9389$

Using these values, equations (12.19) and (12.20) give

$$x_B = 1038.5924 \text{ mm}$$

$$\theta_B = 0.127939 \text{ rad.}$$

With these values p_2 is recalculated and becomes 14 687 347 kN, while $\rho_2 = 0.3717$. Equation (12.20) now gives

$$x_B = 1053.4062 \text{ mm}$$

This locates point E in *Figure 12.15* which is on the non-linear load deflection graph.

Values of x_B calculated at other values of W produces the curve OED.

The condition for elastic instability of the frame is when the determinant $D = 0$, that is

$$hj - g^2 = 0$$

i.e. $\quad (b_1 + 2b_2)(e_1 + e_2) - (\sqrt{2}\,d_2 - d_1)^2 = 0$

Substituting for b, e and d we obtain

$$4(\phi_{5,1} + 5.65684\ \phi_{5,2})(\phi_{3,1} + 1.4142\ \phi_{3,2}) - 3(\phi_{2,1} - 2\phi_{2,2})^2 = 0$$

Assuming that the ratio ρ_2/ρ_1 remains equal to $0.372/0.4$, i.e. $\rho_2 = 0.93\rho_1$, which is reasonable but not strictly accurate, the condition of elastic instability is satisfied when $\rho_1 = \rho_{cr} = 0.987$. Thus $W_{cr} = 0.987 \times 10^9/25.3 = 39.012 \times 10^6$ kN.

Once the critical value of ρ is found, it is helpful to state, using a procedure similar to that given in section 10.16, that the entire nonlinear load deflection curve can be drawn using the simple formula

$$\delta_B = \delta'\delta_c/(\delta_c - \delta') \tag{12.37}$$

where δ_c is the linear deflection at the elastic critical load, AC in *Figure 12.15*, δ' is the linear elastic deflection at a given load, Point G in the figure and δ_B is the non-linear elastic deflection, point E in the figure.

Example 5. Taking the effect of axial forces into consideration calculate the side sway of the frame shown in *Figure 12.6*, example 2 of section 12.6. What is the elastic buckling load of the frame?

Answer: The slope deflection equations for AB and BD are

$$M_{BA} = 4k\,\phi_{3,1}\theta_B + 6k\,\phi_{2,1}x_B/L$$

$$F_{BA} = -6k\,\phi_{2,1}\theta_B/L - 12k\,\phi_{5,1}x_B/L^2$$

$$M_{BD} = 4k\,\phi_{3,2}\theta_B + 2k_2\phi_{4,2}\theta_B$$

$$M_{DB} = 2k\,\phi_{4,2}\theta_B + 4k\,\phi_{3,2}\theta_B$$

The second suffix associated with the stability functions refers to the member number. The joint equilibrium equations are

(i) For horizontal equilibrium at B, neglecting the axial force in the beam

$$-F_{BA} = wL/2$$

$$\therefore \quad 12k\,\phi_{5,1}x_B/L^2 + 6k\,\phi_{2,1}\theta_B/L = wL/2 \tag{12.38}$$

(ii) For the rotational equilibrium at B, see section 10.19,

$$M_{BA} + M_{BD} = wL^2/(12\phi_2)$$

$$\therefore \quad 6k\phi_{2,1}x_B/L + (4k\,\phi_{3,1} + 4k\,\phi_{3,2} + 2k\,\phi_{4,2})\theta_B = wL^2/(12\phi_2) \mid (12.39)$$

These equations can be written in matrix form as

$$\begin{bmatrix} 12k\,\phi_{5,1}/L^2 & 6k\,\phi_{2,1}/L \\ 6k\,\phi_{2,1}/L & 4k\,\phi_{3,1} + 4k\,\phi_{3,2} + 2k\,\phi_{4,2} \end{bmatrix} \begin{bmatrix} x_B \\ \theta_B \end{bmatrix} = \begin{bmatrix} \dfrac{wL}{2} \\ \dfrac{wL^2}{12\phi_2} \end{bmatrix} \tag{12.40}$$

i.e. $K\Delta = W$

With $w = 100$ kN/mm, $W = 1.0975 \times 10^8$ kN;
$wL/2 = 100 \times 3 \times 10^3/2 = 150 \times 10^3$
and $wL^2/12 = 100 \times 9 \times 10^6/12 = 75 \times 10^6$.

Assuming the axial forces in the columns are the same, the value of ρ for these is

$$\rho_1 = \frac{1.0975 \times 10^8 \times 9 \times 10^6}{\pi^2 \times 200 \times 10^{12}} = 0.5$$

$$\therefore \quad \phi_{5,1} = 0.5035, \quad \phi_{2,1} = 0.9147, \quad \phi_{3,1} = 0.8236$$

Stiffness Method for Analysis   333

There is no axial force in member 2, thus $\rho_2 = 0$ and $\phi_{3,2} = \phi_{4,2} = 1$.
Solving equations (12.40)

$$x_B = 6.45 \text{ mm}$$

which is nearly three times the deflection obtained by the linear analysis.
The frame buckles when the loads $\{wL/2 \quad wL^2/12\}$ are equal to $\{0 \quad 0\}$ and instability occurs when

$$|K| = D = (12k/L^2) \phi_{5,1} \times k(4\phi_{3,1} + 6) - 36k^2 \phi_{2,1}^2/L^2 = 0$$

Simplifying, instability takes places when

$$d = \phi_{5,1}(\phi_{3,1} + 1.5) - 0.75 \phi_{2,1}^2 = 0$$

At $\quad \rho_1 = 1, \phi_{5,1} = 0$ and $\phi_{2,1} = 0.8225$, thus
$\quad d = -0.75 (0.8225)^2 = -0.50738$

At $\quad \rho_1 = 0, \phi_{5,1} = \phi_{3,1} = \phi_{2,1} = 1$

∴ $\quad d = 2.5 - 0.75 = 1.75$

Interpolating between these two values $1.75/\rho_1 = 0.50738/(1 - \rho_1)$

∴ $\quad \rho_1 \simeq 0.775$

Try $\rho_1 = 0.76, \phi_{5,1} = 0.2426, \phi_{3,1} = 0.7207, \phi_{2,1} = 0.8677$
∴ $\quad d = 0.2426 (0.7207 + 1.5) - 0.75 (0.8677)^2 = -0.0259356$
while at $\rho_1 = 0.74, \phi_{5,1} = 0.2627, \phi_{3,1} = 0.7289, \phi_{2,1} = 0.8714$
∴ $\quad d = 0.2627 (0.7289 + 1.5) - 0.75 (0.8714)^2 = 0.0160286$
Interpolating again between $\rho_1 = 0.74$ and $\rho_1 = 0.76$ we obtain $\rho_{\text{critical}} = 0.74764$
Thus $W_{\text{cr}} = \rho_1 \pi EI/L^2 = 0.74764 \times \pi^2 \times 200 \times 10^{12}/(9 \times 10^6)$

$$W_{\text{cr}} = 164.099 \times 10^6 \text{ kN}.$$

12.11. Assembly and thermal forces

These forces can also be calculated by the displacement method. Consider the pin ended member AB, shown in *Figure 2.6*, which is λ units of length too short. To connect this member to its support, a tensile force $p = k\lambda$ has to be applied to the member by some external agency. Here k is the stiffness of the member. Once the operation is completed the member sustains the force p.

A similar short member which is to be connected to two points in a frame also requires a force p. However, once the connection is secured and the external agency is removed, the member tends to contract back to its original length. This relieves some of the force in the short member but causes forces to develop in the other members of the frame. Consider that the frame is subsequently subjected to a set of external loads which prevents the contraction of this imperfect member but extends it a further δ units of length. The total force sustained by the member is then

$$p = k\delta + k\lambda = k(\delta + \lambda) \qquad (12.41)$$

The frame may, of course, have other imperfect members, some too short and others too long. The forces in these due to the total effect of the external loads and their own imperfection is written as

$$\left.\begin{array}{l} p_1 = k_1(\delta_1 + \lambda_1) \\ p_2 = k_2(\delta_2 + \lambda_2) \\ \vdots \\ p_i = k_i(\delta_i + \lambda_i) \\ \vdots \\ p_N = k_N(\delta_N + \lambda_N) \end{array}\right\} \qquad (12.42a)$$

where i refers to a typical member and N is the total number of members. Here it is assumed that all the members are imperfect. In these equations the value of λ for a perfect member is considered to be zero, while that for a member which is too long is taken to be negative. In the case of a perfect member which is subsequently heated by $t°C$, the elongation is $\theta L t$ where θ is the linear coefficient of thermal expansion and L is the length. Such a member would become too long to be connected to the joints at its ends and thus its 'lack of fit' is taken as $\lambda = -\theta L t$. Equations (12.42) can therefore be used to deal with assembly as well as thermal forces. These equations can be written in matrix form as

$$\mathbf{P} = \mathbf{k}(\boldsymbol{\delta} + \boldsymbol{\lambda}) \qquad (12.42b)$$

where k is the member stiffness matrix of the unassembled structure.

$\boldsymbol{\delta} = \{\delta_1 \quad \delta_2 --- \delta_i --- \delta_N\}$ is the vector of the member deformations and $\boldsymbol{\lambda} = \{\lambda_1 \quad \lambda_2 --- \lambda_i --- \lambda_N\}$ is the 'lack of fit' vector and may be either due to imperfection or due to temperature changes.

So far we have been dealing with pin ended members that can sustain axial forces only. The problem can however be generalised to include flexural imperfections. These may arise either because members are slightly curved or cranked or because only one surface of the member is heated. This may arise in cases of fire or in structures, such as boilers and furnaces, in which the inside of the structure is at a higher temperature than the outside. In these cases equations (12.42) refer to the direct as well as flexural member stiffness equations.

Now equations (12.4) gives $\boldsymbol{\delta} = \mathbf{D}\boldsymbol{\Delta}$ where $\boldsymbol{\Delta}$ is the vector of joint displacements and $\mathbf{D}$ is the displacement transformation matrix.

Equations (12.42) can thus be expressed as

$$P = k\, D\Delta + k\lambda \tag{12.43}$$

The virtual work equations (7.44b) are again used to derive the overall stiffness matrix of the structure. These equations state that

$$W^T \Delta = P^T \delta$$

where W is the vector of the externally applied loads. Using equations (12.4), the virtual work equations become

$$W^T \Delta = P^T D \Delta$$

and thus $W^T = P^T D$

i.e. $$W = D^T P \tag{12.44}$$

These and equations (12.43) give

$$W = D^T k D \Delta + D^T k \lambda \tag{12.45}$$

Hence the joint displacements Δ are calculated by solving equations (12.45), i.e.

$$\Delta = (D^T k\, D)^{-1}\, (W - D^T k \lambda) \tag{12.46}$$

Before applying the external loads W, the joint deflections due to the assembly and thermal effects are calculated from equations (12.46) by replacing W by a null vector. Once the joint deflections are calculated the member forces are obtained using equations (12.43).

Consider now the case of a horizontal uniform prismatic member of a rigidly jointed frame in which the temperature at the top surface is $t°C$ higher than that at the bottom surface. The top fibres then expand relative to those at the bottom surface and while the length of the member is unaltered, it becomes curved as shown in *Figure 3.16* (Section 3.5). The length of an element measured along the neutral axis is $ab = dx$. The difference in the lengths of nq and mp is $qq' + pp' = \beta t dx$, where β is the coefficient of thermal expansion. Now from *Figure 3.16*

$$d\theta = dx/R = \beta t dx/q'p' = \beta t dx/D$$

thus

$$1/R = \beta t/D \tag{12.47}$$

where $q'p' = D$ is the total depth of the section. But equation (3.38) states that $M = EI/R$ which with equation (12.47) gives

336   Stiffness Method for Analysis

$$M = EI\beta t/D \tag{12.48}$$

In section 7.7, example 2 and *Figure 7.8*, it was shown that when a member AB of length L is in pure bending and subtends an angle θ at its centre of curvature, the end rotations are

$$\theta_{AB} = -\theta_{BA} = 0.5\theta$$

and since $\theta = L/R = ML/EI$, it follows, using equation (12.48) that

$$\theta_{AB} = -\theta_{AB} = \beta Lt/2D \tag{12.49}$$

Just as the linear thermal extension of a member was considered to be a negative 'lack of fit', heating one surface of a member causes a negative rotational lack of fit at the end which ends up having an anticlockwise positive slope. For instance if member AB in *Figure 7.8* is bent by heating its top surface, then thermal rotational lack of fit at A is $\theta_{\lambda AB} = -\beta Lt/2D$ while that at B is $\theta_{\lambda BA} = +\beta Lt/2D$.

12.12. Examples

Example 1. To begin with, the pin ended member BD in the structure shown in *Figure 12.12*, example 2 of section 12.10 is 5 mm too short. This member is forced into its position, the entire structure is then heated by 300°C and loaded as shown in the figure. Calculate the joint deflections and the maximum bending moment in the column. The data given in section 12.10 are unaltered.
Answer: For BD, with $\theta = 4 \times 10^{-6}$ per degree C,

$$\theta Lt = 4 \times 10^{-6} \times 10^3 \times 300 = 1.2 \text{mm}$$

$$\therefore \quad \lambda \text{ for BD} = 5 - 1.2 = 3.8 \text{mm}$$

From section 12.10, the equations $\mathbf{P} = \mathbf{k\delta}$ are

$$\begin{bmatrix} F_{AB} \\ M_{BA} \\ F_{CB} \\ M_{BC} \\ P_{BD} \end{bmatrix} = \begin{bmatrix} b_1 & d_1 & 0 & 0 & 0 \\ d_1 & e_1 & 0 & 0 & 0 \\ 0 & 0 & b_2 & d_2 & 0 \\ 0 & 0 & d_2 & e_2 & 0 \\ 0 & 0 & 0 & 0 & a_3 \end{bmatrix} \begin{bmatrix} v_{AB} \\ \theta_{BA} \\ v_{BC} \\ \theta_{BC} \\ \delta_{BD} \end{bmatrix}, \lambda = \begin{bmatrix} 0 \\ 0 \\ 0 \\ 0 \\ 3.8 \end{bmatrix}$$

where $a_3 = EA/L$ for BD.
The displacement transformation equations $\mathbf{\delta} = \mathbf{D\Delta}$ were obtained as

$$\begin{bmatrix} v_{AB} \\ \theta_{BA} \\ v_{BC} \\ \theta_{BC} \\ \delta_{BD} \end{bmatrix} = \begin{bmatrix} 1 & 0 \\ 0 & 1 \\ -1 & 0 \\ 0 & 1 \\ -1 & 0 \end{bmatrix} \begin{bmatrix} x_B \\ \theta_B \end{bmatrix}, \mathbf{D}^T = \begin{bmatrix} 1 & 0 & -1 & 0 & -1 \\ 0 & 1 & 0 & 1 & 0 \end{bmatrix}$$

Stiffness Method for Analysis

$$\therefore \mathbf{kD} = \begin{bmatrix} b_1 & d_1 \\ d_1 & e_1 \\ -b_2 & d_2 \\ -d_2 & e_2 \\ -a_3 & 0 \end{bmatrix}, \quad \mathbf{D^T k} = \begin{bmatrix} b_1 & d_1 & -b_2 & -d_2 & -a_3 \\ d_1 & e_1 & d_2 & e_2 & 0 \end{bmatrix}$$

$$\therefore \mathbf{K} = \mathbf{D^T k D} = \begin{bmatrix} b_1 + b_2 + a_3 & d_1 - d_2 \\ d_1 - d_2 & e_1 + e_2 \end{bmatrix}, \quad \mathbf{D^T k \lambda} = \begin{bmatrix} 3.8 a_3 \\ 0 \end{bmatrix}$$

Increasing the temperature of the column ABC by 300°C increases its length by

$$\theta \times 3L \times t = 4 \times 10^{-6} \times 3 \times 10^3 \times 300 = 3.6 \text{ mm}$$

Preventing this expansion cause a compressive force in the member. This is

$$p = EA \times 3.6/(3L) = 200 \times 10^4 \times 3.6/(3 \times 10^3) = 2\,400 \text{ kN}$$

Due to the external load

$$p_1 = 2W/(3\sqrt{2}) = 2 \times 210 \times 10^3/(3\sqrt{2}) = 98\,994.95 \text{ kN}$$

∴ The net axial force in member 1 = 98 994.95 − 2 400 = 96 594.95 kN

$$\rho_1 = -p_1 L^2/(\pi^2 EI) = 96\,594.95 \times 10^6/(\pi^2 \times 200 \times 10^9) = -0.049$$

Similarly in member 2 the compressive force is

$$p_2 = W/(3\sqrt{2}) = 210 \times 10^3/(3\sqrt{2}) = 49\,497.47 \text{ kN}$$

and the total force = 49 497.47 + 2 400 = 51 897.47 kN

$$\rho_2 = 51\,897.47 \times 4 \times 10^6/(\pi^2 \times 200 \times 10^9) = 0.105$$

These are slightly different from those given in section 12.10. This difference is due to the assembly and thermal effects. From appendix 1

∴
$$\phi_{2,1} = 1.008, \quad \phi_{3,1} = 1.016, \quad \phi_{5,1} = 1.0484$$

$$\phi_{2,2} = 0.9826, \quad \phi_{3,2} = 0.9650, \quad \phi_{5,2} = 0.8963$$

Equations (12.45) become

$$\begin{bmatrix} b_1 + b_2 + a_3 & d_1 - d_2 \\ d_1 - d_2 & e_1 + e_2 \end{bmatrix} \begin{bmatrix} x_B \\ \theta_B \end{bmatrix} = \begin{bmatrix} 210 \times 10^3/\sqrt{2} \\ 0 \end{bmatrix} + \begin{bmatrix} 3.8a_3 \\ 0 \end{bmatrix}$$

With the data given above and in section 12.10 these equations give

$$x_B = 70.5112 \text{ mm}$$

and

$$\theta_B = 0.0538 \text{ rad}$$

These are noticed to have increased compared to those given in section (12.10). For member 1, $\phi_{4,1} = 0.9921$ and the maximum bending moment is at A given by

$$M_{AB} = \frac{2 \times 200 \times 10^9 \times 0.9921 \times 0.0538}{10^3} - \frac{6 \times 200 \times 70.5112 \times 1.008}{10^6}$$

$$M_{AB} = -63.941 \times 10^3 \text{ kN m}$$

A difference of 0.18%.

Example 2. A fire increased the temperature inside the pitched roof frame shown in *Figure 12.7* (example 3 of section 12.6 and example 4 of section 12.10) by $t°C$. The coefficient of thermal expansion is β and the depth of the section of the frame is D. Derive equations (12.45).
Answer: From section 12.6, the equations $\mathbf{P} = \mathbf{k}\,\boldsymbol{\delta}$ are

$$\begin{bmatrix} F_{AB} \\ M_{BA} \\ F_{BC} \\ M_{BC} \end{bmatrix} = \begin{bmatrix} b_1 & d_1 & 0 & 0 \\ d_1 & e_1 & 0 & 0 \\ 0 & 0 & b_2 & d_2 \\ 0 & 0 & d_2 & e_2 \end{bmatrix} \begin{bmatrix} v_{AB} \\ \theta_{BA} \\ v_{BC} \\ \theta_{BC} \end{bmatrix}$$

The lack of fit vector λ is

$$\begin{bmatrix} v_{\lambda AB} \\ \theta_{\lambda BA} \\ v_{\lambda BC} \\ \theta_{\lambda BC} \end{bmatrix} = \begin{bmatrix} 0 \\ -\beta Lt/2D \\ 0 \\ \beta Lt/2\sqrt{2}D \end{bmatrix}$$

Heating inside the frame causes the rafters to sag; the rotation $\theta_{\lambda BC}$ will be clockwise, thus positive. The inward curving of the columns cause an anticlockwise rotation $\theta_{\lambda BA}$ which is therefore negative. The rotations at A and C are not included in the lack of fit vector because end A is fixed and because of symmetry no rotation takes place at C.

From section 12.6, the equations $\delta = D\Delta$ are

$$\begin{bmatrix} v_{AB} \\ \theta_{BA} \\ v_{BC} \\ \theta_{BC} \end{bmatrix} = \begin{bmatrix} -1 & 0 \\ 0 & 1 \\ \sqrt{2} & 0 \\ 0 & 1 \end{bmatrix} \begin{bmatrix} x_B \\ \theta_B \end{bmatrix}, \quad D^T = \begin{bmatrix} -1 & 0 & \sqrt{2} & 0 \\ 0 & 1 & 0 & 1 \end{bmatrix}$$

$$\therefore D^T k \lambda = \begin{bmatrix} -b_1 & -d_1 & \sqrt{2}b_2 & \sqrt{2}d_2 \\ d_1 & e_1 & d_2 & e_2 \end{bmatrix} \begin{bmatrix} 0 \\ -\beta Lt/2D \\ 0 \\ \beta Lt/2\sqrt{2}D \end{bmatrix}$$

$$D^T k \lambda = \frac{\beta Lt}{2D} \begin{bmatrix} d_1 + d_2 \\ -e_1 + e_2/\sqrt{2} \end{bmatrix}$$

$$D^T k D = \begin{bmatrix} b_1 + 2b_2 & -d_1 + \sqrt{2}d_2 \\ -d_1 + \sqrt{2}d_2 & e_1 + e_2 \end{bmatrix}$$

Equations (12.45) becomes

$$\begin{bmatrix} b_1 + 2b_2 & -d_1 + \sqrt{2}d_2 \\ -d_1 + \sqrt{2}d_2 & e_1 + e_2 \end{bmatrix} \begin{bmatrix} x_B \\ \theta_B \end{bmatrix} = \begin{bmatrix} W/2 \\ 0 \end{bmatrix} - \frac{\beta Lt}{2D} \begin{bmatrix} d_1 + d_2 \\ -e_1 + e_2/\sqrt{2} \end{bmatrix}$$

Before solving these equations the stability ϕ functions, associated with b, d and e for each member, must be calculated in the manner described in example 4 of section 12.10. This problem demonstrates that the thermal stresses are influenced by the stability functions for the members. The problem is therefore highly non-linear.

Exercises on Chapter 12

1. Calculate the shearing forces and the bending moments at A, B, C and D for the continuous beam shown in *Figure 12.17* (page 340). EI is constant throughout.
Ans.

$\{F_{AB} \quad M_{AB} \quad M_{BA} \quad F_{BC} \quad M_{BC} \quad M_{CB} \quad F_{CD} \quad M_{CD} \quad M_{DC}\}$

$= (M/26L) \{-21 \quad 7L \quad 14L \quad -15 \quad 12L \quad 3L \quad 3 \quad -3L \quad 0\}$

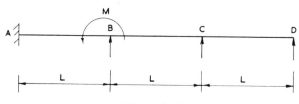

Figure 12.17

2. The beam ABC shown in *Figure 12.18* is built in at A and pinned at C. At point B it is connected to the support at D by a vertical spring of axial stiffness k. Construct the overall stiffness equations $\mathbf{W} = \mathbf{K}\boldsymbol{\Delta}$.

Ans.

$$\begin{bmatrix} M_B \\ M_C \\ V_B \end{bmatrix} = \begin{bmatrix} e_1 + e_2 & f_2 & d_1 - d_2 \\ f_2 & e_2 & -d_2 \\ d_1 - d_2 & -d_2 & b_1 + b_2 + k \end{bmatrix} \begin{bmatrix} \theta_B \\ \theta_C \\ y_B \end{bmatrix}$$

where $f = e/2$.

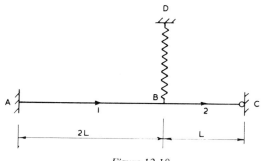

Figure 12.18

3. The symmetrical pitched roof frame of *Figure 12.19* is subject to two moments M at B and D. Sketch the deformed shape of the frame and indicate the joint displacements that are significant. Construct the overall stiffness matrix for this mode of deformation making the order of this matrix as small as possible. If the horizontal sway at B is $0.1L$, calculate the value of M. EI is constant and the axial stiffness EA/L of the members can be neglected.

Ans. $\theta_C = -0.5\,\theta_B$, $\boldsymbol{\Delta} = \{x_B \quad \theta_B\}$

$$\mathbf{K} = \begin{bmatrix} b_1 & -d_1 \\ -d_1 & e_1 + 3e_2/4 \end{bmatrix}$$

$M = -0.2\,EI\,(1 + 6\cos\gamma)/L$

4. The k value of the members of the portal shown in *Figure 12.20* is the same. Calculate the value of the elastic critical load for the frame for the symmetrical

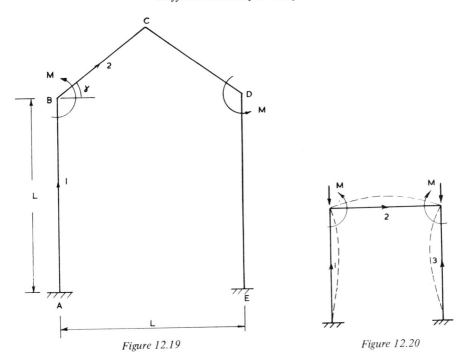

Figure 12.19   Figure 12.20

mode when the side sway is suppressed. Assume $\rho_1 = \rho_3$ and $\rho_2 = 0$.
Ans. $\phi_{3,1} = -0.5$

5. In the rigidly-jointed frame shown in *Figure 12.21* all the members have the same cross section. Taking the effect of axial forces into consideration, calculate the joint rotations when $W/EA = 10^{-5}$. What is the value of ρ_1 when instability takes place?

Ans. $\theta_A = -27.67 \times 10^{-5}$ rad,

$\theta_B = 4.08 \times 10^{-5}$ rad,

$\rho_{1\,cr} = 0.43$

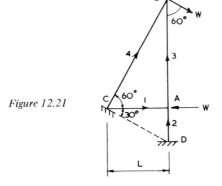

Figure 12.21

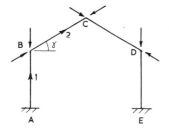
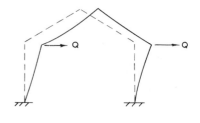

a. The frame b. Antisymmetric mode

Figure 12.22

6. Obtain the condition of elastic instability with an antisymmetric mode for the pitched roof frame shown in *Figure 12.22*.
Ans.

$$\phi_{1,1} \left[e_1 + e_2 - (k_2 \phi_{4,2}^2/\phi_{3,2}) \right] - 3k_1 \phi_{2,1} = 0$$

$$\Delta = \{ x_B \quad \theta_B \}$$

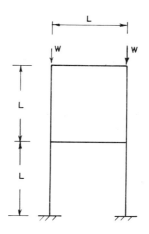

Figure 12.23

7. Neglecting the axial forces in the beams of the two storey symmetrical frame shown in *Figure 12.23* show that if side sway is prevented, the elastic instability conditions are given by

$$(1 + 2\phi_3)(1 + 4\phi_3) = 2\phi_4,$$

the members are identical.

Appendix 1 The stability functions

APPENDIX 1
STABILITY FUNCTIONS FOR COMPRESSIVE FORCES

ρ	ϕ_1	ϕ_2	ϕ_3	ϕ_4	ϕ_5
0.00	1.0000	1.0000	1.0000	1.0000	1.0000
0.02	0.9865	0.9967	0.9934	1.0033	0.9803
0.04	0.9669	0.9934	0.9868	1.0067	0.9605
0.06	0.9592	0.9901	0.9801	1.0101	0.9407
0.08	0.9333	0.9868	0.9734	1.0135	0.9210
0.10	0.9154	0.9834	0.9667	1.0170	0.9012
0.12	0.8993	0.9801	0.9599	1.0205	0.8814
0.14	0.8821	0.9767	0.9531	1.0241	0.8616
0.16	0.8648	0.9734	0.9462	1.0277	0.8418
0.18	0.8474	0.9700	0.9394	1.0313	0.8220
0.20	0.8298	0.9666	0.9324	1.0350	0.8021
0.22	0.8122	0.9632	0.9255	1.0388	0.7823
0.24	0.7943	0.9598	0.9185	1.0426	0.7624
0.26	0.7764	0.9564	0.9114	1.0464	0.7426
0.28	0.7584	0.9530	0.9043	1.0503	0.7227
0.30	0.7402	0.9496	0.8972	1.0543	0.7028
0.32	0.7218	0.9461	0.8901	1.0583	0.6829
0.34	0.7034	0.9427	0.8828	1.0623	0.6630
0.36	0.6848	0.9392	0.8756	1.0665	0.6431
0.38	0.6660	0.9357	0.8683	1.0706	0.6232
0.40	0.6471	0.9323	0.8610	1.0748	0.6033
0.42	0.6281	0.9288	0.8536	1.0791	0.5833
0.44	0.6089	0.9253	0.8462	1.0835	0.5634
0.46	0.5895	0.9217	0.8387	1.0878	0.5434
0.48	0.5701	0.9182	0.8312	1.0923	0.5234
0.50	0.5504	0.9147	0.8236	1.0968	0.5035
0.52	0.5306	0.9111	0.8160	1.1014	0.4835
0.54	0.5106	0.9706	0.8083	1.1060	0.4634
0.56	0.4905	0.9040	0.8006	1.1108	0.4434
0.58	0.4702	0.9004	0.7929	1.1155	0.4234
0.60	0.4498	0.8968	0.7851	1.1204	0.4034
0.62	0.4291	0.8932	0.7772	1.1253	0.3833
0.64	0.4083	0.8896	0.7693	1.1303	0.3632
0.66	0.3873	0.8860	0.7613	1.1353	0.3432
0.68	0.3661	0.8823	0.7533	1.1404	0.3231
0.70	0.3448	0.8787	0.7452	1.1456	0.3030

344 *Appendix*

ρ	ϕ_1	ϕ_2	ϕ_3	ϕ_4	ϕ_5
0.72	0.3233	0.8750	0.7371	1.1509	0.2829
0.74	0.3015	0.8714	0.7289	1.1563	0.2627
0.76	0.2796	0.8677	0.7207	1.1617	0.2426
0.78	0.2575	0.8640	0.7123	1.1672	0.2225
0.80	0.2351	0.8603	0.7040	1.1728	.02023
0.82	0.2126	0.8565	0.6956	1.1785	0.1821
0.84	0.1899	0.8528	0.6871	1.1843	0.1619
0.86	0.1669	0.8491	0.6735	1.1901	0.1417
0.88	0.1438	0.8453	0.6699	1.1961	0.1215
0.90	0.1204	0.8415	0.6612	1.2021	0.1013
0.92	0.0968	0.8377	0.6525	1.2082	0.0811
0.94	0.0729	0.8339	0.6437	1.2144	0.0608
0.96	0.0489	0.8301	0.6348	1.2208	0.0406
0.98	0.0246	0.8263	0.6259	1.2272	0.0203
1.00	−0.0000	0.8225	0.6169	1.2337	−0.0000
1.02	−0.0248	0.8186	0.6078	1.2403	−0.0203
1.04	−0.0498	0.8148	0.5986	1.2471	−0.0406
1.06	−0.0752	0.8109	0.5894	1.2539	−0.0609
1.08	−0.1007	0.8070	0.5801	1.2608	−0.0813
1.10	−0.1266	0.8031	0.5707	1.2679	−0.1016
1.12	−0.1527	0.7992	0.5612	1.2751	−0.1220
1.14	−0.1790	0.7952	0.5517	1.2824	−0.1424
1.16	−0.2057	0.7913	0.5420	1.2898	−0.1628
1.18	−0.2327	0.7873	0.5323	1.2973	−0.1832
1.20	−0.2599	0.7833	0.5225	1.3050	−0.2036
1.22	−0.2875	0.7794	0.5126	1.3128	−0.2241
1.24	−0.3153	0.7754	0.5027	1.3207	−0.2445
1.26	−0.3435	0.7713	0.4926	1.3288	−0.2650
1.28	−0.3720	0.7673	0.4825	1.3370	−0.2855
1.30	−0.4009	0.7633	0.4722	1.3453	−0.3060
1.32	−0.4300	0.7592	0.4619	1.3538	−0.3265
1.34	−0.4595	0.7551	0.4515	1.3624	−0.3470
1.36	−0.4894	0.7510	0.4409	1.3712	−0.3675
1.38	−0.5196	0.7469	0.4303	1.3802	−0.3881
1.40	−0.5502	0.7428	0.4196	1.3893	−0.4087
1.42	−0.5811	0.7387	0.4087	1.3985	−0.4292
1.44	−0.6125	0.7345	0.3978	1.4080	−0.4499
1.46	−0.6442	0.7303	0.3867	1.4176	−0.4705
1.48	−0.6763	0.7261	0.3755	1.4247	−0.4911
1.50	−0.7089	0.7219	0.3642	1.4373	−0.5118
1.52	−0.7418	0.7177	0.3528	1.4475	−0.5324
1.54	−0.7752	0.7135	0.3413	1.4578	−0.5531
1.56	−0.8090	0.7092	0.3297	1.4684	−0.5738
1.58	−0.8433	0.7050	0.3179	1.4791	−0.5945
1.60	−0.8781	0.7007	0.3060	1.4901	−0.6153
1.62	−0.9133	0.6964	0.2940	1.5012	−0.6360
1.64	−0.9490	0.6921	0.2818	1.5126	−0.6568
1.66	−0.9852	0.6877	0.2695	1.5242	−0.6776
1.68	−1.0219	0.6834	0.2571	1.5360	−0.6984
1.70	−1.0592	0.6790	0.2445	1.5481	−0.7192

Appendix

ρ	ϕ_1	ϕ_2	ϕ_3	ϕ_4	ϕ_5
1.72	−1.0969	0.6746	0.2317	1.5604	−0.7400
1.74	−1.1353	0.6702	0.2189	1.5730	−0.7609
1.76	−1.1741	0.6658	0.2058	1.5858	−0.7817
1.78	−1.2136	0.6614	0.1926	1.5988	−0.8026
1.80	−1.2537	0.6569	0.1793	1.6122	−0.8235
1.82	−1.2944	0.6524	0.1657	1.6258	−0.8445
1.84	−1.3357	0.6479	0.1520	1.6397	−0.8654
1.86	−1.3776	0.6434	0.1382	1.6539	−0.8864
1.88	−1.4202	0.6389	0.1241	1.6684	−0.9074
1.90	−1.4635	0.6343	0.1099	1.6833	−0.9284
1.92	−1.5076	0.6298	0.0954	1.6984	−0.9494
1.94	−1.5523	0.6252	0.0808	1.7139	−0.9704
1.96	−1.5978	0.6206	0.0660	1.7297	−0.9915
1.98	−1.6440	0.6159	0.0509	1.7459	−1.0126
2.00	−1.6910	0.6113	0.0357	1.7624	−1.0337
2.02	−1.8388	9.6066	0.0202	1.7793	−1.0548
2.04	−1.7875	0.6019	0.0046	1.7966	−1.0759
2.06	−1.8371	0.5972	−0.0114	1.8143	−1.0971
2.08	−1.8875	0.5925	−0.0275	1.8324	−1.1183
2.10	−1.9388	0.5877	−0.0439	1.8510	−1.1395
2.12	−1.9911	0.5829	−0.0606	1.8700	−1.1607
2.14	−2.0444	0.5781	−0.0775	1.8894	−1.1819
2.16	−2.0986	0.5733	−0.0947	1.9093	−1.2032
2.18	−2.1539	0.5685	−0.1121	1.9297	−1.2245
2.20	−2.2163	0.5636	−0.1299	1.9506	−1.2458
2.22	−2.2678	0.5587	−0.1479	1.9720	−1.2671
2.24	−2.3264	0.5538	−0.1662	1.9940	−1.2885
2.26	−2.3863	0.5489	−0.1849	2.0165	−1.3099
2.28	−2.4473	0.5440	−0.2039	2.0396	−1.3313
2.30	−2.5096	0.5390	−0.2232	2.0633	−1.3527
2.32	−2.5733	0.5340	−0.2428	2.0876	−1.3741
2.34	−2.6383	0.5290	−0.2628	2.1126	−1.3956
2.36	−2.7047	0.5239	−0.2832	2.1382	−1.4171
2.38	−2.7725	0.5189	−0.3040	2.1646	−1.4386
2.40	−2.8419	0.5138	−0.3251	2.1916	−1.4601
2.42	−2.9129	0.5087	−0.3467	2.2195	−1.4817
2.44	−2.9855	0.5035	−0.3687	2.2480	−1.5033
2.46	−3.0598	0.4984	−0.3912	2.2774	−1.5249
2.48	−3.1359	0.4932	−0.4141	2.3077	−1.5465
2.50	−3.2138	0.4880	−0.4375	2.3388	−1.5682
2.52	−3.2936	0.4927	−0.4613	2.3709	−1.5899
2.54	−3.3754	0.4775	−0.4857	2.4039	−1.6116
2.56	−3.4592	0.4722	−0.5107	2.4379	−1.6333
2.58	−3.5452	0.4669	−0.5362	2.4729	−1.6551
2.60	−3.6335	0.4615	−0.5622	2.5090	−1.6769
2.62	−3.7241	0.4561	−0.5889	2.5463	−1.6987
2.64	−3.8172	0.4507	−0.6162	2.5874	−1.7206
2.66	−3.9128	0.4453	−0.6442	2.6244	−1.7424
2.68	−4.0111	0.4399	−0.6729	2.6654	−1.7643
2.70	−4.1122	0.4344	−0.7023	2.7077	−1.7863

ρ	ϕ_1	ϕ_2	ϕ_3	ϕ_4	ϕ_5
2.72	−4.2162	0.4289	−0.7324	2.7514	−1.8082
2.74	−4.3233	0.4233	−0.7633	2.7967	−1.8302
2.76	−4.4336	0.4178	−0.7951	2.8435	−1.8522
2.78	−4.5473	0.4122	−0.8277	2.8919	−1.8743
2.80	−4.6645	0.4066	−0.8612	2.9421	−1.8964
2.82	−4.7854	0.4009	−0.8957	2.9941	−1.9185
2.84	−4.9103	0.3952	−0.9312	3.0480	−1.9406
2.86	−5.0392	0.3895	−0.9677	3.1039	−1.9628
2.88	−5.1725	0.3837	−1.0053	3.1619	−1.9850
2.90	−5.3104	0.3780	−1.0441	3.2222	−2.0072
2.92	−5.4531	0.3722	−1.0841	3.2848	−2.0294
2.94	−5.6009	0.3663	−1.1255	3.3499	−2.0517
2.96	−5.7540	0.3605	−1.1682	3.4177	−2.0740
2.98	−5.9129	0.3545	−1.2123	3.4883	−2.0964
3.00	−6.0778	0.3486	−1.2580	3.5618	−2.1188
3.02	−6.2491	0.3426	−1.3053	3.6358	−2.1412
3.04	−6.4273	0.3366	−1.3543	3.7186	−2.1637
3.06	−6.6127	0.3306	−1.4052	3.8023	−2.1862
3.08	−6.8058	0.3245	−1.4581	3.8897	−2.2087
3.10	−7.0072	0.3184	−1.5130	3.9812	−2.2312
3.12	−7.2174	0.3123	−1.5701	4.0771	−2.2538
3.14	−7.4369	0.3061	−1.6297	4.1776	−2.2764
3.16	−7.6666	0.2999	−1.6917	4.2831	−2.2991
3.18	−7.9071	0.2936	−1.7565	4.3940	−2.3218
3.20	−8.1592	0.2874	−1.8243	4.5106	−2.3445
3.22	−8.4238	0.2810	−1.8952	4.6334	−2.3673
3.24	−8.7019	0.2747	−1.9695	4.7630	−2.3901
3.26	−8.9947	0.2683	−2.0475	4.8997	−2.4130
3.28	−9.3032	0.2618	−2.1294	5.0444	−2.4359
3.30	−9.6290	0.2554	−2.2157	5.1975	−2.4588
3.32	−9.9734	0.2488	−2.3067	5.3600	−2.4818
3.34	−10.3382	0.2423	−2.4028	5.5325	−2.5048
3.36	−10.7253	0.2357	−2.5046	5.7162	−2.5278
3.38	−11.1369	0.2290	−2.6124	5.9120	−2.5509
3.40	−11.5754	0.2224	−2.7271	6.1212	−2.5740
3.42	−12.0435	0.2157	−2.8491	6.3452	−2.5972
3.44	−12.5445	0.2089	−2.9795	6.5856	−2.6204
3.46	−13.0820	0.2021	−3.1189	6.8441	−2.6437
3.48	−13.6602	0.1952	−3.2686	7.1230	−2.6670
3.50	−14.2840	0.1883	−3.4297	7.4245	−2.6903
3.52	−14.9591	0.1814	−3.6037	7.7517	−2.7137
3.54	−15.6922	0.1744	−3.7922	8.1077	−2.7371
3.56	−16.4912	0.1674	−3.9973	8.4967	−2.7606
3.58	−17.3655	0.1603	−4.2211	8.9232	−2.7841
3.60	−18.3264	0.1532	−4.4667	9.3930	−2.8077
3.62	−19.3875	0.1460	−4.7374	9.9128	−2.8313
3.64	−20.5657	0.1388	−5.0373	10.4911	−2.8550
3.66	−21.8814	0.1316	−5.3717	11.1380	−2.8787
3.68	−23.3606	0.1242	−5.7470	11.8667	−2.9024
3.70	−25.0358	0.1169	−6.1713	12.6932	−2.9262

Appendix 347

ρ	ϕ_1	ϕ_2	ϕ_3	ϕ_4	ϕ_5
3.72	−26.9492	0.1095	−6.6552	13.6388	−2.9501
3.74	−29.1556	0.1020	−7.2124	14.7308	−2.9740
3.76	−31.7284	0.0945	−7.8612	16.0059	−2.9980
3.78	−34.7673	0.0869	−8.6266	17.5140	−3.0220
3.80	−38.4124	0.0793	−9.5436	19.3251	−3.0461
3.82	−42.3655	0.0716	−10.6627	21.5402	−3.0702
3.84	−48.4298	0.0639	−12.0595	24.3107	−3.0944
3.86	−55.5813	0.0561	−13.8533	27.8748	−3.1186
3.88	−65.1139	0.0483	−16.2423	32.6293	−3.1429
3.90	−78.4560	0.0404	−19.5837	39.2885	−3.1673
3.92	−98.4647	0.0324	−24.5919	49.2810	−3.1917
3.94	−131.8069	0.0244	−32.9334	65.9400	−3.2161
3.96	−198.4823	0.0163	−49.6083	99.2557	−3.2406
3.98	−398.4912	0.0082	−99.6166	199.2579	−3.2652

STABILITY FUNCTIONS FOR TENSILE FORCES

0.00	1.0000	1.0000	1.0000	1.0000	1.0000
−0.02	1.0164	1.0033	1.0066	0.9967	1.0197
−0.04	1.0327	1.0066	1.0131	0.9935	1.0395
−0.06	1.0489	1.0098	1.0196	0.9903	1.0592
−0.08	1.0649	1.0131	1.0261	0.9872	1.0789
−0.10	1.0809	1.0163	1.0325	0.9840	1.0986
−0.12	1.0968	1.0196	1.0389	0.9810	1.1183
−0.14	1.1126	1.0228	1.0452	0.9779	1.1380
−0.16	1.1283	1.0260	1.0516	0.9749	1.1576
−0.18	1.1438	1.0292	1.0579	0.9719	1.1773
−0.20	1.1593	1.0324	1.0642	0.9690	1.1969
−0.22	1.1747	1.0356	1.0704	0.9661	1.2166
−0.24	1.1900	1.0388	1.0766	0.9632	1.2362
−0.26	1.2052	1.0420	1.0828	0.9604	1.2558
−0.28	1.2203	1.0452	1.0890	0.9576	1.2755
−0.30	1.2354	1.0483	1.0951	0.9548	1.9521
−0.32	1.2503	1.0515	1.1012	0.9521	1.3147
−0.34	1.2652	1.0546	1.1073	0.9494	1.3343
−0.36	1.2799	1.0578	1.1133	0.9467	1.3539
−0.38	1.2946	1.0609	1.1193	0.9441	1.3734
−0.40	1.3092	1.0640	1.1253	0.9414	1.3930
−0.42	1.3237	1.0671	1.1313	0.9388	1.4126
−0.44	1.3381	1.0702	1.1372	0.9363	1.4321
−0.46	1.3525	1.0733	1.1431	0.9338	1.4517
−0.48	1.3668	1.0764	1.1490	0.9312	1.4712
−0.50	1.3809	1.0795	1.1549	0.9288	1.4907
−0.52	1.3951	1.0826	1.1607	0.9263	1.5103
−0.54	1.4091	1.0856	1.1665	0.9239	1.5298
−0.56	1.4231	1.0887	1.1723	0.9215	1.5493
−0.58	1.4369	1.0917	1.1780	0.9191	1.5688
−0.60	1.4508	1.0948	1.1838	0.9168	1.5883

ρ	ϕ_1	ϕ_2	ϕ_3	ϕ_4	ϕ_5
−0.62	1.4645	1.0978	1.1895	0.9145	1.6077
−0.64	1.4782	1.1008	1.1952	0.9122	1.6272
−0.66	1.4918	1.1039	1.2008	0.9099	1.6467
−0.68	1.5053	1.1069	1.2065	0.9077	1.6661
−0.70	1.5187	1.1099	1.2121	0.9054	1.6856
−0.72	1.5321	1.1129	1.2177	0.9032	1.7050
−0.74	1.5454	1.1158	1.2232	0.9011	1.7245
−0.76	1.5587	1.1188	1.2288	0.8989	1.7439
−0.78	1.5719	1.1218	1.2343	0.8968	1.7633
−0.80	1.5850	1.1248	1.2398	0.8947	1.7827
−0.82	1.5980	1.1277	1.2453	0.8926	1.8021
−0.84	1.6110	1.1307	1.2508	0.8905	1.8215
−0.86	1.6239	1.1336	1.2562	0.8885	1.8409
−0.88	1.6368	1.1366	1.2616	0.8864	1.8603
−0.90	1.6496	1.1395	1.2670	0.8844	1.8797
−0.92	1.6623	1.1424	1.2724	0.8824	1.8991
−0.94	1.6750	1.1453	1.2778	0.8805	1.9184
−0.96	1.6876	1.1482	1.2831	0.8785	1.9378
−0.98	1.7002	1.1511	1.2884	0.8766	1.9572
−1.00	1.7127	1.1540	1.2937	0.8747	1.9765
−1.02	1.7251	1.1569	1.2990	0.8728	1.9958
−1.04	1.7375	1.1598	1.3042	0.8710	2.0152
−1.06	1.7498	1.1627	1.3095	0.8691	2.0345
−1.08	1.7621	1.1655	1.3147	0.8673	2.0538
−1.10	1.7743	1.1684	1.3199	0.8655	2.0731
−1.12	1.7865	1.1713	1.3251	0.8637	2.0924
−1.14	1.7986	1.1741	1.3302	0.8619	2.1117
−1.16	1.8106	1.1770	1.3354	0.8601	2.1310
−1.18	1.8226	1.1798	1.3405	0.8584	2.1503
−1.20	1.8346	1.1826	1.3456	0.8566	2.1696
−1.22	1.8464	1.1854	1.3507	0.8549	2.1886
−1.24	1.8583	1.1883	1.3558	0.8532	2.2081
−1.26	1.8701	1.1911	1.3608	0.8516	2.2274
−1.28	1.8818	1.1939	1.3659	0.8499	2.2466
−1.30	1.8935	1.1967	1.3709	0.8483	2.2659
−1.32	1.9051	1.1995	1.3759	0.8466	2.2851
−1.34	1.9167	1.2022	1.3809	0.8450	2.3043
−1.36	1.9282	1.2050	1.3858	0.8434	2.3236
−1.38	1.9397	1.2078	1.3908	0.8418	2.3428
−1.40	1.9512	1.2106	1.3957	0.8402	2.3620
−1.42	1.9626	1.2133	1.4006	1.8387	2.3812
−1.44	1.9739	1.2161	1.4055	0.8371	2.4004
−1.46	1.0852	1.2188	1.4104	0.8356	2.4196
−1.48	1.9965	1.2216	1.4153	0.8341	2.4388
−1.50	2.0077	1.2243	1.4201	0.8326	2.4580
−1.52	2.0188	1.2270	1.4250	0.8311	2.4772
−1.54	2.0300	1.2297	1.4298	0.8296	2.4963
−1.56	2.0410	1.2325	1.4346	0.8282	2.5155
−1.58	2.0521	1.2352	1.4394	0.8267	2.5347
−1.60	2.0631	1.2379	1.4442	0.8253	2.5538

Appendix

ρ	ϕ_1	ϕ_2	ϕ_3	ϕ_4	ϕ_5
−1.62	2.0740	1.2406	1.4489	0.8239	2.5730
−1.64	2.0849	1.2433	1.4537	0.8225	2.5921
−1.66	2.0958	1.2460	1.4584	0.8211	2.6113
−1.68	2.1066	1.2487	1.4631	0.8197	2.6304
−1.70	2.1174	1.2513	1.4678	0.8183	2.6495
−1.72	2.1281	1.2540	1.4725	0.8169	2.6686
−1.74	2.1388	1.2567	1.4772	0.8156	2.6878
−1.76	2.1495	1.2593	1.4819	0.8143	2.7069
−1.78	2.1601	1.2620	1.4865	0.8129	2.7260
−1.80	2.1706	1.2646	1.4911	0.8116	2.7451
−1.82	2.1812	1.2675	1.4958	0.8103	2.7642
−1.84	2.1917	1.2699	1.5004	0.9080	2.7833
−1.86	2.2021	1.2725	1.5049	0.8078	2.8023
−1.88	2.2126	1.2752	1.5095	0.8065	2.8214
−1.90	2.2230	1.2778	1.5141	0.8052	2.8405
−1.92	2.2333	1.2804	1.5186	0.8040	2.8596
−1.94	2.2436	1.2830	1.5232	0.8027	2.8786
−1.96	2.2539	1.2856	1.5277	0.8015	2.8977
−1.98	2.2641	1.2882	1.5322	0.8003	2.9167
−2.00	2.2743	1.2908	1.5367	0.7991	2.2958
−2.02	2.2845	1.2934	1.5412	0.7979	2.9548
−2.04	2.2946	1.2960	1.5457	0.7967	2.9738
−2.06	2.3047	1.2986	1.5501	0.7955	2.9929
−2.08	2.3148	1.3012	1.5546	0.7944	3.0119
−2.10	2.3248	1.3037	1.5590	0.7932	3.0309
−2.12	2.3348	1.3063	1.5634	0.7921	3.0499
−2.14	2.3447	1.3089	1.5678	0.7909	3.0689
−2.16	2.3547	1.3114	1.5722	0.7898	3.0879
−2.18	2.3646	1.3140	1.5766	0.7887	3.1069
−2.20	2.3744	1.3165	1.5810	0.7876	3.1259
−2.22	2.3842	1.3191	1.5853	0.7865	3.1449
−2.24	2.3940	1.3216	1.5897	0.7854	3.1639
−2.26	2.4038	1.3241	1.5940	0.7843	3.1829
−2.28	2.4135	1.3266	1.5948	0.7832	3.2019
−2.30	2.4232	1.3292	1.6027	0.7821	3.2208
−2.32	2.4329	1.3317	1.6070	0.7811	3.2398
−2.34	2.4425	1.3342	1.6113	0.7800	3.2588
−2.36	2.4521	1.3367	1.6155	0.7790	3.2777
−2.38	2.4617	1.3392	1.6198	0.7780	3.2967
−2.40	2.4712	1.3417	1.6241	0.7769	3.3156
−2.42	2.4807	1.3442	1.6283	0.7759	3.3346
−2.44	2.4902	1.3467	1.6326	0.7749	3.3535
−2.46	2.4997	1.3492	1.6368	0.7739	3.3724
−2.48	2.5091	1.3516	1.6410	0.7729	3.3913
−2.50	2.5185	1.3541	1.6452	0.7719	3.4103
−2.52	2.5278	1.3566	1.6494	0.7709	3.4292
−2.54	2.5372	1.3590	1.6536	0.7700	3.4481
−2.56	2.5465	1.3615	1.6577	0.7690	3.4670
−2.58	2.5558	1.3639	1.6619	0.7680	3.4859
−2.60	2.5650	1.3664	1.6660	0.7671	3.5048

ρ	ϕ_1	ϕ_2	ϕ_3	ϕ_4	ϕ_5
−2.62	2.5742	1.3688	1.6702	0.7662	3.5237
−2.64	2.5834	1.3713	1.6743	0.7652	3.5426
−2.66	2.5926	1.3737	1.6784	0.7643	3.5615
−2.68	2.6017	1.3762	1.6825	0.7634	3.5804
−2.70	2.6108	1.3786	1.6866	0.7625	3.5992
−2.72	2.6199	1.3810	1.6907	0.7616	3.6181
−2.74	2.6290	1.3834	1.6948	0.7607	3.6370
−2.76	2.6380	1.3858	1.6989	0.7598	3.6559
−2.78	2.6470	1.3883	1.7029	0.7589	3.6747
−2.80	2.6560	1.3907	1.7070	0.7580	3.6936
−2.82	2.6649	1.3931	1.7110	0.7571	3.7124
−2.84	2.6739	1.3955	1.7151	0.7563	3.7313
−2.86	2.6828	1.3978	1.7191	0.7554	3.7501
−2.88	2.6916	1.4002	1.7231	0.7545	3.7689
−2.90	2.7005	1.4026	1.7271	0.7537	3.7878
−2.92	2.7093	1.4050	1.7311	0.7528	3.8066
−2.94	2.7181	1.4074	1.7351	0.7520	3.8254
−2.96	2.7269	1.4098	1.7390	0.7512	3.8443
−2.98	2.7375	1.4121	1.7430	0.7504	3.8631
−3.00	2.7444	1.4145	1.7470	0.7495	3.8819
−3.02	2.7531	1.4168	1.7509	0.7487	3.9007
−3.04	2.7618	1.4192	1.7548	0.7479	3.9195
−3.06	2.7704	1.4216	1.7588	0.7471	3.9383
−3.08	2.7791	1.4239	1.7627	0.7463	3.9571
−3.10	2.7877	1.4262	1.7666	0.7455	3.9759
−3.12	2.7963	1.4286	1.7705	0.7447	3.9947
−3.14	2.8048	1.4309	1.7744	0.7440	4.0135
−3.16	2.8134	1.4332	1.7783	0.7432	4.0322
−3.18	2.8219	1.4356	1.7822	0.7424	4.0510
−3.20	2.8304	1.4379	1.7860	0.7417	4.0698
−3.22	2.8388	1.4402	1.7899	0.7409	4.0886
−3.24	2.8473	1.4425	1.7937	0.7402	4.1073
−3.26	2.8557	1.4448	1.7976	0.7394	4.1261
−3.28	2.8641	1.4472	1.8014	0.7387	4.1448
−3.30	2.8725	1.4495	1.8052	0.7379	4.1636
−3.32	2.8809	1.4518	1.8090	0.7372	4.1824
−3.34	2.8892	1.4541	1.8128	0.7365	4.2011
−3.36	2.8975	1.4563	1.8166	0.7358	4.2198
−3.38	2.9058	1.4586	1.8204	0.7350	4.2386
−3.40	2.9141	1.4609	1.8242	0.7343	4.2573
−3.42	2.9224	1.4632	1.8280	0.7336	4.2760
−3.44	2.9306	1.4655	1.8318	0.7329	4.2948
−3.46	2.9388	1.4678	1.8355	0.7322	4.3135
−3.48	2.9470	1.4700	1.8393	0.7315	4.3322
−3.50	2.9552	1.4723	1.8430	0.7308	4.3509
−3.52	2.9634	1.4746	1.8468	0.7302	4.3696
−3.54	2.9715	1.4768	1.8505	0.7295	4.3883
−3.56	2.9796	1.4791	1.8542	0.7288	4.4071
−3.58	2.9877	1.4813	1.8579	0.7281	4.4258
−3.60	2.9958	1.4836	1.8616	0.7275	4.4445

Appendix 351

ρ	ϕ_1	ϕ_2	ϕ_3	ϕ_4	ϕ_5
−3.62	3.0038	1.4858	1.8653	0.7268	4.4631
−3.64	3.0119	1.4881	1.8690	0.7261	4.4818
−3.66	3.0199	1.4903	1.8727	0.7255	4.5005
−3.68	3.0279	1.4925	1.8764	0.7248	4.5192
−3.70	3.0359	1.4948	1.8800	0.7242	4.5379
−3.72	3.0438	1.4970	1.8837	0.7236	4.5566
−3.74	3.0518	1.4992	1.8873	0.7229	4.5752
−3.76	3.0597	1.5014	1.8910	0.7223	4.5930
−3.78	3.0676	1.5036	1.8946	0.7217	4.6126
−3.80	3.0755	1.5059	1.8983	1.7210	4.6312
−3.82	3.0834	1.5081	1.9019	0.7204	4.6499
−3.84	3.0912	1.5103	1.9055	0.7198	4.6685
−3.86	3.0990	1.5125	1.9091	0.7192	4.6872
−3.88	3.1068	1.5147	1.9127	0.7186	4.7058
−3.90	3.1146	1.5169	1.9163	0.7180	4.7245
−3.92	3.1224	1.5191	1.9199	0.7174	4.7431
−3.94	3.1302	1.5212	1.9235	0.7168	4.7618
−3.96	3.1379	1.5234	1.9271	0.7162	4.7804
−3.98	3.1456	1.5256	1.9306	0.7156	4.7990

Index

Antisymmetrical loading, 45-49
Arches
  parabolic, 101-105
  three pinned, 97, 100, 106
Assembly effects, 337
Assembly (or lack of fit) forces, 231, 333
Axial forces, 109, 195, 200, 266, 294, 305, 315-317, 322, 325, 330

Beams
  collapse of, *See* Collapse
  compound, 58
  continuous
    collapse load, 188
    collapse mechanism, 189-191, 193
  definition, 58
  deflection, *See* Deflection
  encastré, 58
  fixed ended, carrying control load, 119
  hyperstatic, *See* Hyperstatic
  isostatic, *See* Isostatic
  loading, 58-59
  simply supported, 58, 64
    and loaded at midspan, 116-117
    carrying a number of moving loads, 148
    carrying point load unsymmetrically, 117
    slope of, 117
    uniformly loaded, 120
    with end moments and midspan load, 117
  uniformly loaded, 83, 120, 129
    fixed ended, 121
  with external moments, 84
  with several point loads, 122
  see also Cantilevers
Bending, 58, 111, 161
Bending moment, 155, 168, 239, 243, 248, 294, 316
  at specific point, 127
  beam with end moments and midspan load, 118
  bent member loaded out of plane, 96
  calculation of, 67, 79, 105, 107, 110, 174, 319, 336-338
  cantilever with end load, 59
  cantilever with several loads, 63-64

Bending moment (continued)
  columns, 248, 325
  dangerous section for, 147, 152
  definition, 59
  elastic bending, 111
  fixed ended beam, 121
  generalised Macaulay's method, 81-87
  influence line for, 145
  load and shearing force relationship, 70-73
  sign conventions, 61
  simply supported beam, 64-67, 120
  three-pinned arch, 97-98
  uniformly loaded beam, 120, 121
Bending moment diagram
  beam with end supports and midspan load, 118-119
  cantilever uniformly loaded, 63
  cantilever with end load, 61
  cantilever with inclined force, 62
  cantilever with several loads, 64
  columns, 109, 322
  elastic bending, 112
  examples, 69-70, 72-78
  fixed end beam, 186
  parabolic arch, 102
  plane bent members, 91-94
  rigidly jointed frame with uniformly distributed load, 134-137
  simply supported beam, 65, 66
  three-pinned arch, 100
  uniformly loaded beam, 67, 129
Bending stresses, 78-81
Bent members, 90
  deflection, of 222
  loaded out of plane, 96
  plane, 91
Brittle materials, 26, 200
Buckling load, 239, 253, 317, 321, 332

Cantilever
  carrying inclined force, 62
  circular, 96
  definition, 58
  deflection of, 114-115
  loading, 59

354 *Index*

Cantilever (continued)
  propped, 58, 246, 247, 304
  slope of, 114-115
  uniformly loaded, 62-63
  with end load, 59
  with several loads, 63-64
Cartesian components, 271-272
Centroid, 177
Collapse load of continuous beam, 188
Collapse load factor, 185, 189, 193
Collapse mechanism, 183-189, 200
  example, 191
  in continuous beams, 189-191, 193
Column vector, 9, 24, 25, 40, 280
Columns
  analysis, 239-265
  deformed shape, 239
  design, 261
  double curvature, 248-249
  eccentrically loaded, 255, 257
  initially curved pin-ended, 258, 261
  laterally loaded, 261
  maximum stress in curved pin-ended, 260
  pin-ended, 239, 261
  pure sway, 249-251
  sway, 249-251, 254
Compatibility conditions, 41, 42
Compatibility equations, 172
Compressive forces, 343
Compressive stress, 52, 53, 58
Computers, 51
Concrete, 26
Concurrent forces in space, 266
Conforming function, 243
Coplanar forces, 3-4, 6
Critical load, 52, 239, 246, 249, 253, 257, 321
Curved members
  loaded out of plane, 96
  plane, 91-92

Dangerous section
  for bending moment, 147, 152
  for several loads, 148
  for shearing force, 147
Deflection, 32, 35-43, 61
  at unloaded specific point, 127
  calculation of, 111-140, 167, 170-171, 173, 205, 208, 209, 213, 226, 228, 262, 276, 285, 336
  in beams with several loads, 122
  in fixed end beam carrying central load, 119
  in simply supported beam carrying point load, 117
  in simply supported beam loaded at mid-span, 116-117
  in uniformly loaded beams, 129
  in uniformly loaded fixed ended beam, 121
  in uniformly loaded simply supported beam, 120

Deflection (continued)
  of bent members, 222
  of cantilever, 114-115
  of eccentrically loaded column, 255
  of space frames, 276
  of structures, 205
  of structures with non-linear material characteristics, 210
  of unloaded joints, 210
  relationship between member and joint, 205-208
Deflection diagram, 30
Deflection function, 112-113
Deformation, 32, 39, 200, 212, 217, 218, 221, 234, 278-280, 295-302
Diagonal matrix, 24, 27, 28
Diagonal square matrix, 27
Direct strain, 25
Direct stress, 23
Direction cosines, 269-271, 275
Displacement diagram, 310
Displacement function, 242
Displacement method, 282, 333
Displacement transformation matrix, 33, 297, 303, 334
Displacement transformation vector, 280
Ductile materials, 26, 200

Elastic behaviour, 25
Elastic instability, 315
Elastic-plastic failure load analyses, 196
Elastic section modulus, 179
Elastic structure, 38
Encastré beam, 58
End moments, 117-118
Equilibrium, 5, 9, 60, 61, 267, 272, 312
Equilibrium of forces, 41, 42
Equilibrium equations, 4-9, 280-284, 303-319, 326, 327, 332
Equivalent half frame, 48-51
Equivalent load, 50-51
Euler critical load, 239

Finite element method, 41, 240
Flexibility, 27-28, 158
  definition, 27
  example, 159, 160
Flexibility coefficient, 158
Flexibility equations, 158, 224, 227, 229
Flexibility matrix, 27, 28, 158, 159
Flexural elements, 155
Flexural properties, 155-182
Flexural stiffness, *See* Stiffness
Force deformation diagram, 39
Force method, 205-238
  hyperstatic frames, 213-215
Force-rotation curve, 246-247
Frictionless hinges, 58
Frictionless pins, 1, 3

Index

Homogeneous equations, solution of, 317–318
Hooke's law, 26, 37, 41, 78, 206, 209, 217, 280, 294
Hyperstatic beams, 58, 120, 174, 176
Hyperstatic frame, 12, 13, 18, 19, 106, 108, 137, 213–216, 219, 220, 230, 282
Hyperstatic structures, 34, 120

Incipient collapse, 183
Influence line diagram, 145–154
Influence lines, 141–154
  definition, 141
  examples, 144, 146, 152
  for bending moments, 145
  for pin jointed frames, 149–151
  for shearing force, 143
  for support reactions, 141–142
  significance of, 149
Instability, 196, 251, 318, 319, 321, 328, 331, 333
Isostatic beams, 58–110
Isostatic frames, 12–14, 97, 220, 221, 272, 327
Isostatic structures, 1, 5, 6, 13, 34, 106, 183, 193, 267

Joint deflections and member deformations, 278–280
Joint displacement vector, 302
Joint displacements and member deformations, 295–302

Kinematic theorem, 188, 190, 200

'Lack of fit', 232, 334, 336
'Lack of fit' forces, 231
'Lack of fit' vector, 338, 339
Limit of proportionality, 25, 28, 36, 38, 39, 164
Limiting stress matrix, 24
Linear analysis, 294, 321
Livesley's stability functions, 245
Load deflection diagram, 329
Load factor, 183–184, 189
Local axes, 98
Local coordinates, 98

Matrix equations, 52, 287
Matrix form, 24
Matrix methods, 52
Member forces, 6–9, 24
  calculation of, 10, 15, 21, 44–47, 52, 53, 273, 274, 282
  components of, 302–304
  sign of, 274
Member stresses, 24

Method of joints, 6, 11, 12, 52
Method of sections, 9, 11

Non-conforming function, 243
Non-linear elastic analysis, 315
Non-linear elastic method, 294

Parabolic arch, 101–105
Parabolic displacement function, 243
Parallelogram of forces, 266
Permanent set, 25
Perry-Robertson's formula, 261
Perry's formula, 261
Pin jointed frames, influence lines for, 149–151
Pin jointed isostatic plane structures, 1–22
Pin jointed space structures, 266–293
  forces in, 266–267
  stability of, 267–269
Plane frames, 12
  rigidly jointed, analysis of, 294–342
Plane structures, 1, 2, 6
Plastic behaviour, 25
Plastic hinge moment, 195–196, 201, 254
Plastic hinges, 185–201
Plastic modulus, calculation of, 197
Plastic moment, 177–178
Plastic section modulus, 178–179, 195
Plastic theory, 183–204
  limitation, 200
Plasticity, 176–177
Plates, yield line theory in, 200–201
Portal frames, 190, 196, 226
Principle of superposition, 35–38, 44, 45, 48, 108, 142, 329
  exceptions to, 36, 37
Principle of virtual work, 144
Proportional loading, 184–185
Propped cantilever, 58, 246

Radius of curvature, 111
Radius of gyration, 260
Reaction components, 3, 5–7, 13, 213, 214
Reactive moment, 61
Redundant force, 14, 19
Redundant members, 21
Reinforced concrete slabs, yield line pattern in, 202–203
Reinforced concrete structures, 2
Rigid-plastic theory, 193
Rigidly jointed plane frames, analysis of, 294–342
Roller support, 3

Section modulus, 79, 80
Shape factor, 178, 184
Shear, strain energy in, 168–171

Shear modulus, 169
Shear strain, 169
Shear stress, 58, 169
Shear stress distribution, 170
Shearing force, 59, 155, 168, 239, 294
  absolute maximum, 147
  bent member loaded out of plane, 96
  calculation of, 67-69, 105, 159, 319
  column in double curvature, 248
  dangerous section for, 147
  definition, 59-60
  elastic bending, 111-112
  generalised Macaulay's method, 81-87
  influence line for, 143
  load, and bending moment relationship, 70-73
  simply supported beam, 65-66
  three-pinned arch, 98
Shearing force diagram
  cantilever uniformly loaded, 63
  cantilever wth end load, 61
  cantilever with inclined force, 62
  cantilever with several loads, 64
  examples, 69-70, 72-78
  parabolic arch, 102
  plane bent members, 91
  plane curved members, 92
  simply supported beam, 66
  three-pinned arch, 100
Sign convention, 61
Slenderness ratio, 260
Slope-deflection equation, 113-114
  application of, 116, 120, 122, 123, 127, 155-158, 165, 174, 252, 263, 312, 324
  hyperstatic problems, 176
  initially curved pin-ended columns, 258
  laterally loaded columns, 262
  portal frame, 308
  propped cantilever, 305-306
Space frame, 266
Stability of pin jointed space structures, 267-269
Stability functions, 243, 245, 247, 251, 263, 318, 324, 325, 339, 343
Stable frame, 12, 13
Stiffness, 26-28, 155, 165, 289, 317
  and stress and strain relationship, 40-41
  definition, 26, 156
  example, 157
Stiffness coefficient, 155, 156, 243
Stiffness equation, 280, 289, 307
Stiffness matrix, 27, 28, 41, 155-157, 282, 286, 303, 320, 334, 335
Stiffness method, 294-342
Strain, 37, 78
  stress and stiffness relationship, 40-41

Strain energy, 38-39, 41, 241
  examples, 162-168
  in pure bending, 161
  in pure shear, 168-171
Strain hardening, 25, 176
Stress distribution diagram, 177
Stress, strain and stiffness relationship, 40-41
Stress-strain curves, 25, 28, 36, 39
Stress-strain relationships, 41, 42
Stresses in curved pin-ended column, 260
Stresses in eccentrically loaded column, 257
Structural analysis, principles of, 41-42
Structural properties, 23-57
Structural steel, 26
Structure, definition, 1
Support reaction components, 97
Support reactions, 7
  influence line for, 141-142
Supports, 13, 64
  representation, 2-3
Sway, 155, 157, 249-251, 254, 310, 311, 326
Sway equation, 175
Symmetrical loading, 48, 50
Symmetrical structures, 44-48

Temperature strain, 34
Temperature stresses, 33-34
Tensile forces, 347
Tensile stress, 52, 53, 58, 120
Tension, 61
Thermal effects, 337
Thermal expansion coefficient, 335, 338
Thermal forces, 333
Thermal stresses, 33-34, 339
Three-pinned arch, 97, 100, 106
Trigonometric function, 243
Twisting moment, 97

Unsymmetrical loading, 45, 47, 50

Vector diagrams, 326
Virtual work, 39-40
Virtual work equation, 40, 42, 171-172, 175, 176, 185-188, 191, 192, 199-201, 304, 335

Yield line pattern in reinforced concrete slabs, 201-203
Yield line theory in plates, 200-201
Young's modulus of elasticity, 26, 207